CONCEPTS OF PROBABILITY

Concepts
of
Probability

William C. Guenther

Professor of Statistics
University of Wyoming

McGraw-Hill Book Company
New York, St. Louis, San Francisco, Toronto, London, Sydney

To Clayton A. Guenther

Preface

This book was written for an introductory course in probability at the pre-calculus level. Nearly all the discussion is devoted to discrete probability, but the continuous case is introduced in Chap. 6 and is used in a few later sections to approximate probabilities. The choice of material presented was strongly influenced by applications; nevertheless, an attempt has been made to derive nearly all the results that are within the scope of the book. No claim is made that this is a theoretical text, nor is it intended to be strictly applied. It is hoped that its classification is somewhere between the two and that the book will appeal to those who believe that probability is useful.

A good working knowledge of the second high school algebra course should provide sufficient mathematical background. It is, however, a good idea to precede a course taught from this book with a semester of college mathematics (perhaps college algebra), mainly for the experience and maturity to be gained from it.

Some of the main features of the book are:

1. Applications are stressed to help motivate the material and to give the reader a working knowledge of probability.
2. The topic of derived random variables is given a thorough treatment.
3. A very complete discussion is given for the five most important discrete distributions. It is repeatedly emphasized that one should check the model conditions before using the distribution to compute probabilities for a given situation.
4. Stress is given to the fact that good modern tables are available for the

most important one-dimensional random variables and should be used to evaluate probabilities.

5   A fairly thorough treatment is given for n-dimensional random variables.

6   Probability proofs are demonstrated and advocated. Frequently derivations can be considerably simplified by a probability argument.

7   Two chapters are devoted to statistical applications, including point estimation, interval estimation, hypothesis testing, and sampling.

8   A moderately extensive discussion of Markov chains is presented with the use of matrices.

There is undoubtedly more than enough material for a one-semester course. If possible, an attempt should be made to complete at least the first five chapters. Should further time be available, one or more of the last four chapters could be covered. Chapter 6, the first seven sections of Chap. 7, and Chap. 9 in no way depend upon one another. Section 7-8 and parts of Chap. 8 require knowledge of Chap. 6.

Although the tables of Appendix B are sufficient for working nearly all the problems in the text, it is recommended that the student have access to (and learn to use) good tables of the binomial, hypergeometric, and Poisson distributions, such as those referenced at the end of Chap. 4. These tables take the drudgery out of many problems by eliminating most of the need for tedious calculations with approximations, logarithms, factorials, and combinations.

WILLIAM C. GUENTHER

Contents

CONCEPTS OF PROBABILITY

CHAPTER ONE

Definitions and Interpretation of Probability

1-1 INTRODUCTION

The use of the word "probability" has become so commonplace that almost everyone has a vague notion about its meaning.   One frequently hears statements such as the probability of rain is .3, the probability that a sprinter will win a race is $\frac{2}{3}$, or the probability that two dice will produce a sum of 7 is $\frac{1}{6}$.   Two important considerations arise immediately.   The first concerns the meaning of such a statement, while the second involves the problem of determining the numerical value of a probability.   Both of these important aspects will be considered in detail.

In discussing probability we shall find it convenient to use the term *event*.   Although we shall give a mathematical definition of this term later in the chapter, for the present we can consider an event in its everyday

1

context. Usually we think of an event as something that has occurred, is about to occur, or might conceivably occur. In many important applications we shall be talking about experiments and shall regard any conceptual result of such an experiment as an event. Thus examples of events are rain falling during a given day, a specified sprinter winning a race, and two dice falling so that the sum is 7.

One might consider probability as a measure of likelihood so constructed that it ranges from 0 to 1. If an event is impossible, the weight or measure (or probability) 0 is assigned to it. On the other hand, an event that is certain to happen is assigned weight 1. In between these extremes, we endeavor to assign weights so that the more likely an event, the higher its probability. As we shall see, the weights that are assigned may depend upon prior knowledge of the situation, experimental evidence, or just plain intuition. For example, if a coin is tossed a large number of times, it is reasonable to expect about the same number of heads and tails (or heads about $\frac{1}{2}$ of the time) unless the coin is unbalanced or the tossing procedure is biased. It would seem sensible, therefore, to associate the weight $\frac{1}{2}$ with the head when discussing coin-tossing probabilities. There is no way to prove that this is the correct number to use; nevertheless, it seems like a sensible choice when we consider the symmetry of the two events associated with a balanced coin.

With the coin example in mind there are two points we wish to emphasize. First, as we have already implied, the figure $\frac{1}{2}$ is associated with long-run behavior. Thus when we say that the probability of a head is $\frac{1}{2}$, we do not mean that one out of every two tosses results in a head; we mean, rather, that in the long run we expect heads to show about half the time. If we form the

$$\text{Frequency ratio} = \frac{\text{number of heads}}{\text{number of tosses}}$$

we might suspect that this fraction would tend to get closer to $\frac{1}{2}$ as the number of tosses is increased. That this is indeed what happens has been borne out by actual experience of many investigators. In other words the ratio tends to stabilize near $\frac{1}{2}$ when the experiment is repeated a large number of times. This property of long-run stability of frequency ratios has been found to characterize the results of many experiments.

Second, the coin-tossing example is typical of a very large class of experiments. Even though the experiment is repeated in exactly (or nearly exactly) the same manner, the result is not always the same. In spite of the fact that we make every effort to toss the coin in exactly the same way, the result is sometimes heads and sometimes tails, and which will show on any particular toss cannot be predicted. This is in sharp contrast with what we expect in physical experiments. Suppose we consider an experiment that consists of closing a door. With this act let us associate two events, "the latch catches" and "the latch does not catch." Provided the parts of the latch are all in working order (the experiment is repeated under the same conditions) the result is the same every time since the latch will (or should) catch. In the study of probability we shall be interested in experiments of the coin-tossing type. In fact, from now on, when we use the word *experiment*, we shall mean any act that does not yield the same result every time it is repeated even though the conditions of performance are exactly the same. Experiments of this type are usually called *random experiments*.

Applications of probability have expanded rapidly during the missile and space age. There is hardly a major field of endeavor that does not find some use for this important subject. Some of those who have found probability indispensable include statisticians, geneticists, engineers, physicists, and space scientists in general.

We now turn our attention to the task of defining probability. Although we shall rely almost exclusively upon the mathematical definition given in Sec. 1-4, we shall present two other definitions in chronological order of development. It is usually an aid to understanding to review the thought processes that led to the final result we use.

1-2   THE CLASSICAL DEFINITION OF PROBABILITY

The first definition of probability arose several hundred years ago from consideration of the well-known games of chance with dice, cards, roulette, etc. Let us consider a simple example. Suppose we have a well-balanced (or symmetric) die with six sides numbered 1 to 6. We may feel that any one of the sides is as likely to show as any other when the die is rolled. If our assumption is correct and we wish to form an estimate of our chances of rolling a 4, then it seems fairly logical to use the ratio $\frac{1}{6}$. The fraction is obtained by using for the numerator the number of ways a 4 can be obtained and for the denominator the number of different results that a die can produce. The classical definition of probability was developed from this kind of thought process. That definition is as follows:

CLASSICAL DEFINITION   *If an experiment can produce n dif-*   (1-1)
ferent mutually exclusive results all of which are equally likely, and if f of these results are considered favorable (or result in event A), then the probability of a favorable result (or the probability of event A) is f/n.

Events are mutually exclusive if the occurrence of one prevents the occurrence of any of the others. Thus a die, when rolled, can produce six mutually exclusive results, namely, 1, 2, 3, 4, 5, 6. Only one side can be up when the die comes to rest. The probability that event A happens will be denoted by $\Pr(A)$.

EXAMPLE 1-1

A single die is rolled. What is the probability of rolling either a 4 or a 5?

Solution

We have agreed that the die can produce six equally likely results. Of these, two are favorable (either a 4 or a 5). Hence $n = 6, f = 2$, and the probability of rolling a 4 or a 5 is $\frac{2}{6}$. If we let A be the event that either a 4 or a 5 is rolled, then $\Pr(A) = \frac{2}{6}$.

EXAMPLE 1-2

From an ordinary deck of cards that has been well shuffled one is drawn. What is the probability that the card is a heart?

Solution

An ordinary deck contains 52 cards. When a drawing is made from a well-shuffled deck, then it is not unreasonable to assume that every card has an equal chance of being drawn or that all of the 52 possible results are equally likely. Of these, 13 are hearts. Thus $n = 52, f = 13$, and the probability of drawing a heart is $\frac{13}{52}$. If we let A be the event drawing a heart, then we can write $\Pr(A) = \frac{13}{52}$.

EXAMPLE 1-3

Two coins are tossed. What is the probability that one of the coins shows a head, the other a tail?

Solution

Let H stand for head, T for tail. The possible outcomes can be denoted by HH, HT, TH, and TT where the first letter indicates the result of the first toss, the second letter the result of the second toss. Thus there are four possible outcomes and $n = 4$. Of these, two are favorable to obtaining a head and a tail. Hence, if we can agree that the four results are equally likely, then the desired probability is $\frac{2}{4}$. With balanced coins and a fair tossing procedure we have no reason to think that the four results are not equally likely. Letting A be the event one coin shows a head, the other a tail, we may write $\Pr(A) = \frac{2}{4}$.

If we let $\bar{A}$ denote the event A does not occur, then $n - f$ outcomes are favorable to $\bar{A}$ and the probability that $\bar{A}$ happens is

$$\Pr(\bar{A}) = \frac{n - f}{n}$$

according to definition (1-1). We note that

$$\Pr(A) + \Pr(\bar{A}) = \frac{f}{n} + \frac{n - f}{n} = \frac{f + n - f}{n} = \frac{n}{n} = 1 \qquad (1\text{-}2)$$

Hence, if we know the probability that A occurs, we know the probability that it does not occur, and vice versa.

One objection to the classical definition is that it contains the phrase "equally likely," an undefined term. Attempts to define it usually lead to the use of phrases such as "equally probable," and critics have pointed out that this involves circular reasoning. The objection is not serious, however, since almost everyone has an intuitive feeling of what is meant. We might be hard pressed to define time, yet we use the word frequently with little or no confusion. It is possible that, owing to inexperience, we might label events as equally likely when actually they are not.

A practical objection to the classical definition is that it depends upon being able to classify the outcomes as equally likely. When calculating probabilities associated with a symmetric die, no difficulty arises. Suppose one has a crude homemade wooden die cut in some odd shape with six sides numbered 1 to 6. Due to the lack of symmetry, one would have little faith in regarding the outcomes 1 to 6 as equally likely. What is the probability that the side numbered 4 will show if the die is thrown? The classical definition offers no assistance.

EXERCISES

1-1 Two four-sided symmetric dice with sides numbered 1 to 4 are rolled. What is the probability that the sum of the two numbers is 4?

1-2 An ordinary coin is tossed three times. What is the probability that two out of three times the result is heads?

1-3 From an urn containing four white and two red balls, all of the same size, one ball is drawn. What is the probability of drawing a red ball?

1-4 A card is drawn from an ordinary deck that has been well shuffled. What is the probability that the card is an ace, king, or queen?

1-5 A roulette wheel is divided into 38 equal parts numbered 1 to 36, 0, and 00. Eighteen of the numbers are colored red, eighteen are colored black, and 0, 00 are colored white. A perfectly built wheel is supposed to be symmetric so that all numbers are equally likely. For such a wheel what is the probability that it stops on a black number? What is the probability that it stops on a number greater than 20? What is the probability that it stops on a number that is a multiple of 5?

1-6 A committee consists of five men and two women one of whom must serve as secretary. To determine the unlucky individual a drawing is held. Each person writes his name on a slip of paper, the slips are put in a hat and well mixed, and one is drawn. If this procedure is followed, what is the probability that the secretary is a woman? Suppose each man, instead of writing his own name, wrote the name of a woman on his slip, while each woman wrote her own name as instructed. Now what is the probability of having a woman secretary?

1-3 THE RELATIVE-FREQUENCY DEFINITION

We have illustrated, with the example involving the homemade die, that the classical definition has a serious deficiency. If we had to limit ourselves to situations in which it is possible to divide the outcomes into equally likely cases, the usefulness of probability would be severely restricted. As another example, suppose we wish to know the probability that an individual who is apparently in good health at age 20 will live to be age 70. It is difficult to imagine how we could begin to solve the problem on the basis of the classical definition. Yet the answer to just such a question is highly useful in determining life-insurance rates.

In both of the situations mentioned in the previous paragraph it seems logical, for want of something better to do, to perform the experiment a number of times, say n, count the number of successes f, and use the ratio f/n as the probability. In other words, if we rolled the unbalanced die 1,000 times and counted 250 fours, we could use $250/1{,}000$ as the probability of obtaining a four. Similarly, if we had case histories of 100 individuals that were started when they were age 20, and 46 of them lived to be age 70, we could use $46/100$ as the probability that a healthy individual of age 20 will live to be 70. In each of these cases, if the probability is to be meaningful the conditions of the experiment should be kept as constant as possible. If the die deteriorates as it is thrown, then it is likely that the probability of four will change and the fraction we have obtained will not be too meaningful or useful in predicting future behavior of the die. With the second example it is even more difficult to get a meaningful probability. First, we would have to restrict ourselves to the kind of individual for which we desire to use the probability. Undoubtedly the result would be different for men than for women. It would certainly be different for white male Americans than for black male South Africans. Even if we restrict the study to white male Americans, by the time we collect the data it is possible that chances for increased length of life have improved and the probability we calculate is an underestimate of one applicable to the current situation.

Let us return to the die example and assume that the die does not deteriorate as it is thrown. Some important questions that should occur to us are (1) How many times should the die be rolled to get a number we are willing to use as the probability of a four? (2) How does the fraction behave as the number of throws increases? The answers to these two questions are interrelated and depend upon the stability property mentioned in Sec. 1-1. We know from accumulated experimental evidence that such frequency ratios tend to stabilize for most situations of practical interest. As to the number of repetitions required, it is difficult to give a good answer. Perhaps the best we can say is that the experiment should be repeated until the frequency ratios one desires to use as probabilities appear to be changing very little as n is increased.

The preceding discussion should provide some motivation for the relative-frequency definition of probability. That definition is:

RELATIVE-FREQUENCY DEFINITION *Suppose that an experi-* (1-3)
ment is performed n times with f successes. Assume that the relative frequency f/n approaches a limit as n increases (so that, as n increases, so does f). Then the probability of a success is $\lim_{n \to \infty} f/n$.

Of course, no experiment can be repeated an infinite number of times. The best we can do is to make an estimate based upon large n, a fact that might leave us somewhat dissatisfied with the definition.

EXAMPLE 1-4

The unbalanced homemade die was tossed 100 times. The side labeled 1 occurred 12 times, 2 occurred 18 times, 3 occurred 20 times, 4 occurred 27 times, 5 occurred 13 times, and 6 occurred 10 times. What probabilities should be associated with each of the sides?

Solution

The best we can do is to use the relative frequencies obtained from the 100 throws.   Thus:

Probability of a 1 = $12/100$
Probability of a 2 = $18/100$
Probability of a 3 = $20/100$
Probability of a 4 = $27/100$
Probability of a 5 = $13/100$
Probability of a 6 = $10/100$

From a practical point of view, n is probably too small to expect the frequencies to stabilize.   Certainly $n = 1,000$ or $n = 10,000$ would be much more preferable if we had to use these numbers as probabilities in an important problem.

EXAMPLE 1-5

Table 1-1 is an excerpt from the American Experience Mortality Table. At age 10 a group of 100,000 individuals were alive.   In the original table x runs from 10 to 95.   The column headed by l_x gives the number of this group alive at age x.   The number dying between age x and $x + 1$ is denoted by d_x.   For the probability that a person of age x dies within one year, we use the relative frequency $q_x = d_x/l_x$, and for the probability that he lives we use $p_x = 1 - q_x$.   Find the probability that an individual age 10 will live to be 45.   Find the probability that an individual age 43 lives to be 45.

Table 1-1   Abbreviated American experience mortality table

Age x	Number living l_x	Number of deaths d_x	Yearly probability of dying q_x	Yearly probability of living p_x
10	100,000	749	.007490	.992510
11	99,251	746	.007516	.992484
12	98,505	743	.007543	.992431
. . .	. . .	. . .	. . .	. . .
43	75,782	797	.010517	.989483
44	74,985	812	.010829	.989171
45	74,173	828	.011163	.988837

Solution

Of the 100,000 alive at age 10, there were 74,173 alive at age 45.   Hence,
for the answer to the first question we use $74{,}173/100{,}000 = .74$, approximately.   To answer the second question we note that 74,173 of the
75,782 alive at age 43 are still living.   Hence we get $74{,}173/75{,}782 = .98$,
approximately.

EXERCISES

1-7   Three sprinters A, B, and C race against each other frequently;
they have won 60, 30, and 10 percent of the races, respectively.
With only this information available, what should we use as the
probability that A wins the next race?   That A loses?

1-8   Based upon very extensive study, the records of an insurance
company reveal that the population of the United States can be
classified according to ages as follows: under 20, 35 percent; 20 to
35, 25 percent; 35 to 50, 20 percent; 50 to 65, 15 percent; and over
65, 5 percent.   Suppose that you could select an individual in
such a way that everyone in the United States has an equal chance
of being chosen.   What is the probability that the individual is
over 35?

1-9   Suppose the weather bureau classifies each day according to wind
condition as windy or calm, according to rainfall as moist or dry,
and according to temperature as above normal, normal, or below
normal.   Records are available for the past 100 years.   If you
had to use a number for the probability that July 4 will be calm,
dry, and normal, how would you obtain that number?

1-10   As a class exercise perform the following experiment.   Let each
individual throw two coins ten times, record the number of times
that one head and one tail appear, and then form the frequency
ratio f/n.   Find the range (largest value minus the smallest value)
of all such ratios obtained.   Next let each individual repeat the
experiment with 100 tosses of the two coins and again find the
range of the frequency ratios.   How do the two ranges compare?

1-11   As a class exercise perform the following experiment.   Let each
individual roll an ordinary die ten times, record the number of
1's and 2's, and then form the frequency ratio that would be used
for the probability of obtaining a 1 or a 2.   Find the range of all
such ratios obtained.   Next, let each individual repeat the experiment rolling the die 100 times and again find the range of the
frequency ratios.   How do these two ranges compare?

1-12   From Table 1-1 find the probability that an individual age 12
lives to be 43.

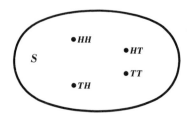

Figure 1-1 A sample space representation for all possible outcomes of two tosses of a coin.

1-4 THE MATHEMATICAL DEFINITION OF PROBABILITY

As we have already observed, both the classical and relative-frequency definitions of probability have limitations. However, one should not jump to the conclusion that the third definition we are about to discuss sweeps away all difficulties. The main reason for the invention of the mathematical definition was to establish a relationship between probability and modern mathematics.

For our next definition of probability, we need the notion of a sample space. To illustrate the concept, consider tossing a coin twice. All possible outcomes can be listed a number of ways. First, they can be denoted by *HH*, *HT*, *TH*, and *TT*, where the first letter indicates the result of the first toss, the second letter the result of the second toss. A second way of characterizing the outcomes is by listing the number of heads, 0, 1, and 2. Third, one of the many other listings of all possible outcomes is obtained by recording the result of the first throw, the result of the second throw, and the temperature at the time of throwing. With any such listing which includes all possible outcomes, each outcome is called a *sample point*, and the set or collection *S* of all sample points for a given listing is called the *sample space*.

We shall frequently use pictorial representations as a visual aid to a discussion. Figure 1-1 shows the sample space when the results of both throws are recorded. Figure 1-2 serves the same purpose when outcomes are given in terms of the number of heads obtained. Thus, if we prefer, we can think of each outcome as a point in space. The oval-shaped

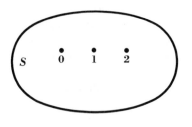

Figure 1-2 A second sample space representation for all possible outcomes of two tosses of a coin using the number of heads obtained.

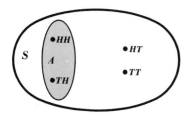

Figure 1-3   A sample space and an event.   The event A, a subset of the sample space, represents the outcome "head on the second toss."

curve around the points serves no purpose other than to indicate that the sample spaces are contained therein.

Having observed that different sample spaces can be used for the same experiment, it is reasonable to ask which one is the more suitable.   The choice will depend upon the features of the experiment one needs to know in order to work the problem at hand.   About the best advice that can be given is to label the points of the sample space so as to include all the necessary information, but no useless information.   If we knew nothing about probabilities associated with the coin example, very likely we would first be interested in the representation of Fig. 1-1.   Our intuition might tell us that these are mutually symmetric results and should be assigned equal probabilities of $\frac{1}{4}$.   Having somehow determined the probabilities that should be associated with the points of the sample space given in Fig. 1-2, perhaps by using the first sample space, we would probably find that the second representation would prove more useful in applied problems.

Any subset A of the sample space is called an *event*.   Figure 1-3 shows an event associated with the sample space of Fig. 1-1.   The event A represents the outcome resulting in a head on the second toss.   To generalize the discussion, suppose that for a particular experiment we have agreed upon a sample space with n outcomes.   Denote these outcomes by $a_1, a_2, \ldots, a_n$ as in Fig. 1-4.   A subset consisting of only one sample point is called a *simple (or elementary) event*.   Hence a_1, $a_2, \ldots, a_n$ are all simple events.   We observe that the terms sample point, simple event, and elementary event are all synonymous.   Any event A will consist of one or more simple events.   For example, the event A of Fig. 1-4 includes the simple events a_2, a_3, and a_5.

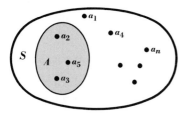

Figure 1-4   A sample space with n simple events.

Now assign to each sample point a_i of S a weight w_i subject to

(a) $w_i \geqq 0$

(b) $w_1 + w_2 + \cdots + w_n = \sum\limits_{i=1}^{n} w_i = 1$ (1-4)

All (1-4) implies is that the weights are positive numbers between 0 and 1 inclusive and that the sum of the weights is equal to 1.   The capital sigma is standard mathematical notation for sum.   (For more on summation notation see Appendix A1.)   We now make the following definition:

MATHEMATICAL DEFINITION   *Let weights satisfying (1-4) be* (1-5)
assigned to points of the sample space.   Then the probability
Pr(A) *of any event A is the sum of the weights of all sample*
points in A.

According to the definition, each weight is itself a probability since it is possible for an event A to contain only one sample point.

   Although (1-5) is well adapted to proving theorems, it is of no assistance in determining the weights.   In choosing the weights, we have to depend upon experience and intuition.   Frequently we recognize from the symmetry of the simple events of the sample space that equal weights are reasonable.   When this situation arises each of the n sample points is assigned weight $1/n$ and definition (1-5) can be modified to read:

MODIFIED MATHEMATICAL DEFINITION   *The probability* Pr(A) (1-6)
that event A will occur is the ratio of the number of sample points
in A to the total number of sample points in S.   That is, Pr(A) =
$n(A)/n$ *where $n(A)$ is the number of sample points in A.*

EXAMPLE 1-6

Four slips of paper numbered 1, 2, 3, 4 are placed in a hat and one slip is drawn out.   What is an appropriate sample space?   What weights would you assign to the sample points?

Solution

In order to simplify the selection of weights, we would probably choose the sample space consisting of four sample points that could be designated 1, 2, 3, 4.   If the slips are the same size so that each has an equal opportunity of being selected, the weights $\frac{1}{4}$, $\frac{1}{4}$, $\frac{1}{4}$, $\frac{1}{4}$ are reasonable.

EXAMPLE 1-7

Suppose that in Example 1-6 the number is recorded, the slip replaced, and a second slip is drawn.   What sample space would probably be selected and what weights would you assign to the sample points?

Solution

When two slips are drawn as described, the sample space we would probably select contains 16 points that could be designated

(1,1) (2,1) (3,1) (4,1)
(1,2) (2,2) (3,2) (4,2)
(1,3) (2,3) (3,3) (4,3)
(1,4) (2,4) (3,4) (4,4)

where the first number is the result of the first draw and the second is the result of the second draw. If the slips are well mixed after each draw, it is reasonable to assign a weight of $\frac{1}{16}$ to each point.

EXAMPLE 1-8

In Example 1-7 let A be the event that includes all the sample points such that the sum of the numbers from the two draws is 6. What is $\Pr(A)$?

Solution

We have already selected a sample space containing 16 points each with weight $\frac{1}{16}$. Of these (2,4), (3,3), and (4,2) belong to A. Thus according to definition (1-5) or definition (1-6), we get $\Pr(A) = \frac{1}{16} + \frac{1}{16} + \frac{1}{16} = \frac{3}{16}$.

EXAMPLE 1-9

Consider again the unbalanced homemade die of Example 1-4. If the die is rolled once, what is the logical sample space? What weights would you assign to the sample points? Let A be the event that an even number shows. What is $\Pr(A)$?

Solution

The sample space that we would probably select consists of six sample points that could be labeled 1, 2, 3, 4, 5, 6. We do not know the exact weights that ought to be assigned. The best we can do is to use our accumulated experience with the die and choose weights $\frac{12}{100}$, $\frac{18}{100}$, $\frac{20}{100}$, $\frac{27}{100}$, $\frac{13}{100}$, $\frac{10}{100}$. Then $\Pr(A) = \frac{18}{100} + \frac{27}{100} + \frac{10}{100} = \frac{55}{100}$.

EXAMPLE 1-10

Suppose that for the roulette wheel described in Exercise 1-5 we are interested in the sample space containing the simple events red, black, and white. What weights should be assigned and how does one get them?

Solution

It is probably obvious that one would select weights $^{18}\!/_{38}$, $^{18}\!/_{38}$, $^{2}\!/_{38}$, respectively. These numbers are obtained by considering the sample space consisting of 38 sample points that could be designated by 1, 2, . . . , 36, 0, 00. Because of the symmetry of the simple events associated with the latter sample space, it is logical to assign weights of $^{1}\!/_{38}$ to each point. The event "red" contains 18 sample points, the event "black" the same number, and the event "white" contains two sample points. Hence the weights $^{18}\!/_{38}$, $^{18}\!/_{38}$, $^{2}\!/_{38}$. This example is a further demonstration of a comment made earlier in this section. That is, we may first work with a sample space that is not particularly interesting but consists of simple events that are mutually symmetric in order to determine the probabilities associated with the simple events in a second sample space that is of interest.

Thus far we have considered only experiments with a *finite sample space*, that is, a sample space consisting of a finite number of points. Although we shall concentrate almost entirely on problems involving a finite number of sample points, in later chapters we shall consider several interesting examples in which the number of sample points is countably infinite. By this we mean that the points of the sample space can be put into one-to-one correspondence with the positive integers. In other words the points can be counted 1, 2, 3, etc., but the counting process never ends. As a simple illustration suppose that a coin is tossed until a head appears. The outcomes could be denoted by H, TH, TTH, $TTTH$, etc., which can be counted by identifying H with 1, TH with 2, TTH with 3, etc. A sample space that includes a finite number of points or an infinite number that is countable is called a *discrete sample space*. The study of probability associated with discrete sample spaces could be referred to as *discrete probability*.

Let us reexamine condition (*b*) of (1-4) when the sample space contains an infinite number of points $a_1, a_2, a_3, \ldots$. Now an infinite number of weights are required satisfying condition (*a*) of (1-4) as before, but condition (*b*) must be changed to read

$$(b') \quad w_1 + w_2 + w_3 + \cdots = \sum_{i=1}^{\infty} w_i = 1 \qquad (1\text{-}7)$$

From a practical point of view no difficulty is encountered since, in most applications of interest, the sum of the weights from a certain point on in the series is practically zero. Thus the probability of the event containing sample points $a_1, a_2, \ldots, a_n$, for some n, is almost 1, and the probability of the event containing sample points $a_{n+1}, a_{n+2}, \ldots$ is nearly zero. Theoretically we are faced with interpreting the meaning of a sum containing an infinite number of terms. Obviously we cannot

14

write down all such terms and form their sum as in the finite case. What is meant by (1-7) is that the sum can be made as close to 1 as we like by taking a sufficiently large number of terms. To illustrate the situation a little more graphically, let us consider the following example. A frog 1 inch away from a pond jumps toward it in such a way that with each jump he reduces the remaining distance by one-half. How many jumps are required to reach the pond? Theoretically, since there is always some distance left to be jumped, the frog never arrives. However, most people would be willing to concede that the frog is at the pond after 50 (or perhaps less) jumps since the remaining distance is hardly worth any consideration. When the nonnegative weights satisfy (1-7), definition (1-5) is still correct without modification. Definition (1-6) is, of course, useless since it is applicable only when the sample space contains n (a finite number) points.

EXERCISES

1-13 An ordinary coin is tossed three times. Select a sample space that is apt to be useful for calculating probabilities. What weights would you assign to the points of the sample space? What is the probability that two out of the three times the result is heads?

1-14 Two four-sided symmetric dice with sides numbered 1 to 4 are rolled. We want to find the probability that the sum of the two numbers is 4. Select a suitable sample space and assign weights. Then find the desired probability.

1-15 From an urn containing four white and two red balls, one ball is drawn. Construct a sample space such that equal weights are reasonable. What is the probability of drawing a red ball?

1-16 A symmetric six-sided die is rolled once. We want to find the probability that the number appearing is greater than 4. Select a suitable sample space, assign weights, and compute the probability.

1-17 A symmetric six-sided die is rolled twice. We are interested in finding the probability that a total of 7 appears on the two rolls. Select a suitable sample space, assign weights, and compute the probability.

1-18 A card is selected from a deck of 52 cards. We would like to know the probability that the card is a heart. Select a suitable sample space, assign weights, and compute the probability.

1-19 Three sprinters A, B, and C race against each other frequently; they have won 60, 30, and 10 percent of the races respectively. What sample space and what weights does this information suggest? What is the probability that A loses the next race?

1-20 What sample space is suggested by the information of Exercise 1-8? What weights should be assigned to the sample points?

What is the probability that the selected individual is under 35? over 35?

1-21 A shipment of paint contains 2,000 one-gallon cans of which 800 are white, 500 are yellow, 300 are red, 300 are green, 100 are blue. During transit the cans are accidentally submerged in water and all labels are lost. Upon arrival the cans are placed upon a platform and one is selected and opened. What sample space is suggested regarding the color of the paint in the selected can? What weights should be assigned to the various sample points? What is the probability that the selected can contains red, white, or blue paint?

1-22 What sample space is suggested by the information given in Exercise 1-9 if one wishes to characterize a day? How might you assign weights to the sample points?

1-23 Suppose we had available the complete table described in Example 1-5. What sample space would make maximum use of this information? What weights should be assigned to the sample points?

1-24 A four-sided die is so constructed that in the long run the side labeled 4 lands on the bottom about twice as frequently as each of the sides labeled 1, 2, 3. If a 4-point sample space is used with points being designated by 1, 2, 3, 4, what weights should be used?

1-25 Suppose that the die in Exercise 1-24 is so constructed that the weights should be chosen proportional to the number appearing on the side. Now what weights should be used?

1-26 The balls for various games of pool are numbered consecutively from 1 to 15. In addition there is a white cue ball which is unnumbered. These balls are all placed in a bag and one is drawn out. We are interested in the probability that the ball has an even number on it. Select a suitable sample space, assign weights, and compute the probability.

1-27 To determine who pays for coffee, three people each toss a coin and the odd man pays. If the coins all show heads or all show tails, they are tossed again. Describe a sample space that characterizes the manner in which a decision is reached.

1-5 CONCLUDING REMARKS

We have discussed three definitions of probability. Although none of the three is wholly satisfactory, the mathematical definition comes the closest to meeting our demands and is the one we shall use. However, it has not been a waste of time to consider the classical and relative-frequency definitions since the experience gained thereby will almost certainly be of assistance in choosing probabilities for the simple events of a sample space.

The construction of the definition of probability and the assignment of weights are parts of a process that can be called model building. We might regard a model as an idealized representation that we hope characterizes the behavior of an experiment and is sufficiently close to reality to be useful in predicting future behavior. For our purposes it is probably more accurate to say that a model is a set of assumptions that hold true for a class of experiments. Frequently, as we shall see in later chapters, these assumptions lead to a formula that is also sometimes referred to as a model. Perhaps the chief objective of probability theory is to construct models suitable for use with random experiments. Having conceived of a model that appears reasonable, we must rely on experience to determine whether use of the model yields satisfactory results. We may be convinced without a shadow of doubt that the probability of obtaining a 2 with a die is $\frac{1}{6}$. However, if after 10,000 rolls, a 2 has been counted 5,000 times, our confidence in the figure $\frac{1}{6}$ may not be very great.

1-2 Classical Definition of Probability (1-1)

If an experiment can produce n different mutually exclusive results all of which are equally likely, and if f of these results are considered favorable (or result in event A), then the probability of a favorable result (or the probability of event A) is f/n.

1-3 Relative-frequency Definition of Probability (1-3)

Suppose that an experiment is performed n times with f successes. Assume that the relative frequency f/n approaches a limit as n increases (so that as n increases, so does f). Then the probability of a success is $\lim\limits_{n \to \infty} \dfrac{f}{n}$.

1-4 Mathematical Definition of Probability (1-5)

Assign to each point a_i of the sample space S a weight w_i subject to the conditions

(a) $w_i \geqq 0$, for all i.

(b) $w_1 + w_2 + \cdots + w_n = \sum\limits_{i=1}^{n} w_i = 1$ with a finite number n of sample points.

(b') $w_1 + w_2 + w_3 + \cdots = \sum\limits_{i=1}^{\infty} w_i = 1$ with a countable infinity of sample points.

Then the probability $\Pr(A)$ of any event A is the sum of the weights of all sample points in A.

Modified Mathematical Definition of Probability (1-6)

Let the sample space S consist of a finite number of points n each assigned weight $1/n$. Then the probability $\Pr(A)$ that event A will occur is the ratio of the number of sample points in A to the total number of sample points in S. That is,

$$\Pr(A) = \frac{n(A)}{n}$$

where $n(A)$ is the number of sample points in A.

CHAPTER TWO

The Calculation of Probabilities

2-1 INTRODUCTION

In many probability problems it is convenient to regard an event as resulting from two or more simpler events. In Example 1-7 the event (2,3) from the sample space with 16 points arises from the event 2 with sample space 1, 2, 3, 4 on the first draw and from the event 3 with sample space 1, 2, 3, 4 on the second draw. Since we know

Pr(2 on first draw) $= \frac{1}{4}$
Pr(3 on second draw) $= \frac{1}{4}$

and

Pr(2 on first draw, 3 on second draw) $= \frac{1}{16}$

these facts might suggest that the latter probability could be obtained by taking the product of the first two. If this is true, then it would seem logical that we should attempt to find ways of evaluating the probability of an event in a sample space with many points by considering the event as resulting from several events in sample spaces with few points. As we shall see, it is often easier to evaluate the probability of an event by first finding the probabilities of the simpler events and then using theorems to compute the probability of the more complicated situation.

2-2 SOME ALGEBRA OF EVENTS

Before discussing some of the devices we shall use to simplify the calculation of probabilities, we shall first introduce some useful terminology. Consider again the sample space of Fig. 1-1 used in connection with tossing a coin twice. Let A be the event "at least one head shows." Another event related to A, say $\bar{A}$, is the event "A does not occur" or "no heads show." The events A and $\bar{A}$ are called complementary events (Fig. 2-1). If A occurs, $\bar{A}$ does not and vice versa. Thus, the *complement* of an event A is the event $\bar{A}$ that includes all the sample points not in A. Since, according to conditions (1-4), the sum of the weights must be 1, it is obvious that

$$\Pr(A) + \Pr(\bar{A}) = 1$$

or

$$\Pr(A) = 1 - \Pr(\bar{A}) \tag{2-1}$$

Since it is frequently easier to calculate $\Pr(\bar{A})$ than $\Pr(A)$, formula (2-1) is itself a useful device. Having already agreed that the sample points in Fig. 2-1 should each be assigned weight $\frac{1}{4}$, we get for this example $\Pr(A) = \frac{3}{4}$, $\Pr(\bar{A}) = \frac{1}{4}$, and the sum of the two probabilities is 1.

Suppose that we throw an ordinary coin and a symmetric die simultaneously. Let A_1 represent the occurrence of a head on the coin and A_2

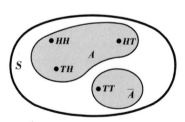

Figure 2-1   Complementary events.

20

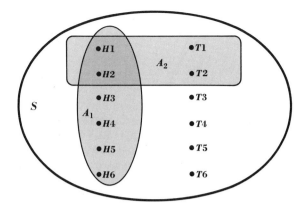

Figure 2-2 Sample space for a coin and die thrown simultaneously.

represent the occurrence of a 1 or a 2 on the die. The 12-point sample space S and events A_1 and A_2 are pictured in Fig. 2-2. The oval includes all the points in A_1, the rectangle all the points in A_2. From the events A_1 and A_2 arise two new events. One of these is "either A_1 or A_2 or both occur" and is called the union of events A_1 and A_2. The *union* of two events A_1 and A_2 is the event that contains all the sample points in A_1 or A_2 or both A_1 and A_2. Three notations for union are in common usage. These are $A_1 \cup A_2$, A_1 or A_2, and $A_1 + A_2$. We shall usually use the first merely because it is the more prevalent in the literature. In Fig. 2-2 we observe with no difficulty that $A_1 \cup A_2$ contains the eight points $H1$, $H2$, $H3$, $H4$, $H5$, $H6$, $T1$, $T2$. The second new event is defined by the condition "both A_1 and A_2 occur" and is called the intersection of events A_1 and A_2. Thus, the *intersection* of two events A_1 and A_2 is the event that contains all the sample points in both A_1 and A_2. Of the three common notations for intersection—$A_1 \cap A_2$, A_1 and A_2, and A_1A_2—we shall prefer the latter. For the events A_1 and A_2 pictured in Fig. 2-2, A_1A_2 contains the points $H1$ and $H2$.

The definitions of union and intersection easily extend to any number k of events. Thus by $A_1 \cup A_2 \cup \cdots \cup A_k$ we mean that at least one of the events $A_1, A_2, \ldots, A_k$ occurs and $A_1A_2 \cdots A_k$ means that all k of the events occur.

Another important term we shall need is *mutually exclusive*. We say that events $A_1, A_2, \ldots, A_k$ are mutually exclusive if the events have no sample points in common. As an illustration let us again use the coin-die example and the three events pictured in Fig. 2-3. Thus, A_1 contains the points $H5$ and $H6$; A_2 contains $T1$, $T2$, and $T3$; and A_3 contains the point $H1$; the three events have no points in common and are, according to the definition, mutually exclusive. In everyday language, if several events are mutually exclusive, the occurrence of one of the events excludes or prevents the occurrence of any of the other events. In other

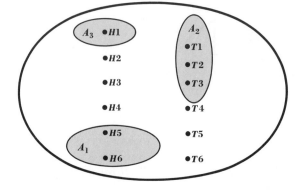

Figure 2-3 Three
mutually exclusive
events resulting from
the simultaneous
tossing of a coin and
a die.

words, two such events cannot occur simultaneously. Stated in more formal manner, if A_1 and A_2 are mutually exclusive, then A_1A_2 contains no points (or forms an empty set). We observe that any event A and its complement $\bar{A}$ are always mutually exclusive.

EXAMPLE 2-1

Consider again the sample space of Example 1-7. Let A_1 be the event "the sum of the two numbers is 5," A_2 be the event "at least one 2 is drawn," and A_3 be the event "the second number is a 3." List the sample points in A_1A_2, A_1A_3, A_2A_3, $A_1A_2A_3$, $A_1 \cup A_2$, $A_1 \cup A_3$, $A_2 \cup A_3$, $A_1 \cup A_2 \cup A_3$, $\overline{A_1 \cup A_2 \cup A_3}$.

Solution

Reproducing the sample space and labeling the events will be of great assistance. This yields Fig. 2-4. We then get the following listing of

Figure 2-4 The events of Example 2-1.

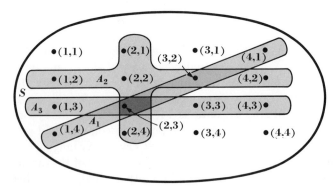

sample points:

For A_1A_2: (2,3), (3,2)
$\quad A_1A_3$: (2,3)
$\quad A_2A_3$: (2,3)
$\quad A_1A_2A_3$: (2,3)
$\quad A_1 \cup A_2$: (1,2), (1,4), (2,1), (2,2), (2,3), (2,4), (3,2), (4,1), (4,2)
$\quad A_1 \cup A_3$: (1,3), (1,4), (2,3), (3,2), (3,3), (4,1), (4,3)
$\quad A_2 \cup A_3$: (1,2), (1,3), (2,1), (2,2), (2,3), (2,4), (3,2), (3,3), (4,2), (4,3)
$\quad A_1 \cup A_2 \cup A_3$: (1,2), (1,3), (1,4), (2,1), (2,2), (2,3), (2,4), (3,2), (3,3), (4,1), (4,2), (4,3)
$\quad \overline{A_1 \cup A_2 \cup A_3}$: (1,1), (3,1), (3,4), (4,4)

EXERCISES

2-1   Repeat Example 2-1 if the sample space is created by drawing the second number without replacing the first.

2-2   In Exercise 1-13 a coin was tossed three times and a sample space was constructed. Using this sample space and letting A_1 be the event "not all three tosses yield the same result," A_2 be the event "at least two heads occur," and A_3 be the event "a head occurs on the first toss," list the points in the sets A_1A_2, A_1A_3, A_2A_3, and $A_1A_2A_3$.

2-3   In Exercise 1-17 we used a 36-point sample space for two rolls of a die. Let A_1 be the event "the sum is 7" and A_2 be the event "both rolls produce the same number." Are A_1 and A_2 mutually exclusive?

2-3   THE ADDITION THEOREM

Consider again the sample space for the coin-die example and the events A_1 and A_2 as defined for Fig. 2-2. It seems reasonable to assign each of the 12 points a weight $1/12$. With this assignment we see by using definition (1-5) that

$$\Pr(A_1A_2) = \frac{1}{12} + \frac{1}{12} = \frac{2}{12}$$
$$\Pr(A_1 \cup A_2) = 8(\frac{1}{12}) = \frac{8}{12}$$

We can also write

$$\Pr(A_1 \cup A_2) = 6(\frac{1}{12}) + 4(\frac{1}{12}) - 2(\frac{1}{12})$$
$$= \Pr(A_1) + \Pr(A_2) - \Pr(A_1A_2) \qquad (2\text{-}2)$$

The reasoning behind (2-2) is simple. We add the weights of the points for events A_1 and A_2 and subtract the weights for A_1A_2 because they have been added to the sum twice. Although we have used a specific example, the argument holds for any two events A_1 and A_2. Formula (2-2) is called the *addition theorem*. An important special case arises when A_1 and A_2 are mutually exclusive so that the event A_1A_2 contains no points. Since there are then no weights to add up, $\Pr(A_1A_2) = 0$ and (2-2) reduces to

$$\Pr(A_1 \cup A_2) = \Pr(A_1) + \Pr(A_2) \tag{2-3}$$

The formulas of the previous paragraph can be extended to handle more than two events. If the events $A_1, A_2, \ldots, A_n$ are mutually exclusive, then with no difficulty we find the generalization of (2-3) to be

$$\Pr(A_1 \cup A_2 \cup \cdots \cup A_n) = \Pr(A_1) + \Pr(A_2) + \cdots + \Pr(A_n) \tag{2-4}$$

Figure 2-3 can be used to illustrate (2-4) for three events. Having assigned equal weights of $\frac{1}{12}$ to each point, we get

$$\Pr(A_1 \cup A_2 \cup A_3) = \frac{2}{12} + \frac{3}{12} + \frac{1}{12} = \frac{6}{12}$$

If the n events are mutually exclusive and exhaust all possibilities—that is, $A_1 \cup A_2 \cup \cdots \cup A_n = S$—then (2-4) yields the useful result

$$\Pr(A_1) + \Pr(A_2) + \cdots + \Pr(A_n) = 1 \tag{2-5}$$

For the situation pictured in Fig. 2-3, let A_4 contain all sample points not in A_1, A_2, or A_3. Then

$$\Pr(A_1) + \Pr(A_2) + \Pr(A_3) + \Pr(A_4) = \frac{2}{12} + \frac{3}{12} + \frac{1}{12} + \frac{6}{12} = 1$$

It is easy to see that we might save considerable work by using (2-5). If we desire $\Pr(A_2) + \Pr(A_3) + \cdots + \Pr(A_n)$, it may be much easier to compute $\Pr(A_1)$ and evaluate the sum as $1 - \Pr(A_1)$.

When $A_1, A_2, \ldots, A_n$ are not mutually exclusive, the generalization is more complicated. We shall consider only the case for three events A_1, A_2, A_3. To get $\Pr(A_1 \cup A_2 \cup A_3)$ we begin by dividing the events formed by A_1, A_2, and A_3 into the seven mutually exclusive events pic-

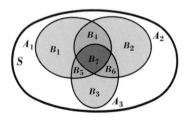

Figure 2-5 Three events, not mutually exclusive, represented as the union of mutually exclusive events.

tured in Fig. 2-5. We then note that

$$
\begin{aligned}
\Pr(A_1) + \Pr(A_2) + \Pr(A_3) &= [\Pr(B_1) + \Pr(B_4) + \Pr(B_5) + \Pr(B_7)] \\
&\quad + [\Pr(B_2) + \Pr(B_4) + \Pr(B_6) + \Pr(B_7)] \\
&\quad + [\Pr(B_3) + \Pr(B_5) + \Pr(B_6) + \Pr(B_7)] \\
&= \Pr(B_1) + \Pr(B_2) + \Pr(B_3) + 2[\Pr(B_4) \\
&\quad + \Pr(B_5) + \Pr(B_6)] + 3\Pr(B_7) \quad (2\text{-}6)
\end{aligned}
$$

and

$$
\begin{aligned}
\Pr(A_1 A_2) + \Pr(A_1 A_3) + \Pr(A_2 A_3) &= [\Pr(B_4) + \Pr(B_7)] + [\Pr(B_5) \\
&\quad + \Pr(B_7)] + [\Pr(B_6) + \Pr(B_7)] \\
&= [\Pr(B_4) + \Pr(B_5) + \Pr(B_6)] \\
&\quad + 3\Pr(B_7) \quad (2\text{-}7)
\end{aligned}
$$

Obviously,

$$
\Pr(B_1) + \Pr(B_2) + \cdots + \Pr(B_7) = \Pr(A_1 \cup A_2 \cup A_3) \quad (2\text{-}8)
$$

The left-hand side of (2-8) can be obtained by subtracting (2-7) from (2-6) and adding $\Pr(B_7)$. Since $B_7 = A_1 A_2 A_3$, we get

$$
\begin{aligned}
\Pr(A_1 \cup A_2 \cup A_3) &= \Pr(A_1) + \Pr(A_2) + \Pr(A_3) - \Pr(A_1 A_2) \\
&\quad - \Pr(A_1 A_3) - \Pr(A_2 A_3) + \Pr(A_1 A_2 A_3) \quad (2\text{-}9)
\end{aligned}
$$

the addition theorem for three events that are not mutually exclusive.

EXAMPLE 2-2

Suppose we select one card from an ordinary deck that has been well shuffled. Use formula (2-2) to find the probability that the card is a heart or a face card.

Solution

Let A_1 and A_2 be the events "a heart is drawn" and "a face card is drawn," respectively. Having selected a 52-point sample space and assigned weights of $\frac{1}{52}$ to each point, we get

$$
\Pr(A_1) = \frac{13}{52} \qquad \Pr(A_2) = \frac{12}{52} \qquad \Pr(A_1 A_2) = \frac{3}{52}
$$

Hence

$$
\Pr(A_1 \cup A_2) = \frac{13}{52} + \frac{12}{52} - \frac{3}{52} = \frac{22}{52}
$$

Of course, since there are 13 hearts and 9 face cards not hearts, we have immediately, using definition (1-5), $\Pr(A_1 \cup A_2) = 22(\frac{1}{52})$.

EXAMPLE 2-3

Verify that formula (2-9) holds for the events defined in Example 2-1. Use equal weights of $\frac{1}{16}$.

25

Solution

Since 12 points fall within at least one of the three events, we have, using definition (1-5),

$$Pr(A_1 \cup A_2 \cup A_3) = {}^{12}\!/_{16}$$

Also

$$Pr(A_1) = {}^4\!/_{16} \qquad Pr(A_2) = {}^7\!/_{16} \qquad Pr(A_3) = {}^4\!/_{16}$$
$$Pr(A_1A_2) = {}^2\!/_{16} \qquad Pr(A_1A_3) = Pr(A_2A_3) = Pr(A_1A_2A_3) = {}^1\!/_{16}$$

Thus

$$Pr(A_1 \cup A_2 \cup A_3) = {}^4\!/_{16} + {}^7\!/_{16} + {}^4\!/_{16} - {}^2\!/_{16} - {}^1\!/_{16} - {}^1\!/_{16} + {}^1\!/_{16} = {}^{12}\!/_{16}$$

EXAMPLE 2-4

Suppose the probability of a cloudy day is .40, the probability of a windy day is .30, and the probability that a day is both cloudy and windy is .18. What is the probability that a day is either cloudy or windy or both?

Solution

Letting A_1 be the event "a day is cloudy," A_2 be the event "a day is windy," and using formula (2-2) we get

$$Pr(A_1 \cup A_2) = .40 + .30 - .18 = .52$$

EXERCISES

2-4   Suppose five coins are thrown simultaneously and it is known that the probability of obtaining three heads is ${}^{10}\!/_{32}$, and the probability of obtaining four heads is ${}^5\!/_{32}$. What is the probability of obtaining either three or four heads? If the probability of obtaining five heads is ${}^1\!/_{32}$, what is the probability of obtaining three or more heads?

2-5   If the probability that a person has blond hair is .23, blue eyes is .17, both blue eyes and blond hair is .11, what is the probability that a person has either blond hair or blue eyes or both?

2-6   Two runners are competing in a 5-mile race. Suppose that the probability that the first runner wins is ${}^3\!/_{10}$ and the probability that the second wins is ${}^3\!/_{20}$. What is the probability that one of these two men wins the race?

2-7   Freshmen at the state university are required to take English, mathematics, and history. It is known that 70 percent pass history, 65 percent pass English, 60 percent pass mathematics,

55 percent pass history and English, 50 percent pass history and mathematics, 45 percent pass English and mathematics, and 40 percent pass all three subjects. What is the probability that a freshman will pass at least one course? at least two courses? (Use Fig. 2-5 to assist in answering the last question.)

2-8 Two cards are drawn from an ordinary deck. The probability of drawing either one or two hearts is $15\!\!\!/_{34}$, the probability of drawing either one or two spades is $15\!\!\!/_{34}$, and the probability of drawing a heart and a spade is $13\!\!\!/_{102}$ What is the probability of drawing a heart or a spade or both?

2-9 Two ordinary symmetric dice are rolled. Let A_1, A_2, and A_3 be the events "the sum is 7," "the sum is 8," and "the sum is 9," respectively. Suppose $\Pr(A_1) = 6\!\!\!/_{36}$, $\Pr(A_2) = 5\!\!\!/_{36}$, and $\Pr(A_3) = 4\!\!\!/_{36}$. Is this enough information to find $\Pr(A_1 \cup A_2 \cup A_3)$ and if so, what is the answer?

2-10 One ordinary die is rolled twice. Let A_1 and A_2 be the events "an ace appears on the first throw" and "an ace appears on the second throw." If $\Pr(A_1) = \Pr(A_2) = 6\!\!\!/_{36}$, is this sufficient information to find $\Pr(A_1 \cup A_2)$? If not, what else would we need to know?

2-11 A student applies for a fellowship at two different universities. The probability is $\frac{1}{3}$ that he will be awarded a fellowship by the first school, $\frac{1}{4}$ that he will be awarded a fellowship by the second, and $\frac{1}{6}$ that he will be awarded a fellowship by both. What is the probability that the student is awarded at least one fellowship?

2-4 CONDITIONAL PROBABILITY AND THE MULTIPLICATION THEOREM

A standard deck of cards has been well shuffled and the card on the top of the deck is to be dealt. Let A_2 be the event "the top card is a heart." To evaluate $\Pr(A_2)$ we would visualize a sample space with 52 points, assign each sample point a weight of $\frac{1}{52}$, observe that 13 points are included in the event A_2, and arrive at $\Pr(A_2) = 13\!\!\!/_{52}$. Now suppose that the dealer slides the top card a fraction of an inch forward from the rest of the stack, far enough so that we see from the reflection in the table that the card is red. The possession of this additional information should change the weights which we are willing to assign to the sample points and the probability that the top card is a heart. Let A_1 be the event "the top card is red." We are now interested in calculating the probability that A_2 happens subject to the condition that A_1 is known to have

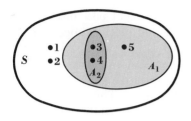

Figure 2-6   Five-point sample space for sprinter example.

occurred.   This probability will be denoted by $\Pr(A_2|A_1)$ and is called a *conditional probability*.   In words, the symbol $\Pr(A_2|A_1)$ will be referred to as the "probability of A_2 given A_1."   Since the 26 points corresponding to a black card are now impossible, the sample space of interest has been reduced to the 26 points corresponding to the red cards.   Undoubtedly we would assign a weight of $\frac{1}{26}$ to each point, observe that 13 points still correspond to the event "the top card is a heart," and conclude that $\Pr(A_2|A_1) = \frac{13}{26}$.   We see that a conditional probability is one calculated under a condition that provides further or new information that was not previously known or available.

Let us consider another example, one in which equal weights are not appropriate.   Suppose that five sprinters are entered in a race.   Sprinters 1 and 2 represent State University, 3 and 4 represent State College, and 5 represents State Teacher's College.   We shall assume on the basis of past experience that the weights which are associated with each individual's winning are respectively $\frac{3}{20}, \frac{2}{20}, \frac{4}{20}, \frac{3}{20}, \frac{8}{20}$.   Let A_2 be the event that State College wins the race.   We easily calculate $\Pr(A_2) = \frac{4}{20} + \frac{3}{20} = \frac{7}{20}$.   The five-point sample space is pictured in Fig. 2-6 with the number of each point corresponding to that particular sprinter's winning.   Now suppose that State University is disqualified and cannot participate because of illegal recruiting practices, thus eliminating sprinters 1 and 2.   Letting A_1 be the event "State College or State Teacher's College wins the race," we observe that A_1 must occur.   Since neither 1 or 2 can win the race, we can use the three-point sample space of Fig. 2-7 with A_1 now playing the roll of S.   The main problem involves the reassignments of weights in a manner that makes sense and is consistent with the information already available.   It is reasonable to

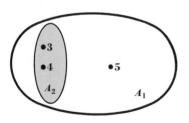

Figure 2-7   Sample space for sprinter example, given that only 3, 4, and 5 can participate.

pick new weights that are proportional to the old ones, $\frac{4}{20}$, $\frac{3}{20}$, $\frac{8}{20}$. According to conditions (1-4) the sum of the weights must be 1. In other words, if k is the constant of proportionality, we must have

$$\frac{4}{20}\,k + \frac{3}{20}\,k + \frac{8}{20}\,k = 1 \qquad k = \frac{20}{15} = \frac{1}{15\!\!/_{20}}$$

We observe that $^{15}\!/_{20} = \Pr(A_1)$ in the original sample space. *Thus, the new weights are obtained by dividing the old weights by* $\Pr(A_1)$ and are $\frac{4}{15}$, $\frac{3}{15}$, $\frac{8}{15}$, respectively. Using these, we find

$$\Pr(A_2|A_1) = \frac{4}{15} + \frac{3}{15} = \frac{7}{15}$$

which can also be written

$$\Pr(A_2|A_1) = \frac{\frac{4}{20}}{\frac{15}{20}} + \frac{\frac{3}{20}}{\frac{15}{20}} = \frac{\frac{7}{20}}{\frac{15}{20}} = \frac{\Pr(A_2)}{\Pr(A_1)}$$

Since A_2 and A_1A_2 contain the same sample points, we can also write the previous equation as

$$\Pr(A_2|A_1) = \frac{\Pr(A_1A_2)}{\Pr(A_1)} \tag{2-10}$$

Another form of (2-10) obtained by multiplying both sides of the equation by $\Pr(A_1)$ is

$$\Pr(A_1A_2) = \Pr(A_1)\Pr(A_2|A_1) \tag{2-11}$$

Formula (2-11) is called the *multiplication* theorem.

In both the card and sprinter examples all the sample points in A_2 were also in A_1 (A_2 is a subset of A_1) so that A_2 and A_1A_2 represented the same event. All hearts are red cards and event "State College wins the race" is included in the event "State College or State Teacher's College wins the race." In most cases where we shall want to use (2-11), A_2 will not be a subset of A_1. To illustrate such a situation, consider again the coin-die example of Fig. 2-2. Suppose we consider the probability that A_2 happens, given A_1 has already occurred. (That is, the probability of a 1 or 2 on the die given that the coin has produced a head.) Now two points in A_2, $T1$ and $T2$, are not in A_1. Knowing that A_1 has already happened, A_1 in Fig. 2-8 plays the role of S, and the number of sample points in A_2 is reduced from four to two, yielding A_2'. The A_2' of Fig. 2-8 is A_1A_2 in the original sample space. Undoubtedly, we would agree that equal weights of $\frac{1}{6}$ are reasonable for the sample space pictured in Fig. 2-8, so that

$$\Pr(A_2|A_1) = \frac{2}{6} = \frac{\frac{2}{12}}{\frac{6}{12}} = \frac{\Pr(A_1A_2)}{\Pr(A_1)}$$

as before.

Although we used specific examples to derive (2-10) and (2-11), the argument is exactly the same with n sample points $a_1, a_2, \ldots, a_n$

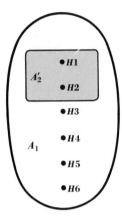

Figure 2-8 *Sample space for coin-die example, given the coin shows a head.*

assigned weights $w_1, w_2, \ldots, w_n$. The new weights, given A_1 is certain to occur, are for $i = 1, 2, \ldots, n$

$$w_i' = \frac{w_i}{\Pr(A_1)} \qquad \text{if } a_i \text{ is in } A_1 \tag{2-12}$$
$$= 0 \qquad \text{if } a_i \text{ is not in } A_1$$

These weights satisfy (1-4) since each is nonnegative and their sum is 1 [because $\Pr(A_1)$ is the sum of the weights in A_1]. With the new assignment of weights definition (1-5) applies as before, and to obtain $\Pr(A_2|A_1)$ we add the new weights in A_1A_2, which yields

$$\Pr(A_2|A_1) = \frac{\Pr(A_1A_2)}{\Pr(A_1)}$$

as before.

Perhaps it should be added that whenever $\Pr(A_1)$ appears in the denominator of a fraction, as in (2-10) and (2-12), we must require $\Pr(A_1) > 0$ in order that the fraction be meaningful.

EXAMPLE 2-5

In Example 1-7 a sample space was constructed for an experiment in which a slip of paper was drawn from a hat containing four numbers, the slip replaced, and a second number drawn. Use that sample space and let A_1 be the event "the sum is 6" and A_2 be the event "one number is a 2." Find $\Pr(A_2|A_1)$.

Solution

Since A_1 contains three points, $\Pr(A_1) = \frac{3}{16}$. Similarly A_1A_2 contains two points and $\Pr(A_1A_2) = \frac{2}{16}$. Thus, according to (2-10),

$$\Pr(A_2|A_1) = \frac{\frac{2}{16}}{\frac{3}{16}} = \frac{2}{3}$$

30

EXAMPLE 2-6

With the same sample space used in Example 2-5 let A_1 be the event "the sum is greater than 5" and A_2 be the event "the sum is 7." Find $\Pr(A_2|A_1)$.

Solution

Since six points yield a sum of either 6, 7, or 8, $\Pr(A_1) = \frac{6}{16}$. Similarly $A_1A_2 = A_2$ contains two points and $\Pr(A_1A_2) = \frac{2}{16}$. Thus

$$\Pr(A_2|A_1) = \frac{\frac{2}{16}}{\frac{6}{16}} = \frac{2}{6}$$

The generalization of (2-11) can be obtained in a straightforward manner. To illustrate the procedure, consider first three events A_1, A_2, A_3. If we replace A_1A_2 by B_1, we can write

$$\Pr(B_1A_3) = \Pr(B_1)\Pr(A_3|B_1)$$

using formula (2-11). Now substituting back and applying (2-11) again, we get

$$\begin{aligned} \Pr(A_1A_2A_3) &= \Pr(A_1A_2)\Pr(A_3|A_1A_2) \\ &= \Pr(A_1)\Pr(A_2|A_1)\Pr(A_3|A_1A_2) \end{aligned} \qquad (2\text{-}13)$$

With n events A_1, A_2, $\ldots$, A_n the same procedure with $B_1 = A_1A_2 \cdots A_{n-1}$, $B_2 = A_1A_2 \cdots A_{n-2}$, etc., can be used to get

$$\Pr(A_1A_2 \cdots A_n) = \Pr(A_1)\Pr(A_2|A_1)\Pr(A_3|A_1A_2) \cdots \\ \Pr(A_n|A_1A_2 \cdots A_{n-1}) \qquad (2\text{-}14)$$

In deriving formula (2-10) our objective was to get an expression for a conditional probability in terms of probabilities of events associated with the original given sample space S. From this result we immediately obtained (2-11), which tells us that the probability that both A_1 and A_2 occur can be obtained from the product of the probability that A_1 occurs and the probability that A_2 occurs, given A_1 has already happened. It is this latter form, and the generalization (2-14), which is particularly useful in many practical applications. If the sample space one selects contains a large number of points, it may be much more difficult to evaluate $\Pr(A_1A_2 \cdots A_n)$ directly than to find $\Pr(A_1)$, $\Pr(A_2|A_1)$, $\Pr(A_3|A_1A_2)$, etc., and take their product.

EXAMPLE 2-7

We already know by using definition (1-5) that $\Pr(A_1A_2) = \frac{2}{12}$ for the coin-die example pictured in Fig. 2-2. Obtain this result by using formula (2-11).

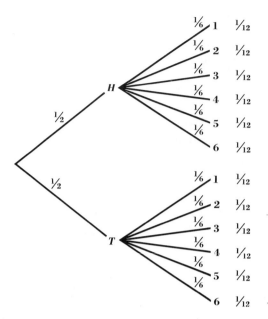

Figure 2-9 Tree diagram for coin-die example.

Solution

In computing $\Pr(A_1)$ we need consider only a two-point sample space, H and T with weights each $\frac{1}{2}$, since A_1 is the event "a head shows on the coin." Hence $\Pr(A_1) = \frac{1}{2}$. To find $\Pr(A_2|A_1)$ only a six-point sample space, 1, 2, 3, 4, 5, 6 with weights each $\frac{1}{6}$, is necessary since we want the probability that the die shows a 1 or a 2, given that the coin shows a head. (The result on the coin, of course, has no influence upon the result produced by the die.) Thus

$$\Pr(A_2|A_1) = \tfrac{2}{6}$$

and

$$\Pr(A_1A_2) = (\tfrac{1}{2})(\tfrac{2}{6}) = \tfrac{2}{12}$$

When a problem is analyzed by considering a sequence of two or more experiments, a tree diagram, such as that pictured in Fig. 2-9 for the coin-die example, can be helpful for summarizing the possible results and for suggesting certain probability formulas (i.e., Bayes' formula considered in Sec. 2-7). Starting from the left, a set of branches is connected to a point on the left side of the diagram. One branch is drawn for each possible outcome of the first experiment and labeled correspondingly, here H for heads and T for tails. Alongside each path is written the probability associated with that particular event. Hence the tree indicates that head and tail each occur with probability $\frac{1}{2}$. The second column of branches shows what can happen on the second experiment, given the result of the first experiment. Thus after a head is obtained, a 1, 2, 3, 4, 5, or 6 can be obtained on the die and six branches emanate from the point labeled H. Similarly, the same six results are possible after a tail is obtained and six paths are connected to the point labeled T. The probabilities (here, each $\frac{1}{6}$) associated with the branches of the second column are conditional probabilities, given the result of the first experiment. Moving from left to right the two columns of branches form 12 paths, each corresponding to a possible result in the sequence of two experiments. By formula (2-11) the probability associated with each path is obtained by multiplication. For Fig. 2-9 each product is $\frac{1}{12}$, the number which appears at the right-hand end of the path. Since the branches coming from a common point represent all possible outcomes of an experiment, the sum of the probabilities associated with these branches necessarily has to be 1. Similarly, the paths represent all possible outcomes of a sequence of experiments and the probabilities appearing at the end of the paths must also sum to 1.

For a sequence of three or more experiments the tree diagram is extended to three or more columns of branches. In each column all probabilities are conditional, given the results of experiments considered earlier in the sequence. Then, the probability associated with each path is the product given by formula (2-14).

EXAMPLE 2-8 ───

Suppose that from an urn containing three white balls and two black balls, two draws are to be made successively without replacement. Let A_1 be the event "getting a white ball on the first draw" and A_2 be the event "getting a black ball on the second draw." Find $\Pr(A_1A_2)$.

Figure 2-10 Appropriate sample space for urn example when both draws are considered.

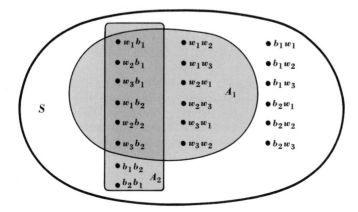

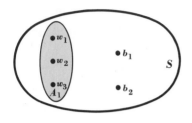

Figure 2-11   A sample space for the first draw. A_1 is the event "a white ball is drawn on first draw."

Solution

Let w_1, w_2, w_3 represent the three white balls and b_1, b_2 represent the black ones.   An appropriate sample space is composed of the 20 points in Fig. 2-10.   The first letter and subscript denote the outcome of the first draw and the second letter and subscript the result of the second draw.   Suppose that we assign a weight of $\frac{1}{20}$ to each point, which is a reasonable choice if the balls are drawn without looking.   By definition (1-5) we have $\Pr(A_1A_2) = \frac{6}{20}$.

Now let us evaluate $\Pr(A_1A_2)$ by using the multiplication theorem.   If we consider the draws individually, a reasonable sample space for the first draw consists of the five points of Fig. 2-11.   If each point is assigned a weight $\frac{1}{5}$, then $\Pr(A_1) = \frac{3}{5}$.   Given that A_1 has happened and, renaming the two remaining white balls W_1, W_2, an appropriate sample space for the second draw consists of the four points of Fig. 2-12.   Assigning a weight of $\frac{1}{4}$ to each point yields $\Pr(A_2|A_1) = \frac{2}{4}$.   Consequently, the theorem gives $\Pr(A_1A_2) = (\frac{3}{5})(\frac{2}{4}) = \frac{6}{20}$ as before.

EXAMPLE 2-9

For the urn problem of Example 2-8, compute $\Pr(A_2)$ and $\Pr(A_1 \cup A_2)$ by using theorems.

Solution

From Fig. 2-10 it is obvious that $\Pr(A_2) = \frac{8}{20}$ and $\Pr(A_1 \cup A_2) = \frac{14}{20}$. To compute $\Pr(A_2)$ by using theorems, we observe that A_2 can happen in two ways, that is, by drawing a white ball first and a black ball second or a black ball first and a black ball second.   The first order of drawing is the event A_1A_2.   The second order can be designated by $\bar{A}_1A_2$.   Obviously

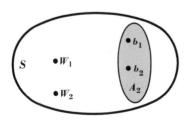

Figure 2-12   Sample space for second draw given that first draw is white. A_2 is the event "black ball is drawn on second draw," given white ball is drawn on first draw.

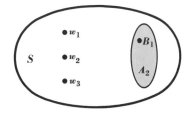

Figure 2-13 Sample space for second draw given first draw is black. A_2 is the event "black ball is drawn on second draw," given black ball is drawn on first draw.

A_1A_2 and $\bar{A}_1A_2$ are mutually exclusive, not only by the definition of complement but also due to the fact that only one order can occur on any series of drawings. Thus, by the addition theorem,

$$\mathrm{Pr}(A_2) = \mathrm{Pr}(A_1A_2) + \mathrm{Pr}(\bar{A}_1A_2)$$

Hence we must compute

$$\mathrm{Pr}(\bar{A}_1A_2) = \mathrm{Pr}(\bar{A}_1)\mathrm{Pr}(A_2|\bar{A}_1)$$

From Fig. 2-11 we see $\mathrm{Pr}(\bar{A}_1) = \frac{2}{5}$. Given that the first draw is black, an appropriate sample space for the second draw consists of the four points in Fig. 2-13 (renaming the remaining black ball B_1). Again using equal weights of $\frac{1}{4}$, we get $\mathrm{Pr}(A_2|\bar{A}_1) = \frac{1}{4}$. Hence $\mathrm{Pr}(\bar{A}_1A_2) = (\frac{2}{5})(\frac{1}{4}) = \frac{2}{20}$ and $\mathrm{Pr}(A_2) = \frac{6}{20} + \frac{2}{20} = \frac{8}{20}$ as it should.

EXAMPLE 2-10

Draw a tree diagram for the urn problem discussed in Examples 2-8 and 2-9, labeling the branches according to the color obtained on the draw. Then find the probability that the first draw was a white ball, given the second draw yields a black ball.

Solution

Let W stand for white, B for black. From Fig. 2-11 we see that the probabilities for the first draw are $\frac{3}{5}$, $\frac{2}{5}$. Having drawn a white ball first, Fig. 2-12 tells us that the probabilities for the second draw are $\frac{2}{4}$, $\frac{2}{4}$. If a black ball is drawn first, we see from Fig. 2-13 that the probabilities for the second draw are $\frac{3}{4}$, $\frac{1}{4}$. The tree diagram appears in Fig. 2-14.

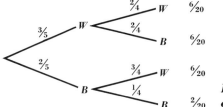

Figure 2-14 Tree diagram for urn example.

Now let A_1 be the event "a black ball is drawn on the second draw" and A_2 be the event "a white ball is drawn on the first draw, a black ball is drawn on the second." Since two paths correspond to drawing a black ball on the second draw, there are two mutually exclusive ways to achieve A_1. With these paths we associate probabilities $\frac{6}{20}$ and $\frac{2}{20}$. Hence $\Pr(A_1) = \frac{6}{20} + \frac{2}{20} = \frac{8}{20}$. Only one path corresponds to A_2 and $\Pr(A_2) = \frac{6}{20}$. Thus $\Pr(A_2|A_1) = (\frac{6}{20})/(\frac{8}{20}) = \frac{3}{4}$, a result in which the denominator is the sum of the probabilities associated with the paths corresponding to A_1, while the numerator is the sum of the probabilities associated with the paths corresponding to A_2, a subset of A_1.

For Examples 2-8 and 2-9 it may seem that it is simpler to use the definition than the theorems. With only 20 sample points this may be true. If the urn contains 35 white and 15 black balls, then the sample space that is the counterpart of the one appearing in Fig. 2-10 contains 2,450 points, and it might be extremely tedious to track down points belonging to specific events. However, use of the multiplication and addition theorems yields probabilities rather quickly.

EXAMPLE 2-11

A symmetric die is thrown twice. What is the probability of getting a 2 and a 3 in that order? In any order?

Solution

Let A_1 be the event "a 2 is obtained on the first roll" and A_2 be the event "a 3 is obtained on the second roll." To compute $\Pr(A_1)$ we select a sample space with six points, each with associated weight $\frac{1}{6}$, and find $\Pr(A_1) = \frac{1}{6}$. After the first roll, we conclude that the same sample space is appropriate for the second roll and $\Pr(A_2|A_1) = \frac{1}{6}$. Hence $\Pr(A_1A_2) = (\frac{1}{6})(\frac{1}{6}) = \frac{1}{36}$. Letting A_3 and A_4 be "a 3 is obtained on the first roll" and "a 2 is obtained on the second roll," respectively, the same argument yields $\Pr(A_3A_4) = \frac{1}{36}$. Now, if we let $B_1 = A_1A_2$ and $B_2 = A_3A_4$, we recognize that B_1 and B_2 are mutually exclusive since only one order of throws can occur with two rolls. Further, these are the only two orders which yield a 2 and a 3. Hence

$$\Pr(B_1 \cup B_2) = \Pr(B_1) + \Pr(B_2) = \frac{1}{36} + \frac{1}{36} = \frac{2}{36}$$

by the addition theorem for mutually exclusive events.

EXAMPLE 2-12

If an ordinary symmetric die is thrown twice, what is the probability of getting at least one 2?

Solution

Let A_1 and A_2 be the events "a 2 appears on the first roll" and "a 2 appears on the second roll." Then, using the same argument as in Example 2-11, we get

$$\Pr(A_1 A_2) = \Pr(A_1)\Pr(A_2|A_1) = (\tfrac{1}{6})(\tfrac{1}{6}) = \tfrac{1}{36}$$
$$\Pr(A_1 \bar{A}_2) = \Pr(A_1)\Pr(\bar{A}_2|A_1) = (\tfrac{1}{6})(\tfrac{5}{6}) = \tfrac{5}{36}$$
$$\Pr(\bar{A}_1 A_2) = \Pr(\bar{A}_1)\Pr(A_2|\bar{A}_1) = (\tfrac{5}{6})(\tfrac{1}{6}) = \tfrac{5}{36}$$

Now we observe that the three events $B_1 = A_1 A_2$, $B_2 = A_1 \bar{A}_2$, $B_3 = \bar{A}_1 A_2$ are the only ways to get at least one 2, and since only one order can be obtained on two rolls, B_1, B_2, and B_3 are mutually exclusive. Thus

$$\Pr(B_1 \cup B_2 \cup B_3) = \tfrac{1}{36} + \tfrac{5}{36} + \tfrac{5}{36} = \tfrac{11}{36}$$

by the addition theorem for mutually exclusive events.

EXAMPLE 2-13

Two cards are dealt from an ordinary deck. What is the probability that at least one is a heart?

Solution

Let A_1 and A_2 be the events "a heart is obtained on the first draw" and "a heart is obtained on the second draw." Then the three mutually exclusive events that yield at least one heart are $B_1 = A_1 A_2$, $B_2 = A_1 \bar{A}_2$, $B_3 = \bar{A}_1 A_2$. For the first draw we would probably select a sample space of 52 points each with weight $\tfrac{1}{52}$. On the second draw there are 51 cards remaining, indicating a space of 51 sample points and a weight of $\tfrac{1}{51}$ for each point. We have, by using the multiplication theorem,

$$\Pr(A_1 A_2) = \Pr(A_1)\Pr(A_2|A_1) = (\tfrac{13}{52})(\tfrac{12}{52}) = (\tfrac{1}{4})(\tfrac{12}{51})$$
$$\Pr(A_1 \bar{A}_2) = \Pr(A_1)\Pr(\bar{A}_2|A_1) = (\tfrac{13}{52})(\tfrac{39}{51}) = (\tfrac{1}{4})(\tfrac{39}{51})$$
$$\Pr(\bar{A}_1 A_2) = \Pr(\bar{A}_1)\Pr(A_2|\bar{A}_1) = (\tfrac{39}{52})(\tfrac{13}{51}) = (\tfrac{1}{4})(\tfrac{39}{51})$$

As contrasted to Example 2-12 the sample space is different for the second event in the series than it is for the first. Finally, by the addition theorem for mutually exclusive events,

$$\Pr(B_1 \cup B_2 \cup B_2) = (\tfrac{1}{4})(\tfrac{12}{51}) + (\tfrac{1}{4})(\tfrac{39}{51}) + (\tfrac{1}{4})(\tfrac{39}{51}) = \tfrac{15}{34}$$

EXAMPLE 2-14

Two cards are dealt from an ordinary deck. What is the probability of drawing a heart or a spade?

Solution

Let A_1 and A_2 be the events "at least one heart is drawn" and "at least one spade is drawn." We want

$$\Pr(A_1 \cup A_2) = \Pr(A_1) + \Pr(A_2) - \Pr(A_1 A_2)$$

From Example 2-13, $\Pr(A_1) = {}^{15}\!\!/_{34}$. Similarly, $\Pr(A_2) = {}^{15}\!\!/_{34}$. We also need $\Pr(A_1 A_2)$, which is the probability of drawing one heart and one spade. If we let B_1, B_2, B_3, B_4 be the events "the first card is a heart," "the first card is a spade," "the second card is a heart," and "the second card is a spade," then the mutually exclusive events $B_1 B_4$ and $B_2 B_3$ are the only ways to get one heart and one spade. Hence

$$\Pr(A_1 A_2) = \Pr(B_1 B_4) + \Pr(B_2 B_3)$$

Using the same type of argument as in Example 2-13 we find

$$\Pr(B_1 B_4) = ({}^{13}\!\!/_{52})({}^{13}\!\!/_{51}) = \Pr(B_2 B_3)$$

so that

$$\Pr(A_1 A_2) = 2({}^{13}\!\!/_{52})({}^{13}\!\!/_{51}) = {}^{13}\!\!/_{102}$$

and

$$\Pr(A_1 \cup A_2) = {}^{15}\!\!/_{34} + {}^{15}\!\!/_{34} - {}^{13}\!\!/_{102} = {}^{77}\!\!/_{102}$$

Alternative Solution

Let A_1 be the event "a heart or spade is drawn." Since we know that $\Pr(A_1) + \Pr(\bar{A}_1) = 1$ or $\Pr(A_1) = 1 - \Pr(\bar{A}_1)$, we can just as well work with the complement of the event. If $\bar{A}_1$ is to happen, then a club or diamond must be drawn each time. Now let B_1 and B_2 be the events "the first card is a club or diamond" and "the second card is a club or diamond" and use the same sample space as in Example 2-13. We get

$$\Pr(B_1 B_2) = \Pr(B_1)\Pr(B_2|B_1) = ({}^{26}\!\!/_{52})({}^{25}\!\!/_{51}) = {}^{25}\!\!/_{102} = \Pr(\bar{A}_1)$$

so that

$$\Pr(A_1) = 1 - {}^{25}\!\!/_{102} = {}^{77}\!\!/_{102}$$

as before.

EXERCISES

Use the addition and multiplication theorems in preference to the definition for evaluating probabilities.

2-12　In Exercise 1-17 we constructed a sample space for the outcomes produced by two rolls of a symmetric six-sided die. Using that sample space and equal weights, find the probability that the sum is greater than 9, given that the sum is greater than 7. Also find the probability that the sum is an even number, given that the sum is greater than 7.

2-13 Use Table 1-1 and find the probability that a person age 43 lives to be 45 by using the formula for a conditional probability.

2-14 Two coins are tossed and you are told that one of them shows a head. What is the probability that the other is also a head?

2-15 Using the sample space and weights selected in Exercise 1-21, find the probability that the selected can contains red paint, given that it does not contain white paint.

2-16 In the urn problem of Example 2-8 find the probability that both a first and a second draw yield black balls, given that one of the balls drawn is black.

2-17 Suppose 80 percent of those who enroll in college are males and 20 percent are females. If half the males and 40 percent of the females finally graduate, what is the probability that a graduate is a male?

2-18 A coin is tossed three times. What is the probability that two out of the three times the result is heads?

2-19 Two cards are dealt from a well-shuffled deck. What is the probability that one is an ace and one is a king?

2-20 An urn contains six red and four black balls. Two balls are drawn out in succession without replacement. What is the probability that the first is red, the second black? What is the probability of getting a red ball and a black ball in either order? What is the probability of drawing a white ball? If five balls are drawn out, what is the probability that at least one is red?

2-21 An ordinary symmetric die is rolled until a 6 appears. What is the probability that exactly four rolls are required?

2-22 Three cards are dealt from a deck. What is the probability that two are red and one is black?

2-23 Suppose it is known that three hunters, say A, B, and C, can kill a pheasant on $\frac{1}{2}$, $\frac{2}{3}$, and $\frac{3}{4}$ of their shots, respectively. What is the probability that they can kill a bird if all three shoot at it simultaneously?

2-24 A hat contains 100 slips of paper numbered from 1 to 100. Three are drawn out successively without replacement. What is the probability that two of the numbers are larger than 90?

2-25 An ordinary die is rolled until a 1 appears, or until it has been rolled three times. What is the probability that the 1 appears on the first roll, given it does not appear on the second?

2-26 In the World Series the two teams continue to play until one team wins at least 4 games. Suppose that it has been determined that the probabilities of a series ending in 4, 5, 6, and 7 games are respectively .20, .26, .24, .30. If the first two games are split, what is the probability that a series lasts 7 games?

2-27 Three cards are dealt from an ordinary deck. Find the probability that the third card is a heart.

2-28 An urn contains three red and two black balls. A second urn contains four red balls and one black ball. One ball is drawn from the first urn and placed in the second. Then a ball is drawn from the second urn. What is the probability that the ball drawn from the second urn is red? Given the second ball is red, what is the probability that the ball drawn from the first urn was red?

2-29 To determine which of two individuals pays for coffee a coin is tossed until one participant loses twice. If one individual loses on the first throw, what is the probability that he will buy the coffee? Suppose the tossing is continued until one individual loses three times. Having lost on the first throw, what is that individual's probability of paying for the coffee?

2-30 A county in Minnesota has two small towns which we shall call towns A and B. The population of town A is 100, the population of town B is 200, and 300 people live in the county but outside of the two towns. In town A 40 percent are women, 60 percent men; in town B 50 percent are women, 50 percent men; and rurally 60 percent are women and 40 percent men. One name is to be selected from all the individuals living in the county. To do this a die is to be thrown and town A selected if a 1 shows, town B selected if a 2 or 3 shows, and the rural area selected if a 4, 5, or 6 shows. Following this, an individual is to be chosen from the selected area in such a way that every individual has an equal chance of being selected. What is the probability of picking a woman? Suppose that instead of using the described procedure, all 600 names are placed in a hat and one is drawn. Now what is the probability of selecting a woman?

2-5 INDEPENDENT EVENTS

In the coin-die problem of Example 2-7 we found that $\Pr(A_2|A_1) = \frac{2}{6}$. From Fig. 2-2 it is easy to see that $\Pr(A_2) = \frac{4}{12} = \frac{2}{6}$. Thus, in this example

$$\Pr(A_2|A_1) = \Pr(A_2) \tag{2-15}$$

In other words, the fact that the coin produces a head does not change the probability of a 1 or a 2 on the die. The latter figure remains $\frac{2}{6}$ irrespective of what happens with the coin. We would probably be surprised if this were not the case, since the throwing of a coin and the rolling of a die are unrelated acts and it is difficult to visualize how the result on the coin could possibly affect the result on the die.

For the coin-die example we also note that $\Pr(A_1|A_2) = \frac{1}{2}$, $\Pr(A_1) = \frac{1}{2}$, so that

$$\Pr(A_1|A_2) = \Pr(A_1) \tag{2-16}$$

It is easy to see that (2-15) implies (2-16) and vice versa since according to (2-11) we can write

$$\Pr(A_1A_2) = \Pr(A_1)\Pr(A_2|A_1) = \Pr(A_2)\Pr(A_1|A_2) \tag{2-17}$$

If (2-15) is true, then (2-17) yields

$$\Pr(A_1)\Pr(A_2) = \Pr(A_2)\Pr(A_1|A_2)$$

which, upon division by $\Pr(A_2)$, reduces to

$$\Pr(A_1) = \Pr(A_1|A_2)$$

If (2-15) is satisfied the multiplication theorem reduces to

$$\Pr(A_1A_2) = \Pr(A_1)\Pr(A_2) \tag{2-18}$$

Whenever Eq. (2-18) holds for two events A_1 and A_2, then A_1 and A_2 are said to be *independent* events. The only way to verify that A_1 and A_2 are independent is to compute $\Pr(A_1)$, $\Pr(A_2)$, and $\Pr(A_1A_2)$, using the sample space and weights selected as suitable for the given experiment, and then substitute in (2-18) and check for equality. However, as was the case with (2-11), formula (2-18) is used primarily to find $\Pr(A_1A_2)$ from the probabilities that appear on the right-hand side of the equation.

In order to use this form of multiplication theorem in this way, we must be able to judge on the basis of our knowledge of the experiment that there is no connection or relationship between the events A_1 and A_2. That is, we must believe that the occurrence or nonoccurrence of A_1 in no way affects the occurrence or nonoccurrence of A_2 and vice versa. Sometimes it will be perfectly clear that events are independent, while in other cases it may be difficult or impossible to make the judgment. Of course, the chief advantage of (2-18) over (2-11) is that no thought or consideration must be given to the other event when calculating either probability.

We would like to generalize (2-18) to any number of events. Looking at the definition for two events it is logical to suspect that we would say that events A_1, A_2, . . . , A_n are independent if

$$\Pr(A_1A_2 \cdots A_n) = \Pr(A_1)\Pr(A_2) \cdots \Pr(A_n) \tag{2-19}$$

41

However, more is necessary, as we shall demonstrate with three events A_1, A_2, A_3. In addition to requiring

$$\Pr(A_1 A_2 A_3) = \Pr(A_1)\Pr(A_2)\Pr(A_3) \tag{2-20}$$

we would also like to have

$$\begin{aligned}
\Pr(A_1 A_2) &= \Pr(A_1)\Pr(A_2) \\
\Pr(A_1 A_3) &= \Pr(A_1)\Pr(A_3) \\
\Pr(A_2 A_3) &= \Pr(A_2)\Pr(A_3)
\end{aligned} \tag{2-21}$$

One might be tempted to infer that (2-20) implies (2-21) and (2-21) implies (2-20) but unfortunately this is not the case.

To show that (2-20) can hold when (2-21) does not, consider a sample space with four points 1, 2, 3, 4 and weights

$$w_1 = \frac{\sqrt{2}}{2} - \frac{1}{4} \qquad w_2 = \frac{1}{4}$$
$$w_3 = \frac{3}{4} - \frac{\sqrt{2}}{2} \qquad w_4 = \frac{1}{4}$$

Let A_1 contain points 1 and 3, A_2 contain points 2 and 3, A_3 contain points 3 and 4 so that $A_1 A_2 A_3$ contains only the point 3, and $A_1 A_2$ also contains the point 3. Then

$$\Pr(A_1 A_2 A_3) = \frac{3}{4} - \frac{\sqrt{2}}{2}$$

and

$$\Pr(A_1)\Pr(A_2)\Pr(A_3) = \frac{1}{2}\left(1 - \frac{\sqrt{2}}{2}\right)\left(1 - \frac{\sqrt{2}}{2}\right) = \frac{3}{4} - \frac{\sqrt{2}}{2}$$

while

$$\Pr(A_1)\Pr(A_2) = \frac{1}{2}\left(1 - \frac{\sqrt{2}}{2}\right) \neq \Pr(A_1 A_2) = \frac{3}{4} - \frac{\sqrt{2}}{2}$$

(The originator of this example is unknown to the author.)

To show that (2-21) can hold while (2-20) does not, consider a sample space generated by throwing a coin three times. Let the points be HHH, HHT, HTH, HTT, THH, THT, TTH, TTT each with weight $\frac{1}{8}$ and define A_1, A_2, A_3 as the events "the first two throws yield heads or the first two throws yield tails," "the last two throws yield heads or the last two throws yield tails," "the first and last throw yield the same result." We see that A_1 contains the points HHH, HHT, TTH, TTT; A_2 contains HHH, HTT, THH, TTT; A_3 contains HHH, HTH, THT, TTT; while $A_1 A_2$, $A_1 A_3$, $A_2 A_3$, $A_1 A_2 A_3$ all contain HHH, TTT. Thus

$$\Pr(A_1) = \Pr(A_2) = \Pr(A_3) = \frac{4}{8} = \frac{1}{2}$$
$$\Pr(A_1 A_2) = \Pr(A_1 A_3) = \Pr(A_2 A_3) = \Pr(A_1 A_2 A_3) = \frac{2}{8} = \frac{1}{4}$$

so that

$$\Pr(A_1A_2) = \tfrac{1}{4} = \Pr(A_1)\Pr(A_2)$$
$$\Pr(A_1A_3) = \tfrac{1}{4} = \Pr(A_1)\Pr(A_3)$$
$$\Pr(A_2A_3) = \tfrac{1}{4} = \Pr(A_2)\Pr(A_3)$$

but

$$\Pr(A_1A_2A_3) = \tfrac{1}{4} \neq \tfrac{1}{8} = \Pr(A_1)\Pr(A_2)\Pr(A_3)$$

With the two preceding examples in mind, the events A_1, A_2, and A_3 are said to be independent if, and only if, both (2-20) and (2-21) hold. In the case of n events A_1, A_2, . . . , A_n we call the events independent if, and only if, Eq. (2-19) holds and similar equations hold for all combinations of two or more events. It can be shown that for n events it is necessary that $2^n - n - 1$ equations be satisfied. It would appear to be a hopeless task to check all these equations if n is even moderately large. However, as we commented when discussing only two events, in practical situations it is quite often easy to judge that events are independent. Many times the experimenter will attempt to perform a series of experiments in a manner that makes the assumption of independence seem realistic. If we believe that there is no connection or relationship among the events, then our main use of this information arises from the fact that we can calculate the left side of (2-19) from the right side after evaluating the n probabilities individually in such a way that for each evaluation no thought need be given to the other $n - 1$ events.

Events that are not independent are called *dependent* events. The urn problem discussed in Example 2-8 is a good illustration. Any event associated with the second draw will depend upon the result of the first draw. This is obviously true since the composition of the urn for the second draw will be different after a white ball has been drawn than after a black ball is drawn. In that example we had

$$\Pr(A_2|A_1) = \tfrac{6}{20} \neq \tfrac{8}{20} = \Pr(A_2)$$

In particular, we note that mutually exclusive events are not independent (unless one of the events occurs with probability 0). This is true since if A_1 and A_2 are mutually exclusive $\Pr(A_1A_2) = 0 \neq \Pr(A_1)\Pr(A_2)$.

EXAMPLE 2-15

Two cards are drawn without replacement from an ordinary deck. Would you judge that an event associated with the first draw would be independent of an event associated with the second draw?

Solution

No, because probabilities associated with any outcome occurring on the second draw depend upon the result of the first draw.

EXAMPLE 2-16

Is an event associated with a particular roll of a die independent of an event associated with any other particular roll?

Solution

Yes, because the result on any one roll can in no way influence the result on any other roll.

EXAMPLE 2-17

Is an event associated with the weather on a particular day independent of an event associated with the weather on any other particular day?

Solution

To be specific, suppose that we talk about the events "rain on day 1" and "rain on day 2." If the second day immediately follows the first, we would not be inclined to regard the events as independent since rain frequently accompanies a front that can influence weather in a locality for several days. If the days were 100 days apart, then it is quite likely that we would be safe in assuming independence. However, it may still pay to check weather records to see if the assumption is realistic. The real test of any assumption is obtained by checking the results derived therefrom against the actual results we are attempting to characterize.

EXAMPLE 2-18

Toward the end of Sec. 1-4 we considered an experiment that consisted of tossing a coin until a head appears. We observed that a sample space that might be of interest consisted of sample points designated by $H, TH, TTH, TTTH, \ldots$ where the ith sample point can be identified with the event "the first head occurs on the ith toss." Assign weights to these points.

Solution

Each event of the sample space can be regarded as being generated by a series of independent events, each with two sample points with weights $\frac{1}{2}$. Hence we find with no difficulty

$\Pr(H) = \frac{1}{2}$
$\Pr(TH) = \Pr(T)\Pr(H) = (\frac{1}{2})(\frac{1}{2}) = \frac{1}{4}$
$\Pr(TTH) = \Pr(T)\Pr(T)\Pr(H) = (\frac{1}{2})(\frac{1}{2})(\frac{1}{2}) = \frac{1}{8}$
$\Pr(TTTH) = \Pr(T)\Pr(T)\Pr(T)\Pr(H) = (\frac{1}{2})(\frac{1}{2})(\frac{1}{2})(\frac{1}{2}) = \frac{1}{16}$
. . . .

The weights form a geometric progression with an infinite number of terms. The first term is $a = \frac{1}{2}$, the common ratio is $r = \frac{1}{2}$, and the sum is $a/(1 - r) = (\frac{1}{2})/(1 - \frac{1}{2}) = 1$, as required by (1-7).

EXERCISES

2-31 A coin is tossed twice. Let A_1, A_2, A_3 be the events "a head occurs on the first toss," "a tail occurs on the second toss," "the two tosses produce different results." Are A_1, A_2 independent? A_1, A_3? A_2, A_3? A_1, A_2, A_3?

2-32 An ordinary symmetric die is thrown twice. Let A_1, A_2, A_3 be the events "a 1 appears on the first throw," "a 1 appears on the second throw," "both throws produce the same result." Which events are independent?

2-33 An ordinary symmetric die is thrown. Let A_1, A_2, A_3 be the events "a 1 or a 2 appears on the first throw," "a 1 appears on the first throw," "both 1's appear or a 4 appears on first throw or a 5 appears on first throw or a 6 appears on first throw but not on second." Does either (2-20) or (2-21) hold?

2-34 In the urn problem of Example 2-8 we decided that A_1 and A_2 are dependent. How might we change the experiment to make A_1 and A_2 independent?

2-35 Suppose that two successive true-false questions in an examination are (1) Columbus discovered America in 1492 and (2) Columbus discovered America in 1500. If A_1 and A_2 are the events "question (1) is answered correctly" and "question (2) is answered correctly," would you expect A_1 and A_2 to be independent?

2-36 A baseball player has a lifetime batting average of .300. During his last 20 trips to the plate he has no hits (0 for 20 in baseball language). Would you expect that for his next time at bat the probability of a hit is .300?

2-37 Suppose that 20 percent of the students get A in Calculus I and 25 percent get A in Calculus II. For an individual student do you think that the event "getting an A in Calculus I" and the event "getting an A in Calculus II" are independent? In other words, is the probability of getting A in both $(.20)(.25) = .05$?

2-38 Suppose 100 students are enrolled in Calculus I. Each name is written on a slip of paper, the slips placed in a hat, and two names are drawn. If, in the long run, 20 percent get A's in Calculus I, is the probability that both the students whose names were drawn get A's equal to $(.20)(.20) = .04$?

2-39 A company makes $\frac{1}{2}$-inch bolts on the first floor of its plant and $\frac{1}{2}$-inch nuts on the second floor. Later a nut is placed on a bolt by an individual who grabs a bolt from one conveyor and a nut from another. If 1 percent of the bolts are defective and 2 percent of the nuts are defective, what is the probability of getting a good bolt and a good nut?

2-40 In Example 2-18 assume that the probability of a head on each toss is p (instead of $\frac{1}{2}$). Let $q = 1 - p$. Now assign weights to the points of the sample space and verify that the sum of the weights is 1.

2-6 PERMUTATIONS AND COMBINATIONS

We have seen how it is sometimes simpler to compute probabilities by using the addition and multiplication theorems rather than the definition. In this section we shall discuss some counting formulas that can be used to simplify the calculation of probabilities, especially in cases where enumeration of the sample points is excessive. The counting formulas can be used either with or without the theorems.

Suppose that there are three highway routes between towns A and B and two routes between towns B and C. Then there is a total of six possible routes from A to C traveling through B. With each of the three routes for the first lap, there are two choices for the second lap, giving the total of six. This illustrates a fundamental principle of counting. If a first thing can be done in n_1 ways after which a second thing can be done in n_2 ways, then the two things can be done in n_1n_2 ways. The truth of the principle is readily demonstrated by making a list of the ways the two things can be done. For example, using the numbering of Fig. 2-15, the routes are 1-4, 2-4, 3-4, 1-5, 2-5, 3-5. The principle is readily extended to r things. If a first thing can be done in n_1 ways, after which a second thing can be done in n_2 ways, . . . , after which an rth thing can be done in n_r ways, then all r things can be done in $n_1n_2 \cdots n_r$ ways.

Consider three things which we shall call A, B, and C. They can be arranged in six ways if we use all three. The arrangements, called *permutations*, are ABC, ACB, BAC, BCA, CAB, CBA. We would like

Figure 2-15 Highway routes and fundamental counting principle.

46

to have a formula for counting the number of permutations of n things using all n at a time. The counting principle provides the answer. The first thing can be chosen from any of the total of n, after which the second thing can be selected from any of the remaining $n - 1$. Consequently, the first two places in the permutation can be chosen in $n(n - 1)$ ways. With each choice for the first two things, the third can be chosen in $n - 2$ ways. By continuing this line of reasoning, we find that the number of permutations of n things taken n at a time is

$$_nP_n = n(n - 1)(n - 2) \cdots (3)(2)(1) = n! \tag{2-22}$$

where the symbol $n!$ is called n factorial. When $n = 3$, $3! = 3 \cdot 2 \cdot 1 = 6$, agreeing with the result obtained by enumeration. Sometimes we wish to count the number of permutations of n things taken r at a time where r is less than n. With the three letters A, B, C the permutations taken two at a time are AB, AC, BA, BC, CA, CB. We proceed as before, but since there are only r positions to fill in the permutation, the product replacing (2-22) contains only r numbers. The number of permutations of n things taken r at a time is

$$\begin{aligned}_nP_r &= n(n - 1) \cdots [n - (r - 1)] \\ &= n(n - 1) \cdots (n - r + 1)\end{aligned} \tag{2-23}$$

Multiplying and dividing the right-hand side of (2-23) by $(n - r)!$ results in the more convenient form

$$_nP_r = \frac{n!}{(n - r)!} \tag{2-24}$$

If $r = n$, (2-24) reduces to (2-22), since $0!$ is defined to be 1.

EXAMPLE 2-19

Use formula (2-24) to obtain the number of permutations of ten things taken three at a time.

Solution

$n = 10$, $r = 3$, $n - r = 7$, and

$$_{10}P_3 = \frac{10!}{7!} = \frac{10 \cdot 9 \cdot 8 \cdot 7 \cdot 6 \cdot 5 \cdot 4 \cdot 3 \cdot 2 \cdot 1}{7 \cdot 6 \cdot 5 \cdot 4 \cdot 3 \cdot 2 \cdot 1} = 10 \cdot 9 \cdot 8 = 720$$

It is probably just as easy to argue directly as we did when deriving this formula. We immediately get $10 \cdot 9 \cdot 8 = 720$ choices.

Next, suppose that we want to count the number of permutations of n things taken n at a time but that not all n things are distinguishable. To take a simple example, consider the permutations of the letters of the word "book." Four different letters yield $4! = 24$ different permutations. However, this is too large a number, since some of the permutations look exactly alike because the o's are indistinguishable. To clarify this point, assume that the o's have subscripts (denote them o_1 and o_2) so that we can tell them apart. Then bo_1o_2k and bo_2o_1k are different permutations; but if the subscripts are dropped, both look exactly alike. Let P be the number of distinguishable permutations. If, for each of the P permutations, subscripts are added to the o's and the o's are permuted, then the total number of permutations of n distinct things taken n at a time is obtained. By the fundamental counting principle this total is $2!P$. But we already know that the total is $4!$, so consequently $2!P = 4!$, or $P = (4!)/(2!)$. The argument can be readily generalized. Let n_1 things be alike, n_2 alike, . . . , n_k alike such that $n_1 + n_2 + \cdots + n_k = n$. The number of distinct permutations P gives rise to $Pn_1! \, n_2! \cdots n_k!$ permutations if subscripts are added to the like things and if each of the like things is permuted in all possible ways. But this product must be the same as the total number of permutations of n things taken n at a time, which is $n!$. Hence

$$Pn_1! n_2! \cdots n_k! = n!$$

$$P = \frac{n!}{n_1! n_2! \cdots n_k!} \tag{2-25}$$

EXAMPLE 2-20

How many distinct permutations can be constructed from the word "Mississippi"?

Solution

The $n = 11$ letters are composed of $n_1 = 1$ m, $n_2 = 4$ i's, $n_3 = 4$ s's, $n_4 = 2$ p's. Thus

$$P = \frac{11!}{1! \, 4! \, 4! \, 2!} = \frac{11 \cdot 10 \cdot 9 \cdot 8 \cdot 7 \cdot 6 \cdot 5 \cdot 4 \cdot 3 \cdot 2 \cdot 1}{1 \cdot 4 \cdot 3 \cdot 2 \cdot 1 \cdot 4 \cdot 3 \cdot 2 \cdot 1 \cdot 2 \cdot 1}$$
$$= 11 \cdot 10 \cdot 9 \cdot 7 \cdot 5 = 34,650$$

In some counting problems order is not important. For example, we may want to know the number of five-card hands that can be constructed from an ordinary deck. A hand dealt A K Q J 10 of hearts is the same as one with K Q J 10 A of hearts or one with Q J 10 A K of hearts. Consequently, many permutations lead to the same hand or combination of cards. By *combination*, we mean an unordered grouping of things. A formula for the number of combinations of n things taken r at a time can be obtained by counting the number of permutations two ways. We already know from (2-24) that the number of permutations of n things

48

taken r at a time is $n!/(n - r)!$. Let $\binom{n}{r}$ denote the number of combinations of n things taken r at a time. Another procedure for constructing all permutations of n things taken r at a time would be to write down all $\binom{n}{r}$ combinations and then permute all letters of each. This yields $\binom{n}{r} r!$ permutations which must equal $n!/(n - r)!$. Consequently,

$$\binom{n}{r} = \frac{n!}{r!(n - r)!} \tag{2-26}$$

Let us illustrate this construction process with three letters A, B, C. It is easy to recognize that there are three combinations when the letters are taken two at a time. These are AB, AC, and BC. Now permuting all letters of each we see that combination AB produces permutations AB, BA; combination AC produces permutations AC, CA; combination BC produces permutations BC, CB so that we get $3 \cdot 2! = $ six permutations, the same as we obtained earlier. Similarly, the general result is $\binom{n}{r} r!$ as previously stated.

EXAMPLE 2-21

How many five-card hands can be constructed from a deck of 52 cards?

Solution

The answer is the number of combinations of 52 things taken 5 at a time or

$$\binom{52}{5} = \frac{52!}{5!47!} = \frac{52 \cdot 51 \cdot 50 \cdot 49 \cdot 48}{1 \cdot 2 \cdot 3 \cdot 4 \cdot 5} = 2{,}598{,}960$$

EXAMPLE 2-22

What is the probability that a five-card poker hand contains three aces?

Solution

The total number of sample points is $\binom{52}{5}$ from Example 2-21. If a hand is dealt fairly, then equal weights of $1 \Big/ \binom{52}{5}$ for each point are reasonable. If we let A be the event "the hand contains three aces," then to evaluate $\Pr(A)$ we need to count the number of sample points that fall in A. Now three aces can be chosen from four aces in $\binom{4}{3}$ ways. With each choice, the hand can be filled out by choosing two cards from the remaining 48 non-aces in $\binom{48}{2}$ ways. Then, by the fundamental counting principle, the total number of hands containing three aces is $\binom{4}{3}\binom{48}{2}$ and

$$\Pr(A) = \frac{\binom{4}{3}\binom{48}{2}}{\binom{52}{5}} = \frac{94}{54,145} = .0017$$

EXAMPLE 2-23

A committee of college faculty members consists of three full professors, five associate professors, and two assistant professors. A subcommittee of six is selected by drawing names out of a hat. What is the probability that the subcommittee is composed of two full professors, three associate professors, and one assistant professor?

Solution

There are $\binom{10}{6}$ possible subcommittees. Hence a reasonable sample space for the problem contains $\binom{10}{6}$ points, one for each subcommittee. The method used to make the selection suggests equal weights of $1 \Big/ \binom{10}{6}$. If we let A be the event "the subcommittee is composed of two full professors, three associate professors, and one assistant professor," then to evaluate $\Pr(A)$ we need to count the number of sample points which fall in A. Since two full professors can be chosen in $\binom{3}{2}$ ways, three associate professors in $\binom{5}{3}$ ways, and one assistant professor in $\binom{2}{1}$ ways, there are $\binom{3}{2}\binom{5}{3}\binom{2}{1}$ possible subcommittees and

$$\Pr(A) = \frac{\binom{3}{2}\binom{5}{3}\binom{2}{1}}{\binom{10}{6}} = \frac{2}{7}$$

EXERCISES

2-41 In how many ways can five people be seated in a row? Suppose that three are men and two are women and that their positions are determined by drawing numbers out of a hat. What is the probability that men occupy the end positions?

2-42 How many distinct permutations can be formed from the letters *AAABBCCCC*? What would be the answer if the letters were all different?

2-43 Eight men are available to form a basketball team. How many ways can the team be formed if each man can play any of the five positions? If five men are chosen from the eight, how many ways can this be done?

2-44 Suppose that of the eight men available in Exercise 2-43, three are over 6 feet tall. Assuming that height is not considered in picking the team, what is the probability that all three are on the starting five?

2-45 In how many ways can a five-card hand drawn from an ordinary deck be constructed so that it contains two aces and two kings?

2-46 How many different sets of answers are possible for a 10-question true-false examination?

2-47 How many distinct permutations can be made from the word "statistics"?

2-48 A hat contains 25 slips of paper numbered 1 to 25. If three are drawn without replacement, what is the probability that all three numbers are less than 10?

2-49 Five cards are dealt from a well-shuffled deck. What is the probability that the cards form a full house (three of one denomination, two of another denomination, such as three aces and two kings)?

2-50 In a bridge game the declarer holds 5 trump and the dummy has 3. The opponents passed throughout the bidding so that no clue to their distribution is available. What is the probability that the outstanding trump are split 3 and 2? 4 and 1? 5 and 0?

2-51 Suppose that in Exercise 2-50 one round of trump is played and both opponents follow suit. Now what is the probability that trumps were originally split 3 and 2? 4 and 1?

2-7 BAYES' THEOREM

Suppose we have three urns containing black and white balls. The exact composition of the urns is

urn 1: four white and two black balls
urn 2: two white and six black balls
urn 3: three white and no black balls

An urn is to be selected by throwing a coin twice, after which a ball is to be drawn. If no heads are obtained, urn 1 will be chosen. If one head

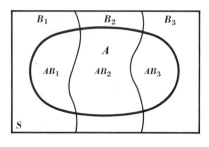

Figure 2-16   *The event A composed of three mutually exclusive events.*

is obtained, urn 2 will be selected, and if two heads show we shall take urn 3.   We already know from Sec. 1-4 that the probabilities associated with the selection of an urn are respectively $\frac{1}{4}$, $\frac{2}{4}$, $\frac{1}{4}$.   Now let us compute the probability that the above procedure produces a white ball.   Let A be the event "a white ball is drawn" and B_1, B_2, B_3 be the events "urn 1 is selected," "urn 2 is selected," "urn 3 is selected."   Obviously B_1, B_2, and B_3 are mutually exclusive and consequently so are AB_1, AB_2, and AB_3, the three events that lead to the selection of a white ball.   In fact

$$A = AB_1 \cup AB_2 \cup AB_3$$

as pictured in Fig. 2-16.   Thus, according to formula (2-4),

$$\Pr(A) = \Pr(AB_1) + \Pr(AB_2) + \Pr(AB_3)$$

and, using the multiplication theorem (2-11), we get finally

$$\begin{aligned}
\Pr(A) &= \Pr(B_1)\Pr(A|B_1) + \Pr(B_2)\Pr(A|B_2) + \Pr(B_3)\Pr(A|B_3) \\
&= (\tfrac{1}{4})(\tfrac{4}{6}) + (\tfrac{2}{4})(\tfrac{2}{8}) + (\tfrac{1}{4})(\tfrac{3}{3}) = \tfrac{13}{24}
\end{aligned} \tag{2-27}$$

The calculations are summarized in the tree diagram shown in Fig. 2-17.

The result (2-27) can be readily generalized.   If the event A can occur in combination with any one of k mutually exclusive events B_1, B_2,

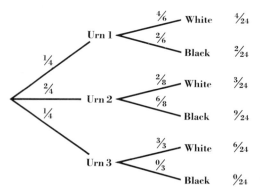

Figure 2-17   *Tree diagram for three-urn example.*

52

$\ldots$, B_k, then

$$\Pr(A) = \Pr(B_1)\Pr(A|B_1) + \Pr(B_2)\Pr(A|B_2) + \cdots + \Pr(B_k)\Pr(A|B_k)$$
$$(2\text{-}28)$$

Now suppose that we are told that a white ball was drawn but not the number of the urn from which it was selected. Let us calculate the probability that it was urn 3. That is, we seek $\Pr(B_3|A)$. By formulas (2-10) and (2-11) we have

$$\Pr(B_3|A) = \frac{\Pr(AB_3)}{\Pr(A)} = \frac{\Pr(B_3)\Pr(A|B_3)}{\Pr(A)} \qquad (2\text{-}29)$$

But all the probabilities on the right-hand side of (2-29) are known so that we obtain

$$\Pr(B_3|A) = \frac{(\frac{1}{4})(\frac{3}{3})}{13\!\!\!/_{24}} = \frac{6}{13}$$

Similarly, we find that $\Pr(B_1|A) = \frac{4}{13}$, $\Pr(B_2|A) = \frac{3}{13}$. The tree diagram can be extremely helpful in evaluating these conditional probabilities. For example, we see from Fig. 2-17 that three paths correspond to the selection of a white ball and the sum of the probabilities associated with these paths is $\Pr(A)$, the denominator of (2-29). Similarly, only one path corresponds to the selection of a white ball from urn 3 and the probability associated with this path is $\Pr(AB_3)$, the numerator of (2-29). Hence, the tree diagram suggests a result that we know to be true by conditional probability formulas. That is,

$$\Pr(B_3|A) = \frac{\text{probability associated with path corresponding to } B_3A}{\text{sum of probabilities associated with paths corresponding to } A}$$

Formula (2-29) is a special case of *Bayes' theorem* (or Bayes' formula). The more general statement is as follows: Let B_1, B_2, $\ldots$, B_k be k mutually exclusive events that exhaust all possibilities, and let A be any event with positive probability. Then

$$\Pr(B_i|A)$$
$$= \frac{\Pr(B_i)\Pr(A|B_i)}{\Pr(B_1)\Pr(A|B_1) + \Pr(B_2)\Pr(A|B_2) + \cdots + \Pr(B_k)\Pr(A|B_k)}$$
$$(2\text{-}30)$$

for $i = 1, 2, \ldots , k$. This result is easily obtained by using the same procedure followed to get (2-29) and then replacing $\Pr(A)$ by (2-28). The tree diagram that suggests the result (2-30) appears in Fig. 2-18, and it is readily apparent that

$$\Pr(B_i|A) = \frac{\text{probability associated with path corresponding to } B_iA}{\text{sum of probabilities associated with paths corresponding to } A}$$

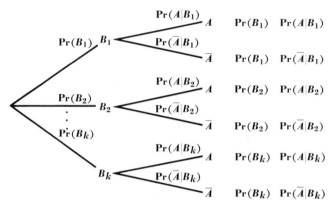

Figure 2-18   General tree diagram that suggests Bayes' formula.

EXAMPLE 2-24 ▭

There are 52 cards in an ordinary deck.   To play the game of sheepshead all cards of denominations 2, 3, 4, 5, and 6 are removed leaving a total of 32 cards.   Suppose that we have available both an ordinary deck and a sheepshead deck.   A coin is tossed to select a deck.   With a head the ordinary deck is taken, with a tail the sheepshead deck.   Then a card is selected after the chosen deck has been well shuffled.   If the selected card is a face card (a jack, queen, or king), what is the probability that the ordinary deck was chosen?

Solution

If we let B_1 be the event "the ordinary deck is chosen," B_2 be the event "the sheepshead deck is chosen," and A be the event "a face card is selected," then with no difficulty we get $\Pr(B_1) = \frac{1}{2}$, $\Pr(B_2) = \frac{1}{2}$,

Figure 2-19   Tree diagram for card example.

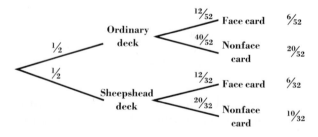

54

$Pr(A|B_1) = 12\!\!/_{52}$, $Pr(A|B_2) = 12\!\!/_{32}$, so that

$$Pr(B_1|A) = \frac{Pr(B_1)Pr(A|B_1)}{Pr(B_1)Pr(A|B_1) + Pr(B_2)Pr(A|B_2)}$$

$$= \frac{(\frac{1}{2})(12\!\!/_{52})}{(\frac{1}{2})(12\!\!/_{52}) + (\frac{1}{2})(12\!\!/_{32})} = \frac{32}{84}$$

Again it may be helpful to draw the tree diagram (Fig. 2-19). Then

$$Pr(B_1|A) = \frac{6\!\!/_{52}}{6\!\!/_{52} + 6\!\!/_{32}} = \frac{32}{84}$$

as before.

EXERCISES

2-52   In Example 2-24 suppose the ordinary deck is chosen with probability p and the sheepshead deck with probability $1 - p$. What is p if $P(B_1|A) = \frac{1}{2}$?

2-53   Two slot machines each pay off a constant amount when the player wins. To select a machine to play, a coin is tossed. If a head shows, machine 1 is played, if a tail shows, machine 2 is played. One of the machines pays off with probability .3 and the other with probability .2 but the player does not know which is which. Draw a tree diagram and find the probability that a player wins on an individual play. If a player wins, what is the probability that he played the machine that pays off with probability .3?

2-54   Three secretaries who work in an office type many things. A carbon copy is always placed in the files. Suppose that it is known that secretaries 1, 2, and 3 type $\frac{1}{6}$, $\frac{2}{6}$, and $\frac{3}{6}$ of the material, respectively, and that the probabilities that a sheet of their work contains at least one undetected error are respectively .03, .02, and .01. One sheet of paper is arbitrarily selected from the files. What is the probability that it contains an error? Suppose that after examining the sheet of paper an error is found. What is the probability that it was the work of secretary 2?

2-55   A student takes a five-answer multiple-choice examination. Suppose the student's knowledge of the material is such that the probability is .6 that he knows the answer to any given question, in which case he answers the question correctly with probability 1. If he does not know the answer, he guesses but his knowledge of the material is sufficient to make the probability of a correct guess $\frac{1}{2}$. What is the probability that a student correctly answers a given question? If the student correctly answers a given question, what is the probability that he knew the answer?

2-56   Suppose that the probability that an electric circuit is overloaded is .01. The current passes through a fuse that breaks with probability .99 when an overload occurs. When no overload occurs, the probability that the fuse does not break is also .99. Given that a fuse has broken, find the probability that the circuit has been overloaded. If the fuse has not broken, what is the probability of an overload?

2-57   The State University gives an entrance examination to all prospective students with .6 of them passing. Of those who do pass, .8 eventually get their bachelor's degree. The University is required by law to admit those who fail the entrance examination. Of these, .1 eventually graduate. What is the probability that a student who takes the entrance examination will graduate? Suppose we know that a given student has graduated. What is the probability that he passed the entrance examination?

2-58   A first urn contains four white balls and two black balls. The balls are well mixed and one is selected without looking at the color and transferred to a second urn that contains two white balls and three black balls. Then, after mixing the balls in the second urn, one is drawn and is black. What is the probability that the transferred ball was black?

2-2 If $\bar{A}$ is the complement of A,

$$Pr(A) = 1 - Pr(\bar{A}) \tag{2-1}$$

2-3 *The addition theorem*

For any two events A_1 and A_2,

$$Pr(A_1 \cup A_2) = Pr(A_1) + Pr(A_2) - Pr(A_1A_2) \tag{2-2}$$

For any three events A_1, A_2, and A_3,

$$Pr(A_1 \cup A_2 \cup A_3) = Pr(A_1) + Pr(A_2) + Pr(A_3) - Pr(A_1A_2)$$
$$- Pr(A_1A_3) - Pr(A_2A_3) + Pr(A_1A_2A_3) \tag{2-9}$$

For two mutually exclusive events A_1 and A_2,

$$Pr(A_1 \cup A_2) = Pr(A_1) + Pr(A_2) \tag{2-3}$$

For n mutually exclusive events $A_1, A_2, \ldots, A_n$,

$$Pr(A_1 \cup A_2 \cup \cdots \cup A_n) = Pr(A_1) + Pr(A_2) + \cdots + Pr(A_n) \tag{2-4}$$

If n events are mutually exclusive and exhaustive, then (2-4) becomes

$$Pr(S) = Pr(A_1) + Pr(A_2) + \cdots + Pr(A_n) = 1 \tag{2-5}$$

2-4 *The multiplication theorem*

For any two events A_1 and A_2,

$$Pr(A_1A_2) = Pr(A_1)Pr(A_2|A_1) \tag{2-11}$$

For any n events $A_1, A_2, \ldots, A_n$,

$$Pr(A_1A_2 \cdots A_n) = Pr(A_1)Pr(A_2|A_1)Pr(A_3|A_1A_2) \cdots$$
$$Pr(A_n|A_1A_2 \cdots A_{n-1}) \tag{2-14}$$

If $Pr(A_1) \neq 0$, then

$$Pr(A_2|A_1) = \frac{Pr(A_1A_2)}{Pr(A_1)} \tag{2-10}$$

2-5 For two independent events A_1, A_2,

$$Pr(A_1A_2) = Pr(A_1)Pr(A_2) \tag{2-18}$$

For n independent events $A_1, A_2, \ldots, A_n$,

$$Pr(A_1A_2 \cdots A_n) = Pr(A_1)Pr(A_2) \cdots Pr(A_n) \tag{2-19}$$

57

2-6   The number of permutations of n things taken n at a time is

$$_nP_n = n! \tag{2-22}$$

The number of permutations of n things taken r at a time, $r \leq n$, is

$$_nP_r = \frac{n!}{(n-r)!} \tag{2-24}$$

The number of permutations of n things taken n at a time if n_1 are alike, n_2 are alike, ... , n_k are alike is

$$P = \frac{n!}{n_1!n_2! \cdots n_k!} \tag{2-25}$$

The number of combinations of n things taken r at a time is

$$\binom{n}{r} = \frac{n!}{r!(n-r)!} \tag{2-26}$$

2-7   If the event A can occur in combination with any one of k mutually exclusive events $B_1, B_2, \ldots , B_k$, then

$$\Pr(A) = \Pr(B_1)\Pr(A|B_1) + \Pr(B_2)\Pr(A|B_2) + \cdots \\ + \Pr(B_k)\Pr(A|B_k) \tag{2-28}$$

If $B_1, B_2, \ldots , B_k$ are k mutually exclusive events that exhaust all possibilities, and A is any event with positive probabilities, then (Bayes' theorem)

$$\Pr(B_i|A) \\ = \frac{\Pr(B_i)\Pr(A|B_i)}{\Pr(B_1)\Pr(A|B_1) + \Pr(B_2)\Pr(A|B_2) + \cdots + \Pr(B_k)\Pr(A|B_k)} \tag{2-30}$$

CHAPTER THREE

Random Variables, Probability Distributions, and Expectation

3-1 RANDOM VARIABLES AND PROBABILITY DISTRIBUTIONS

In the first two chapters we have encountered a number of random experiments. Our examples have included dice throwing, coin tossing, drawing numbers from a hat, drawing balls from an urn, and dealing hands of cards. Sometimes, as in the case of rolling a single die, the points of the sample space we selected were identified by a number. In other situations the sample points were not identified by numbers but by a description. The sample space we have usually selected to describe the results when a coin is thrown twice is of the latter type. Here we named the points of the sample space *HH*, *HT*, *TH*, and *TT*. In applications of probability we are often interested in a number associated with the outcome of a random experiment. Such a quantity whose value is determined by the outcome of a random experiment is called a *random variable*.

59

Table 3-1 Probability distribution for a single die

x	1	2	3	4	5	6
$\Pr(X = x)$	$\frac{1}{6}$	$\frac{1}{6}$	$\frac{1}{6}$	$\frac{1}{6}$	$\frac{1}{6}$	$\frac{1}{6}$

Let us consider some examples of random variables. We have already cited the die example. If a single ordinary die is thrown and the number of spots showing is X, then X is a random variable that can take on the values 1, 2, 3, 4, 5, 6. In addition, X^2, $X + 1$, or any other function of X is a random variable. If two dice are thrown, then the random variable that we would most likely consider is the sum of the spots, say Y. The random variable Y, which is of interest in the game of craps, can assume any of the values 2, 3, . . . , 12. There is, of course, no reason why we could not just as well consider the difference of the spots. If we let Z be the largest minus the smallest number showing, then Z is a random variable that can assume any of the values 0, 1, 2, 3, 4, 5. Other examples of random variables are the number of heads showing in ten tosses of a coin, the number of aces in a five-card hand, and the number of white balls obtained if three balls are drawn from an urn.

A table listing all possible values that a random variable can take on together with the associated probabilities is called a *probability distribution*. The probability distribution for the random variable X, where X is the outcome produced by a single throw of a standard balanced die, is given in Table 3-1. We note that the sum of all probabilities in a probability distribution is 1. The same information can be conveniently expressed by the formula

$$\Pr(X = x) = f(x) = \tfrac{1}{6} \qquad x = 1, 2, 3, 4, 5, 6$$

A formula, or function $f(x)$, from which probabilities associated with various values of a random variable can be obtained, is called a *probability function*. As we shall see, many important probability functions are tabulated, but quite often in another form. Usually tables contain

$$F(x) = \Pr(X \le x) = \sum_{X \le x} f(x)$$

which is known as the *cumulative distribution* or *distribution function*. Table 3-2 gives the cumulative distribution for the random variable X of Table 3-1. Thus, for example,

$$F(3) = f(1) + f(2) + f(3) = \tfrac{1}{6} + \tfrac{1}{6} + \tfrac{1}{6} = \tfrac{3}{6}$$

60

Table 3-2 Cumulative distribution for a die

x	1	2	3	4	5	6
$F(x)$	$1/6$	$2/6$	$3/6$	$4/6$	$5/6$	1

Perhaps a word about notation is worthwhile. We shall use capital letters, usually those letters near the end of the alphabet, for random variables. As we have done in the above paragraph, we let x (or, in general, small letters) denote a specific though unspecified value of X. Small letters will be used for probability functions, large letters for distribution functions. Hence distribution function $F(x)$ goes with probability function $f(x)$, $G(y)$ goes with $g(y)$.

EXAMPLE 3-1

Suppose that two ordinary symmetric dice are to be thrown. Let Y be the sum of the spots, Z be the largest number minus the smallest. Find the probability distribution and the cumulative distribution for each random variable.

Solution

In Table 3-3 we have reproduced the 36-point sample space that was used in Exercise 1-17. The first number below the description of the point is the value of Y, the second the value of Z. We have already decided that equal probabilities are reasonable for this sample space. To obtain $\Pr(Y = y)$ we add the probabilities for all sample points corresponding to the event $Y = y$. Since the event $Y = 2$ contains one point, $Y = 3$ contains two points, etc., we easily get the first two rows of Table 3-4. From these probabilities the cumulative distribution of Y is obtained and appears in the third row of the table. Similarly, we construct Table 3-5 for random variable Z. (Tables 3-3, 3-4, and 3-5 appear on page 62.)

Let us summarize what we have done to obtain the probability distribution of a random variable. First, we chose a sample space to describe the outcomes of the random experiment in such a way as to make computations of probabilities as easy as possible. Frequently this is accomplished by selecting a sample space for which it is reasonable to assign equal probabilities to all sample points. Second, we computed a value of the random variable for each sample point. Third, we listed all possible values that the random variable can assume, each value corresponding

Table 3-3 Sample space for two dice and
values of random variables Y and Z

(1,1)	(2,1)	(3,1)	(4,1)	(5,1)	(6,1)
2	3	4	5	6	7
0	1	2	3	4	5
(1,2)	(2,2)	(3,2)	(4,2)	(5,2)	(6,2)
3	4	5	6	7	8
1	0	1	2	3	4
(1,3)	(2,3)	(3,3)	(4,3)	(5,3)	(6,3)
4	5	6	7	8	9
2	1	0	1	2	3
(1,4)	(2,4)	(3,4)	(4,4)	(5,4)	(6,4)
5	6	7	8	9	10
3	2	1	0	1	2
(1,5)	(2,5)	(3,5)	(4,5)	(5,5)	(6,5)
6	7	8	9	10	11
4	3	2	1	0	1
(1,6)	(2,6)	(3,6)	(4,6)	(5,6)	(6,6)
7	8	9	10	11	12
5	4	3	2	1	0

Table 3-4 Probability distribution and cumulative distribution
for sum of spots on two dice

y	2	3	4	5	6	7	8	9	10	11	12
$f(y)$	$\frac{1}{36}$	$\frac{2}{36}$	$\frac{3}{36}$	$\frac{4}{36}$	$\frac{5}{36}$	$\frac{6}{36}$	$\frac{5}{36}$	$\frac{4}{36}$	$\frac{3}{36}$	$\frac{2}{36}$	$\frac{1}{36}$
$F(y)$	$\frac{1}{36}$	$\frac{3}{36}$	$\frac{6}{36}$	$\frac{10}{36}$	$\frac{15}{36}$	$\frac{21}{36}$	$\frac{26}{36}$	$\frac{30}{36}$	$\frac{33}{36}$	$\frac{35}{36}$	1

Table 3-5 Probability distribution and cumulative distribution for largest number minus smallest number when two dice are thrown

z	0	1	2	3	4	5
$f(z)$	$\frac{6}{36}$	$\frac{10}{36}$	$\frac{8}{36}$	$\frac{6}{36}$	$\frac{4}{36}$	$\frac{2}{36}$
$F(z)$	$\frac{6}{36}$	$\frac{16}{36}$	$\frac{24}{36}$	$\frac{30}{36}$	$\frac{34}{36}$	1

to an event in the original sample space. Finally, we found the probability of each event by adding the probabilities associated with each sample point in the event. The table listing these events and their probabilities of occurrence is then the desired probability distribution. We note that since the assignment of values for the random variable to points of the sample space divides the sample into a set of mutually exclusive and exhaustive events, the sum of the probabilities in the probability distribution necessarily must be 1.

It might occur to us to use the values of the random variable for the points in our sample space. For example, in the game of craps we are not interested in how a specific sum can be obtained, merely in the probability that a given sum will be obtained. This suggests that *the* sample space ought to be 2, 3, 4, 5, 6, 7, 8, 9, 10, 11, 12. If we are clever enough to assign the right probabilities to these 11 sample points, then there is not much need for the sample space of Table 3-3. Our main reason for starting with the sample space of Table 3-3, one with equal probabilities, is to simplify the calculation of probabilities of other events. Once we have the probability distribution of the random variable, we can regard the list of possible values as the sample points and the associated probabilities as the weights to be attached to these points. It is then easy to see that these weights satisfy conditions (1-4) so that definition (1-5) is applicable to the new sample space.

If a random variable X can take on only a finite number of values or an infinite number of values that are countable, then we call X a *discrete random variable*. In this book we shall be concerned almost entirely with discrete random variables.

EXAMPLE 3-2

Suppose the probability function for a random variable X is $f(x) = x/10$, $x = 1, 2, 3, 4$. List the values that would be included in a probability distribution and the values that would be listed in the cumulative distribution.

Solution

$$f(1) = \tfrac{1}{10} \qquad f(2) = \tfrac{2}{10} \qquad f(3) = \tfrac{3}{10} \qquad f(4) = \tfrac{4}{10}$$
$$F(1) = \tfrac{1}{10}$$
$$F(2) = \tfrac{1}{10} + \tfrac{2}{10} = \tfrac{3}{10}$$
$$F(3) = \tfrac{1}{10} + \tfrac{2}{10} + \tfrac{3}{10} = \tfrac{6}{10}$$
$$F(4) = \tfrac{1}{10} + \tfrac{2}{10} + \tfrac{3}{10} + \tfrac{4}{10} = 1$$

EXAMPLE 3-3

A gambling game that was played at small-town celebrations a few years back was known as "chuckaluck." The player bets an amount, say \$1, on a number from 1 to 6. To be specific, let us take 2. Then a cage containing two dice is spun. If two 2's appear, the player gets his dollar back with two additional dollars. If only a single 2 appears, the player receives one dollar in addition to having his bet returned. If no 2's appear, the house wins. Find the probability distribution of the number of 2's and the probability distribution of the player's winnings for a single play.

Solution

In Example 2-12 we found that the probability of getting two 2's is $\frac{1}{36}$ and the probability of getting one 2 is $\frac{10}{36}$. By subtraction, the probability of getting no 2's is $1 - \frac{1}{36} - \frac{10}{36} = \frac{25}{36}$. Letting $X =$ the number of 2's, the probability distribution of X appears in Table 3-6. Let $Y =$ the player's winnings. Since $Y = -\$1$ if $X = 0$, $Y = \$1$ if $X = 1$, and $Y = \$2$ if $X = 2$. The probability distribution of Y is as given in Table 3-7.

Table 3-6 Probability distribution of X

x	0	1	2
$f(x)$	$\frac{25}{36}$	$\frac{10}{36}$	$\frac{1}{36}$

Table 3-7 Probability distribution of Y

y	$-\$1$	$\$1$	$\$2$
$f(y)$	$\frac{25}{36}$	$\frac{10}{36}$	$\frac{1}{36}$

EXERCISES

3-1 Find the probability distribution for the number of heads appearing when two coins are tossed.

3-2 Find the probability distribution for the number of heads appearing when three coins are tossed.

3-3 The probability function for a random variable X is $f(x) = x/21$, $x = 1, 2, 3, 4, 5, 6$. Make a table of the cumulative distribution.

3-4 A hat contains 10 slips of paper numbered 1 to 10. One number is to be drawn. Let X denote the number. Find the probability distribution of X.

3-5 A blindfolded individual is given four different brands of cigarettes to smoke, told the names of the brands, and asked to identify each. Let X equal the number of correct identifications. Under the assumption that the individual possesses no discriminatory ability, find the probability distribution of X.

3-6 In Example 3-1 redefine Z to be the result on the first die minus the result on the second. Find the probability distribution of this new Z.

3-7 Suppose that the cumulative distribution of random variable X is

x	0	1	2	3	4	5
$F(x)$	.13	.27	.53	.84	.92	1

(a) Find $\Pr(X \leq 2)$, $\Pr(X = 2)$, and $\Pr(2 \leq X \leq 4)$.
(b) Find the probability distribution of X.

3-8 Suppose that the cumulative distribution of random variable X is

x	0	1	2	3	4
$F(x)$	$\frac{1}{16}$	$\frac{5}{16}$	$\frac{11}{16}$	$\frac{15}{16}$	1

(a) Find $\Pr(X \leq 3)$, $\Pr(X = 3)$, $\Pr(X = 2 \text{ or } 3)$.
(b) Find the probability distribution of X.

3-9 Find the probability distribution of the sum of the numbers appearing in the two drawings described in Example 1-7.

3-10 The four-sided die described in Exercise 1-24 is rolled twice. If X is the sum of the spots that land on the bottom in the two rolls, find the probability distribution of X.

3-11 Suppose that the die described in Exercise 1-25 is rolled twice. If X is the sum of the spots that land on the bottom in two rolls, find the probability distribution of X. Make a table of the cumulative distribution.

3-12 A coin is tossed until a head appears. Let N be the random variable representing the number of trials required. Find the probability function of N. Show that the sum of the probabilities is 1. (*Hint:* the probabilities form an infinite geometric progression with common ratio less than 1.)

3-13 An ordinary symmetric die is rolled until a 6 appears. Let N be the random variable representing the number of trials required.

Find the probability function of N. Show that the sum of the probabilities is 1.

3-14   A box contains eight screws of the same size. Three of the screws are colored silver and five are colored black. If three screws are selected from the box without observing the color, find the probability distribution of the number of silver screws contained in the three selected.

3-2   DERIVED RANDOM VARIABLES

In Example 3-3 we first found the probability distribution of X, the number of 2's showing when two dice are tossed. Although we used the multiplication and addition theorems, we would have arrived at the same result by using the sample space of Table 3-3 and the definition of probability. We next found the probability distribution of Y, the player's winnings. This also could have been obtained by using the original sample space, assigning $-\$1$, $\$1$, or $\$2$ to each point, and adding the probabilities corresponding to each event. However, having the distribution of X available, we were able to write down the distribution of Y immediately by observing that $X = 0$ and $Y = -\$1$ represent the same event in the original sample space as do $X = 1$, $Y = \$1$ and $X = 2$, $Y = \$2$.

The problem we wish to consider in this section is that of obtaining the distribution of a second random variable Y, given only the probability distribution of a first random variable X and some functional relationship between X and Y. This relationship can be in the form of a table, as in the chuckaluck example. Usually, however, X and Y will be related through a formula $Y = u(X)$; that is, $Y = 2X + 3$ or $Y = 3X^2$ (in the chuckaluck example $Y = -1 + 5X/2 - X^2/2$ would serve as well as the table).

First let us consider the case in which there is a one-to-one correspondence between the values of X and the values of Y. To be specific, let us take $f(x) = x/6$, $x = 1, 2, 3$, and $Y = X^2$. The only values of Y with nonzero probability are 1, 4, and 9 since $X = 1$ makes $Y = 1$, $X = 2$ makes $Y = 4$, and $X = 3$ makes $Y = 9$. We easily compute

$\Pr(Y = 1) = \Pr(X^2 = 1) = \Pr(X = 1) = \frac{1}{6}$
$\Pr(Y = 4) = \Pr(X^2 = 4) = \Pr(X = 2) = \frac{2}{6}$
$\Pr(Y = 9) = \Pr(X^2 = 9) = \Pr(X = 3) = \frac{3}{6}$

or in general

$g(y) = \Pr(Y = y) = \Pr(X^2 = y) = \Pr(X = \sqrt{y}) = f(\sqrt{y}) = \sqrt{y}/6$

Now let $Y = u(X)$ be any function of X that yields one value of X—say $X = w(Y)$—for each Y, and one Y for each X, and let $f(x)$ be a probability function with nonzero probabilities at $X = x_1, x_2, x_3, \ldots$. Then we calculate, as in the above specific example,

$$g(y) = \Pr(Y = y) = \Pr[u(X) = y] = \Pr[X = w(y)] = f[w(y)] \qquad (3\text{-}1)$$

for $Y = u(x_1), u(x_2), u(x_3), \ldots$. We see that Y and X have the same probability distribution. That is, when there is a one-to-one correspondence between X and Y, $f(x)$ and $g(y)$ yield exactly the same probabilities. Only the random variable and the set of values the variable can assume with nonzero probability have changed.

EXAMPLE 3-4

Let X have the probability distribution of Table 3-1. Find the probability function of $Y = X - 3$ and of $Z = 2X^2$.

Solution

$X = 1, 2, 3, 4, 5, 6$ correspond to $Y = -2, -1, 0, 1, 2, 3$, respectively. Hence we immediately get

$$g(y) = \tfrac{1}{6} \qquad y = -2, -1, 0, 1, 2, 3$$

Similarly $X = 1, 2, 3, 4, 5, 6$ correspond to $Z = 2, 8, 18, 32, 50, 72$ and

$$h(z) = \tfrac{1}{6} \qquad z = 2, 8, 18, 32, 50, 72$$

EXAMPLE 3-5

Let X have the probability function

$$f(x) = \frac{x^2 + 1}{18} \qquad x = 0, 1, 2, 3$$

Find the probability function of $Y = X^2 + 1$

Solution

It is easy to see that $X = 0, 1, 2, 3$ correspond to $Y = 1, 2, 5, 10$, respectively. Since $X = \sqrt{Y-1}$, $X^2 = Y - 1$, $X^2 + 1 = Y$, direct substitution in (3-1) yields

$$g(y) = y/18 \qquad y = 1, 2, 5, 10$$

In some interesting problems several values of the first random variable X will give rise to the same value of the second random variable Y. The procedure for finding the probability distribution of Y is almost the same as before, the difference being that it is now necessary to add the several probabilities that are associated with each value of X that produces the

Table 3-8 Probability distribution of $Y = (X - 2)^2$

y	0	1	4
$g(y)$	$3\!/_{15}$	$6\!/_{15}$	$6\!/_{15}$

same value of Y. For example, suppose that the probability function of X is

$$f(x) = \frac{x+1}{15} \qquad x = 0, 1, 2, 3, 4$$

and we seek the probability distribution of $Y = (X - 2)^2$. Now $Y = 0$ if $X = 2$ so that

$$\Pr(Y = 0) = \Pr(X = 2) = 3\!/_{15}$$

and $Y = 1$ if $X = 1$ or 3, so that

$$\Pr(Y = 1) = \Pr(X = 1) + \Pr(X = 3) = 2\!/_{15} + 4\!/_{15} = 6\!/_{15}$$

and $Y = 4$ if $X = 0$ or 4, so that

$$\Pr(Y = 4) = \Pr(X = 0) + \Pr(X = 4) = 1\!/_{15} + 5\!/_{15} = 6\!/_{15}$$

Hence, we have the $g(y)$ that appears in Table 3-8. In summary, if $X = x_1, x_2, \ldots, x_k$ all yield $Y = y$, then

$$\begin{aligned} g(y) &= \Pr(Y = y) \\ &= \Pr(X = x_1) + \Pr(X = x_2) + \cdots + \Pr(X = x_k) \end{aligned} \qquad (3\text{-}2)$$

EXAMPLE 3-6

If the probability function of X is

$$f(x) = x/10 \qquad x = 1, 2, 3, 4$$

find the probability distribution of $Y = |X - 3|$.

68

Table 3-9 *Probability distribution of* $Y = |X - 3|$

y	0	1	2
$g(y)$	$\frac{3}{10}$	$\frac{6}{10}$	$\frac{1}{10}$

Solution

We observe that if

$X = 1$	$Y = |1 - 3| = 2$
$X = 2$	$Y = |2 - 3| = 1$
$X = 3$	$Y = |3 - 3| = 0$
$X = 4$	$Y = |4 - 3| = 1$

Thus

$$\Pr(Y = 0) = \Pr(X = 3) = \tfrac{3}{10}$$
$$\Pr(Y = 1) = \Pr(X = 2) + \Pr(X = 4) = \tfrac{2}{10} + \tfrac{4}{10} = \tfrac{6}{10}$$
$$\Pr(Y = 2) = \Pr(X = 1) = \tfrac{1}{10}$$

which, in table form, yields Table 3-9.

The procedure for finding the derived distribution of Y is the same if X takes on a countable infinity of values. Usually Y will also take on a countable infinity of values, but this need not be the case as is demonstrated by Example 3-7.

EXAMPLE 3-7

If the random variable X has the probability function

$$f(x) = (\tfrac{1}{2})^x \qquad x = 1, 2, 3, \ldots$$

find the probability distribution of Y where

$$Y = 0 \qquad \text{if } X \text{ is odd}$$
$$ = 1 \qquad \text{if } X \text{ is even}$$

Solution

The probability that Y is 0 is

$$f(1) + f(3) + f(5) + \cdots = \frac{1}{2} + \left(\frac{1}{2}\right)^3 + \left(\frac{1}{2}\right)^5 + \cdots = \frac{\frac{1}{2}}{1 - (\frac{1}{2})^2} = \frac{2}{3}$$

and the probability that Y is 1 is

$$f(2) + f(4) + f(6) + \cdots = \left(\frac{1}{2}\right)^2 + \left(\frac{1}{2}\right)^4 + \left(\frac{1}{2}\right)^6 + \cdots$$

$$= \frac{(\frac{1}{2})^2}{1 - (\frac{1}{2})^2} = \frac{1}{3}$$

Hence,

$$g(y) = \frac{2}{3} \quad \text{if } y = 0$$
$$= \frac{1}{3} \quad \text{if } y = 1$$

EXERCISES

3-15 If the probability function of X is $f(x) = x/15$, $x = 1, 2, 3, 4, 5$, find the probability function of $Y = X - 3$. Find the probability distribution of $Z = |X - 3|$ and $\Pr(Z \leq 1)$.

3-16 If the random variable X has the probability function $f(x) = \frac{1}{10}$, $x = 1, 2, \ldots, 10$, find the probability function of $Y = X^2 - 1$ and $Z = 11 - X$.

3-17 If the random variable X has the probability function $f(x) = (\frac{1}{2})^x$, $x = 1, 2, 3, \ldots$, find the probability function of $Y = X^2$ and of $Z = \sqrt{X}$.

3-18 If the random variable X has the probability function $f(x) = (\frac{1}{6})(\frac{5}{6})^{x-1}$, $x = 1, 2, 3, \ldots$, find the probability function of $Y = X - 1$.

3-19 Suppose that Y has the probability distribution given in Table 3-4. A game is played with the following payoffs:

If $Y = 7$ or 11, the player wins \$2
$Y = 2, 4,$ or 8, the player wins \$1
$Y = 3, 6, 9,$ or 12, the player loses \$1
$Y = 5$ or 10, the player loses \$2

Use the probability distribution of Y to find the probability distribution of Z, the player's winnings on a single play.

3-20 Suppose that the chuckaluck game of Example 3-3 is changed by increasing the winning payoffs by \$1. Starting with the probability distribution of X given in Table 3-6, find the probability distribution of Y = the player's winnings.

3-21 Consider again the blindfolded individual already described in Exercise 3-5. If the individual makes 0 or 1 correct identifications he gets nothing. However, with 2, 3, or 4 correct identifications he receives a number of packs of cigarettes, the number being equal to the square of the number of correct identifications. Let Y = the number of packs of cigarettes the individual receives.

Starting with the probability distribution of X, find the probability distribution of Y.

3-22 Let X have the cumulative distribution of Exercise 3-8. Find the probability distribution of $Y = F(X)$ and $Z = f(X)$.

3-3 EXPECTED VALUE OF A RANDOM VARIABLE

Another important concept in probability theory that had its origin in gambling is expected value. As a simple illustration, suppose that 100 tickets are sold for $1 each to raffle off a watch worth $80. Since one ticket is worth $80 and the other 99 will pay off nothing, the average value of a ticket is

$$\frac{\$80 + \$0 + \$0 + \cdots + \$0}{100} = \$.80$$

Another way to write this fraction is

$$\$80(\tfrac{1}{100}) + \$0(\tfrac{1}{100}) + \cdots + \$0(\tfrac{1}{100}) = \$80(\tfrac{1}{100}) + \$0(\tfrac{99}{100}) \tag{3-3}$$

If we let the random variable X be the amount a ticket is worth, then the probability distribution of X is easily found to be the one appearing in Table 3-10. We observe that the right-hand side of Eq. (3-3) can be obtained from Table 3-10 by multiplying each value of X by its probability of occurrence and summing the resulting numbers.

If an individual participated in only one raffle and with a single ticket, he would receive either $0 or $80. The average amount received by all ticketholders is, as we have already seen, $.80. Now suppose that an individual buys one ticket for many such raffles of the type described. After a few thousand raffles it is quite likely that he would win a few times. To be specific, suppose that after 10,000 tries he has won 110

Table 3-10 *Probability distribution of X, the amount a ticket is worth*

x	$0	$80
$g(x)$	$\tfrac{99}{100}$	$\tfrac{1}{100}$

71

times. On the average he has received

$$\frac{\$80(110) + \$0(9,890)}{10,000} = \$80\left(\frac{110}{10,000}\right) + \$0\left(\frac{9,890}{10,000}\right)$$
$$= \$.88$$

Since $110/10,000$ and $9,890/10,000$ are relative frequencies, we would undoubtedly expect, after our discussion in Sec. 1-3, that these numbers should be close to the probabilities appearing in Table 3-10. The closer the relative frequencies are to the probabilities, the closer the average amount received after a number of raffles is to $.80. Hence, if one participated in many such raffles, in the long run one would expect to get about $.80 back for each dollar invested.

In general, we define expected value as follows: If a random variable X has possible outcomes $x_1, x_2, \ldots, x_k$ that occur with probabilities $f(x_1), f(x_2), \ldots, f(x_k)$, respectively, then the *expected value* of X (sometimes called the expectation of X or the mean of the probability distribution) is

$$E(X) = x_1 f(x_1) + x_2 f(x_2) + \cdots + x_k f(x_k)$$
$$= \sum_{i=1}^{k} x_i f(x_i) = \sum_{i=1}^{k} x_i \Pr(X = x_i) \tag{3-4}$$

If the random variable X has an infinite number of outcomes $x_1, x_2, x_3, \ldots$, which can be put into one-to-one correspondence with the positive integers, then

$$E(X) = x_1 f(x_1) + x_2 f(x_2) + x_3 f(x_3) + \cdots$$
$$= \sum_{i=1}^{\infty} x_i f(x_i) = \sum_{i=1}^{\infty} x_i \Pr(X = x_i) \tag{3-5}$$

To cover both cases (3-4) and (3-5) we could also write

$$E(X) = \sum_x x f(x) = \sum_x x \Pr(X = x)$$

with the understanding that the summation is taken over all values of X that occur with nonzero probability. As we have seen using the raffle example, $E(X)$ does not have to be equal to a value the random variable can assume nor is it the value that we expect to get on any one performance of the experiment. If the experiment is repeated many times and the observed values of the random variable are averaged (the sum is divided by the number of repetitions of the experiment), then we do expect that this average will be close to $E(X)$. It is this latter interpretation that justifies the suggestive name "expected value" for $E(X)$.

The Greek letter μ (mu) is often used to denote the expected value of a random variable. Thus $E(X)$ and μ_X are notations that stand for the same quantity. When there is no possibility for confusion (that is, when it is clear which random variable is under consideration), the subscript is sometimes omitted from μ. Hence

$$E(X) = \mu_X = \mu$$

EXAMPLE 3-8

Find the expected value of the random variable whose probability distribution is given in Table 3-1.

Solution

By definition (3-4) we have

$$E(X) = 1(\tfrac{1}{6}) + 2(\tfrac{1}{6}) + 3(\tfrac{1}{6}) + 4(\tfrac{1}{6}) + 5(\tfrac{1}{6}) + 6(\tfrac{1}{6}) = {}^{21}\!/_6 = \tfrac{7}{2}$$

EXAMPLE 3-9

Find the expected value of the random variable whose probability distribution appears in Table 3-4.

Solution

We have

$$\begin{aligned}
E(Y) = {}&2(\tfrac{1}{36}) + 3(\tfrac{2}{36}) + 4(\tfrac{3}{36}) + 5(\tfrac{4}{36}) + 6(\tfrac{5}{36}) + 7(\tfrac{6}{36}) \\
&+ 8(\tfrac{5}{36}) + 9(\tfrac{4}{36}) + 10(\tfrac{3}{36}) + 11(\tfrac{2}{36}) + 12(\tfrac{1}{36}) \\
&\hspace{6cm} = {}^{252}\!/_{36} = 7
\end{aligned}$$

EXAMPLE 3-10

For the chuckaluck game of Example 3-3 find $E(Y)$, the expected winnings on a single play.

Solution

We have

$$E(Y) = -\$1\left(\frac{25}{36}\right) + \$1\left(\frac{10}{36}\right) + \$2\left(\frac{1}{36}\right) = -\frac{\$13}{36} = -\$.36$$

In other words, if the game is played a large number of times, the player would expect to lose on the average about 36 cents for each play. For most games played in gambling casinos, this figure is about 2 to 4 cents on the dollar.

EXAMPLE 3-11

If the random variable X has the probability function $f(x) = (\tfrac{1}{2})^x$, $x = 1, 2, 3, \ldots$, find $E(X)$.

73

Solution

By definition (3-5) we have

$$E(X) = 1(\tfrac{1}{2}) + 2(\tfrac{1}{2})^2 + 3(\tfrac{1}{2})^3 + 4(\tfrac{1}{2})^4 + \cdots$$
$$= \tfrac{1}{2}[1 + 2(\tfrac{1}{2}) + 3(\tfrac{1}{2})^2 + 4(\tfrac{1}{2})^3 + \cdots]$$

According to formula (A2-5) of Appendix A2 we know that

$$\frac{1}{(1-a)^n} = 1 + \binom{n}{1}a + \binom{n+1}{2}a^2 + \binom{n+2}{3}a^3 + \cdots$$

provided that $-1 < a < 1$. In particular, if $n = 2$ this becomes

$$\frac{1}{(1-a)^2} = 1 + 2a + 3a^2 + 4a^3 + \cdots$$

Hence

$$1 + 2(\tfrac{1}{2}) + 3(\tfrac{1}{2})^2 + 4(\tfrac{1}{2})^3 + \cdots = \frac{1}{(1-\tfrac{1}{2})^2} = 4$$

and

$$E(X) = \tfrac{1}{2}(4) = 2$$

EXERCISES

3-23 If the probability function of X is $f(x) = x/15$, $x = 1, 2, 3, 4, 5$, find $E(X)$.

3-24 If the random variable X has the probability function $f(x) = \tfrac{1}{10}$, $x = 1, 2, 3, \ldots, 10$, find $E(X)$.

3-25 In Exercise 3-19 find $E(Z)$.

3-26 In Exercise 3-20 find $E(Y)$.

3-27 In Exercise 3-5 find $E(X)$.

3-28 In Exercise 3-6 find $E(Z)$.

3-29 In Exercise 3-7 find $E(X)$.

3-30 If the random variable X has the probability function $f(x) = (\tfrac{1}{6})(\tfrac{5}{6})^{x-1}$, $x = 1, 2, 3, \ldots$, find $E(X)$. What is the answer if $f(x) = p(1-p)^{x-1}$, $x = 1, 2, 3, \ldots$, and $0 < p < 1$?

3-31 In Exercise 3-21 find $E(Y)$.

3-32 Suppose that the chuckaluck game of Example 3-3 is played with three dice. If the player's number shows on all three dice, he gets his dollar back with 3 additional dollars. If the number shows on two dice, he gets his dollar back with 2 additional dollars. If his number appears once he gets his dollar back plus 1 additional dollar. When none of the dice show his number, he loses his money. Find the expected winnings for one play of the game.

3-33 In Exercise 2-26 find the expected number of games required to terminate the series.

3-34 A box contains two silver screws and three black screws, all the same size. They are drawn from the box one at a time without looking into the box. What is the expected number of drawings required to remove both silver screws?

3-4 EXPECTED VALUE OF A FUNCTION OF A RANDOM VARIABLE; THE VARIANCE

Suppose we consider again the random variable X with the probability function

$$f(x) = \frac{x+1}{15} \qquad x = 0, 1, 2, 3, 4$$

In Sec. 3-2 we found the probability distribution of $Y = (X - 2)^2$ to be the one that appears in Table 3-8. If we seek $E[(X - 2)^2] = E(Y)$, we easily find from the probability distribution of Y

$$E(Y) = 0(\tfrac{3}{15}) + 1(\tfrac{6}{15}) + 4(\tfrac{6}{15}) = 2$$

It seems natural to inquire as to whether $E[(X - 2)^2]$ could have been found directly from the probability distribution of X, making it unnecessary to find the probability distribution of Y as an intermediate step. If we define $E[u(X)]$ by

$$E[u(X)] = \sum_x u(x)\Pr(X = x)$$

then we can demonstrate and prove that $E[u(X)]$ so defined and $E(Y)$ are equal. For example, using the random variables X and Y defined above we can write

$$
\begin{aligned}
E[u(X)] &= E[(X - 2)^2] \\
&= (0 - 2)^2(\tfrac{1}{15}) + (1 - 2)^2(\tfrac{2}{15}) + (2 - 2)^2(\tfrac{3}{15}) \\
&\quad + (3 - 2)^2(\tfrac{4}{15}) + (4 - 2)^2(\tfrac{5}{15}) \\
&= 0(\tfrac{3}{15}) + 1[(\tfrac{2}{15}) + (\tfrac{4}{15})] + 4[(\tfrac{1}{15}) + (\tfrac{5}{15})] \\
&= 0(\tfrac{3}{15}) + 1(\tfrac{6}{15}) + 4(\tfrac{6}{15}) \\
&= E(Y)
\end{aligned}
$$

The generalization of this result is the following theorem:

If a random variable X has possible outcomes $x_1, x_2, \ldots , x_k$, which

75

occur with probabilities $f(x_1)$, $f(x_2)$, . . . , $f(x_k)$, respectively, and $Y = u(X)$, then the expected value of $Y = u(X)$ is

$$E(Y) = E[u(X)] = u(x_1)f(x_1) + u(x_2)f(x_2) + \cdots + u(x_k)f(x_k)$$
$$= \sum_{i=1}^{k} u(x_i)f(x_i) \qquad (3\text{-}6)$$

In constructing a proof of the theorem, let us use the above example as a guide. If we let $x_1 = 0$, $x_2 = 1$, $x_3 = 2$, $x_4 = 3$, $x_5 = 4$, $y_1 = 0$, $y_2 = 1$, $y_3 = 4$, we can write

$$E[u(X)] = u(x_1)f(x_1) + u(x_2)f(x_2) + u(x_3)f(x_3) + u(x_4)f(x_4) + u(x_5)f(x_5)$$
$$= y_1 f(x_3) + y_2 [f(x_2) + f(x_4)] + y_3 [f(x_1) + f(x_5)]$$
$$= y_1 g(y_1) + y_2 g(y_2) + y_3 g(y_3)$$
$$= E(Y)$$

Hence, in general we have

$$E[u(X)] = \sum_{i=1}^{k} u(x_i)f(x_i)$$
$$= y_1 \text{ [sum of } f(x_i) \text{ for which } u(x_i) = y_1]$$
$$+ y_2 \text{ [sum of } f(x_i) \text{ for which } u(x_i) = y_2]$$
$$\cdot \ \cdot \ \cdot \ \cdot \ \cdot \ \cdot \ \cdot \ \cdot \ \cdot \ \cdot \ \cdot \ \cdot \ \cdot \ \cdot \ \cdot \ \cdot$$
$$+ y_j \text{ [sum of } f(x_i) \text{ for which } u(x_i) = y_j]$$

where y_j represents the last value of y and $j \leq k$. But from our discussion of derived distributions we know that

[sum of $f(x_i)$ for which $u(x_i) = y_1$] = [probability $Y = y_1$] = $g(y_1)$
[sum of $f(x_i)$ for which $u(x_i) = y_2$] = [probability $Y = y_2$] = $g(y_2)$
$\cdot \ \cdot$
[sum of $f(x_i)$ for which $u(x_i) = y_j$] = [probability $Y = y_j$] = $g(y_j)$

Thus

$$E[u(X)] = y_1 g(y_1) + y_2 g(y_2) + \cdots + y_j g(y_j)$$
$$= E[Y]$$

If X can take on a countable infinity of values x_1, x_2, x_3, . . . and $Y = u(X)$, then minor modifications in the above argument will still yield $E[u(X)] = E(Y)$.

EXAMPLE 3-12

If the random variable X has the probability function $f(x) = \frac{1}{5}$, $x = 1$, 2, 3, 4, 5, find $E(X)$, $E(X - 3)$, $E(|X - 3|)$, and $E[(X - 3)^2]$.

Solution

We have

$$E(X) = 1(\tfrac{1}{5}) + 2(\tfrac{1}{5}) + 3(\tfrac{1}{5}) + 4(\tfrac{1}{5}) + 5(\tfrac{1}{5}) = {}^{15}\!\!/_{5} = 3$$

$$\begin{aligned}
E(X - 3) &= (1 - 3)(\tfrac{1}{5}) + (2 - 3)(\tfrac{1}{5}) + (3 - 3)(\tfrac{1}{5}) + (4 - 3)(\tfrac{1}{5}) \\
&\quad + (5 - 3)(\tfrac{1}{5}) \\
&= (-2)(\tfrac{1}{5}) + (-1)(\tfrac{1}{5}) + 0(\tfrac{1}{5}) + 1(\tfrac{1}{5}) + 2(\tfrac{1}{5}) = 0
\end{aligned}$$

$$\begin{aligned}
E(|X - 3|) &= |1 - 3|(\tfrac{1}{5}) + |2 - 3|(\tfrac{1}{5}) + |3 - 3|(\tfrac{1}{5}) + |4 - 3|(\tfrac{1}{5}) \\
&\quad + |5 - 3|(\tfrac{1}{5}) \\
&= 2(\tfrac{1}{5}) + 1(\tfrac{1}{5}) + 0(\tfrac{1}{5}) + 1(\tfrac{1}{5}) + 2(\tfrac{1}{5}) = {}^{6}\!\!/_{5} = 1.2
\end{aligned}$$

$$\begin{aligned}
E[(X - 3)^2] &= (1 - 3)^2(\tfrac{1}{5}) + (2 - 3)^2(\tfrac{1}{5}) + (3 - 3)^2(\tfrac{1}{5}) \\
&\quad + (4 - 3)^2(\tfrac{1}{5}) + (5 - 3)^2(\tfrac{1}{5}) \\
&= 4(\tfrac{1}{5}) + 1(\tfrac{1}{5}) + 0(\tfrac{1}{5}) + 1(\tfrac{1}{5}) + 4(\tfrac{1}{5}) = {}^{10}\!\!/_{5} = 2
\end{aligned}$$

EXAMPLE 3-13

If the random variable X has the probability function $f(x) = x/10$, $x = 1$, 2, 3, 4, find $E(X)$, $E(X - 3)$, $E(|X - 3|)$, and $E[(X - 3)^2]$.

Solution

We have

$$E(X) = 1(\tfrac{1}{10}) + 2(\tfrac{2}{10}) + 3(\tfrac{3}{10}) + 4(\tfrac{4}{10}) = {}^{30}\!\!/_{10} = 3$$

$$\begin{aligned}
E(X - 3) &= (1 - 3)(\tfrac{1}{10}) + (2 - 3)(\tfrac{2}{10}) + (3 - 3)(\tfrac{3}{10}) \\
&\quad + (4 - 3)(\tfrac{4}{10}) \\
&= (-2)(\tfrac{1}{10}) + (-1)(\tfrac{2}{10}) + 0(\tfrac{3}{10}) + 1(\tfrac{4}{10}) = 0
\end{aligned}$$

$$\begin{aligned}
E(|X - 3|) &= |1 - 3|(\tfrac{1}{10}) + |2 - 3|(\tfrac{2}{10}) + |3 - 3|(\tfrac{3}{10}) + |4 - 3|(\tfrac{4}{10}) \\
&= 2(\tfrac{1}{10}) + 1(\tfrac{2}{10}) + 0(\tfrac{3}{10}) + 1(\tfrac{4}{10}) = {}^{8}\!\!/_{10} = .8
\end{aligned}$$

$$\begin{aligned}
E[(X - 3)^2] &= (1 - 3)^2(\tfrac{1}{10}) + (2 - 3)^2(\tfrac{2}{10}) + (3 - 3)^2(\tfrac{3}{10}) \\
&\quad + (4 - 3)^2(\tfrac{4}{10}) \\
&= 4(\tfrac{1}{10}) + 1(\tfrac{2}{10}) + 0(\tfrac{3}{10}) + 1(\tfrac{4}{10}) = {}^{10}\!\!/_{10} = 1
\end{aligned}$$

In Examples 3-12 and 3-13 we found that $E(X - \mu) = 0$ [recall $E(X) = \mu$]. It is easy to see that this property is true for any discrete random variable since

$$E(X - \mu) = \sum_{i=1}^{k} (x_i - \mu)f(x_i)$$

$$\begin{aligned}
&= (x_1 - \mu)f(x_1) + (x_2 - \mu)f(x_2) + \cdots + (x_k - \mu)f(x_k) \\
&= [x_1 f(x_1) + x_2 f(x_2) + \cdots + x_k f(x_k)] \\
&\quad - \mu[f(x_1) + f(x_2) + \cdots + f(x_k)] \\
&= E(X) - \mu[1] = 0
\end{aligned}$$

Another constant which is sometimes used to help characterize a probability distribution is

$$E[(X - \mu)^2] = \sum_{i=1}^{k} (x_i - \mu)^2 f(x_i) \tag{3-7}$$

called the *variance*. We evaluated this expected value in both Examples 3-12 and 3-13, getting 2 and 1 respectively. The variance can be interpreted as a measure of spread or dispersion about μ, the mean of the probability distribution of X. In Example 3-13 those values of X near μ (2 and 4) occur with higher probabilities than in the probability distribution of Example 3-12. Conversely, those values of X further away from the mean occur with higher probability in Example 3-12 than in Example 3-13. If large values of $(X - \mu)^2$ occur with low probability and small values with high probability, we would expect to get a smaller variance than if the reverse were true. The two examples illustrate this fact. It is in this sense that we can look at the variance as a measure of dispersion.

Several notations are used for the variance, the most common of which are $\text{Var}(X)$ and σ_X^2. When there is no possibility of confusion, the subscript is sometimes omitted from the sigma symbol. Hence

$$E[(X - \mu)^2] = \text{Var}(X) = \sigma_X^2 = \sigma^2$$

The square root of the variance is called the *standard deviation*. This, too, can be regarded as a measure of dispersion since it is large or small depending upon whether the variance is large or small. For the random variable of Example 3-12, $\sigma = \sqrt{2} \cong 1.414$; and for the random variable of Example 3-13, $\sigma^2 = \sigma = 1$.

We note that the variance cannot be negative since it is the sum of nonnegative numbers. It could take on the value 0 if the probability distribution is *degenerate*, that is, if it concentrates all its probability at one point. Thus if

$$f(x) = 1 \qquad x = 3$$
$$ = 0 \qquad \text{otherwise}$$

we find $E(X) = 3(1) = 3, E[(X - 3)^2] = 0(1) = 0$. Hence the inequality

$$E[(X - \mu)^2] \geqq 0 \tag{3-8}$$

is always satisfied.

The variance can be expressed in another form that is more adaptable to computation in cases for which the quantities $(x_i - \mu)$ are not integers. We shall show that

$$E[(X - \mu)^2] = E(X^2) - \mu^2 \tag{3-9}$$

78

To get this result start with the definition

$$E[(X - \mu)^2] = \sum_{i=1}^{k} (x_i - \mu)^2 f(x_i)$$

$$= (x_1 - \mu)^2 f(x_1) + \cdots + (x_k - \mu)^2 f(x_k)$$

and expand each of the k binomial quantities. This yields

$$
\begin{aligned}
E[(X - \mu)^2] &= (x_1^2 - 2\mu x_1 + \mu^2) f(x_1) + (x_2^2 - 2\mu x_2 + \mu^2) f(x_2) \\
&\quad + \cdots + (x_k^2 - 2\mu x_k + \mu^2) f(x_k) \\
&= [x_1^2 f(x_1) + x_2^2 f(x_2) + \cdots + x_k^2 f(x_k)] \\
&\quad - 2\mu[x_1 f(x_1) + x_2 f(x_2) + \cdots + x_k f(x_k)] \\
&\quad + \mu^2[f(x_1) + f(x_2) + \cdots + f(x_k)] \\
&= E(X^2) - 2\mu[\mu] + \mu^2[1] \\
&= E(X^2) - \mu^2
\end{aligned}
$$

as we wanted to show.

EXAMPLE 3-14

For the random variable X whose probability function is given in Example 3-13, find the variance using formula (3-9).

Solution

We already found $E(X) = \mu = 3$. Next

$$E(X^2) = 1^2(\tfrac{1}{10}) + 2^2(\tfrac{2}{10}) + 3^2(\tfrac{3}{10}) + 4^2(\tfrac{4}{10}) = {}^{100}\!\!/_{10} = 10$$

so that

$$\sigma^2 = E(X^2) - \mu^2 = 10 - 3^2 = 1$$

as before.

In deriving formula (3-9) we have demonstrated that the following more general result is true:

$$E[u_1(X) + u_2(X) + u_3(X)] = E[u_1(X)] + E[u_2(X)] + E[u_3(X)] \quad (3\text{-}10)$$

Since

$$
\begin{aligned}
E[u_1(X) + u_2(X) + u_3(X)] &= \sum_{i=1}^{k} [u_1(x_i) + u_2(x_i) + u_3(x_i)] f(x_i) \\
&= \sum_{i=1}^{k} u_1(x_i) f(x_i) + \sum_{i=1}^{k} u_2(x_i) f(x_i) \\
&\quad + \sum_{i=1}^{k} u_3(x_i) f(x_i)
\end{aligned}
$$

by writing out the first summation and regrouping terms, the result follows. Hence, for example, $E(X^2 + X) = E(X^2) + E(X)$.

Three other useful formulas for expected values are

$$E(a) = a \tag{3-11}$$
$$E[au(X)] = aE[u(X)] \tag{3-12}$$

and

$$E[au(X) + b] = aE[u(X)] + b \tag{3-13}$$

where a and b are constants. All three results follow from (3-6) and the properties of summation.

We could also use $E(|X - \mu|)$, evaluated in Examples 3-12 and 3-13, as a measure of dispersion. That is, this quantity measures dispersion in the same sense as does the variance (or the standard deviation). The variance, however, has better mathematical properties.

EXERCISES

3-35 Find the variance of the random variable X of Exercise 3-23. Find $E(|X - \mu|)$.

3-36 Find the variance of the random variable X of Exercise 3-24. Find $E(|X - \mu|)$.

3-37 Find the variance of the random variable Z of Exercise 3-19. The mean was found in Exercise 3-25.

3-38 Find the variance of the random variable X of Exercise 3-5. The mean was found in Exercise 3-27.

3-39 Find the variance of the random variable Z of Exercise 3-6. The mean was found in Exercise 3-28.

3-40 Find the variance of the random variable X of Exercise 3-7. The mean was found in Exercise 3-29.

3-41 Find the mean and variance of the random variable X of Exercise 3-9.

3-42 Find the variance of the random variable Y of Table 3-7. The mean was found in Example 3-10.

3-43 Find the variance of the random variable of Exercise 3-34.

3-44 In Exercise 3-30 we considered the random variable X whose probability function is $f(x) = p(1 - p)^{x-1}$, $x = 1, 2, 3, \ldots$ and found $E(X) = 1/p$. To find $E(X^2)$ we write

$$
\begin{aligned}
E(X^2) &= 1^2 p + 2^2 p(1 - p) + 3^2 p(1 - p)^2 + 4^2 p(1 - p)^3 + \cdots \\
&= p[1 + 4(1 - p) + 9(1 - p)^2 + 16(1 - p)^3 + \cdots] \\
&= p[1 + 3(1 - p) + 6(1 - p)^2 + 10(1 - p)^3 + \cdots] \\
&\quad + (1 - p) + 3(1 - p)^2 + 6(1 - p)^3 + \cdots \\
&= p(1 + 1 - p)[1 + 3(1 - p) + 6(1 - p)^2 \\
&\quad + 10(1 - p)^3 + \cdots]
\end{aligned}
$$

Compare the sum in the latter bracket with the expansion of $1/(1 - a)^3$ [see formula (A2-10) of Appendix A2]. Hence find $E(X^2)$ and the variance of X. What is the variance of X if $p = \frac{1}{6}$?

3-45 Let X be a random variable with probability function $f(x) = 1/k$, $x = 1, 2, 3, \ldots, k$, where k is any integer. Given that

$$1 + 2 + 3 + \cdots + k = \frac{k(k + 1)}{2}$$

$$1^2 + 2^2 + 3^2 + \cdots + k^2 = \frac{k(k + 1)(2k + 1)}{6}$$

find the mean and variance of X.

3-46 Let X be a random variable with probability function $f(x) = 2x/k(k + 1)$, $x = 1, 2, 3, \ldots, k$, where k is any integer. Given the sums in Exercise 3-45 and also that

$$1^3 + 2^3 + 3^3 + \cdots + k^3 = \left[\frac{k(k + 1)}{2}\right]^2$$

verify that the sum of the probabilities is 1 and find the mean and variance of X.

3-47 Let X be a random variable with probability function $f(x) = x/10$, $x = 1, 2, 3, 4$, and let $Y = X^2$. Find the mean and variance of Y without finding the probability distribution of Y.

3-5 CHEBYSHEV'S INEQUALITY

Although the concept of variance is probably more useful in statistical applications than in probability problems, the inequality we are going to discuss in this section demonstrates the use of variance in a probability theorem. That theorem is as follows:

Let the random variable X have mean μ and variance $\sigma^2 > 0$ and let a be any positive number. Then

$$\Pr(|X - \mu| \geq a\sigma) \leq \frac{1}{a^2} \tag{3-14}$$

That is, the probability that the random variable X differs from μ by a times the standard deviation or more is, at most, $1/a^2$. The inequality (3-14) is known as Chebyshev's inequality.

Before attempting to prove this theorem let us do a few calculations. Suppose we know that X has the probability function $f(x) = x/10$, $x = 1$, 2, 3, 4. We have already found that $\mu = E(X) = 3$, $\sigma^2 = \sigma = 1$. Then we easily verify that

$$\Pr(|X - 3| \geq \sigma) = \Pr(|X - 3| \geq 1) = \Pr(X = 1) + \Pr(X = 2)$$
$$+ \Pr(X = 4)$$
$$= \tfrac{1}{10} + \tfrac{2}{10} + \tfrac{4}{10} = \tfrac{7}{10}$$
$$\Pr(|X - 3| \geq 2\sigma) = \Pr(|X - 3| \geq 2) = \Pr(X = 1) = \tfrac{1}{10}$$
$$\Pr(|X - 3| \geq 3\sigma) = \Pr(|X - 3| \geq 3) = 0$$
$$\Pr\left(|X - 3| \geq \frac{\sigma}{2}\right) = \Pr\left(|X - 3| \geq \frac{1}{2}\right) = \Pr(X = 1) + \Pr(X = 2)$$
$$+ \Pr(X = 4)$$
$$= \frac{7}{10}$$
$$\Pr\left(|X - 3| \geq \frac{3\sigma}{2}\right) = \Pr\left(|X - 3| \geq \frac{3}{2}\right) = \Pr(X = 1) = \frac{1}{10}$$

Since the probability distribution is known we can calculate any probabilities we desire, and inequality (3-14) serves no useful purpose. However, let us compare some exact results with this information furnished by Chebyshev's inequality.

We found	Chebyshev's inequality yields
$\Pr(|X - 3| \geq \sigma) = \tfrac{7}{10}$	$\Pr(|X - 3| \geq \sigma) \leq 1$
$\Pr(|X - 3| \geq 2\sigma) = \tfrac{1}{10}$	$\Pr(|X - 3| \geq 2\sigma) \leq \tfrac{1}{4}$
$\Pr(|X - 3| \geq 3\sigma) = 0$	$\Pr(|X - 3| \geq 3\sigma) \leq \tfrac{1}{9}$
$\Pr\left(|X - 3| \geq \dfrac{\sigma}{2}\right) = \tfrac{7}{10}$	$\Pr\left(|X - 3| \geq \dfrac{\sigma}{2}\right) \leq 4$
$\Pr\left(|X - 3| \geq \dfrac{3\sigma}{2}\right) = \tfrac{1}{10}$	$\Pr\left(|X - 3| \geq \dfrac{3\sigma}{2}\right) \leq \tfrac{4}{9}$

It is easy to see that, for all values of a that we tried, (3-14) is true. Further, if $a \leq 1$, $1/a^2 \geq 1$, and the inequality tells us absolutely nothing since any probability is less than or equal to 1. The important point to observe is that the inequality (3-14) holds not just for the random variable X that we considered but for every random variable that has a mean and a variance.

Since some of the bounds furnished by the inequality were quite different from actual results obtained in the above example, we might be inclined to suspect that a better inequality could be constructed. To show that is impossible consider the random variable X whose probability

First interval	Second interval	Third interval
$X \leqq \mu - a\sigma$	$\mu - a\sigma < X < \mu + a\sigma$	$X \geqq \mu + a\sigma$

Figure 3-1   Intervals used in derivation of Chebyshev's inequality.

distribution is

$$f(x) = \frac{1}{18} \qquad x = 1, 3$$
$$= \frac{16}{18} \qquad x = 2$$

It is easy to verify that $E(X) = 2, \sigma^2 = \frac{1}{9}, \sigma = \frac{1}{3}$, and $\Pr(|X - 2| \geqq 3\sigma)$ $= \Pr(|X - 2| \geqq 1) = \frac{1}{9}$. Thus, we have exhibited a random variable and an a for which equality is actually achieved, and hence this inequality cannot be improved in general without further assumptions about the distribution of X.

To prove the theorem we start with the definition of the variance. That is, $\sigma^2 = E[(X - \mu)^2] = \sum_{i=1}^{k} (x_i - \mu)^2 f(x_i)$. Now each value of X is in one of the three intervals (see Fig. 3-1), $X \leqq \mu - a\sigma, \mu - a\sigma < X < \mu + a\sigma, X \geqq \mu + a\sigma$, since this exhausts all possibilities. We can write

$$\sigma^2 = \Sigma_1(x_i - \mu)^2 f(x_i) + \Sigma_2(x_i - \mu)^2 f(x_i) + \Sigma_3(x_i - \mu)^2 f(x_i)$$

where the first sum is the contribution from those values of X in the first interval, the second sum from those values of X in the second interval, and the third sum from those values of X in the third interval. If we discard the second sum, which cannot be negative, we can write

$$\sigma^2 \geqq \Sigma_1(x_i - \mu)^2 f(x_i) + \Sigma_3(x_i - \mu)^2 f(x_i)$$

Further, in the first interval $x_i \leqq \mu - a\sigma, x_i - \mu \leqq -a\sigma$, so that $(x_i - \mu)^2 \geqq a^2\sigma^2$ (since $a\sigma \geqq 0$), and in the third interval $x_i \geqq \mu + a\sigma$, $x_i - \mu \geqq a\sigma$ and again $(x_i - \mu)^2 \geqq a^2\sigma^2$. Each sum is therefore decreased if each $(x_i - \mu)^2$ is replaced by $a\sigma$, and hence

$$\sigma^2 \geqq \Sigma_1(x_i - \mu)^2 f(x_i) + \Sigma_3(x_i - \mu)^2 f(x_i)$$
$$\geqq \Sigma_1 a^2\sigma^2 f(x_i) + \Sigma_3 a^2\sigma^2 f(x_i)$$
$$= a^2\sigma^2 \Sigma_1 f(x_i) + a^2\sigma^2 \Sigma_3 f(x_i)$$
$$= a^2\sigma^2 [\Sigma_1 f(x_i) + \Sigma_3 f(x_i)]$$

But the quantity in the bracket is the sum of the probabilities over interval 1 and interval 3. This is the sum over those X such that $|X - \mu| \geqq a\sigma$ and hence the quantity in the bracket is $\Pr(|X - \mu| \geqq a\sigma)$. Thus

$$\sigma^2 \geqq a^2\sigma^2 \Pr(|X - \mu| \geqq a\sigma)$$

or

$$\Pr(|X - \mu| \geq a\sigma) \leq \frac{1}{a^2}$$

Several alternative forms of (3-14) can be obtained immediately. First, if we let $a\sigma = b$ so that $a = b/\sigma$, then we get

$$\Pr(|X - \mu| \geq b) \leq \frac{\sigma^2}{b^2} \tag{3-15}$$

From formula (3-15) we confirm what our intuition has already told us. That is, the smaller the variance, the smaller is the probability that this random variable differs from its mean by a fixed amount b or more. We get another form of Chebyshev's inequality if we multiply each side of (3-14) by -1 getting

$$-\Pr(|X - \mu| \geq a\sigma) \geq -\frac{1}{a^2}$$

followed by adding 1 to each side of the inequality, which yields

$$1 - \Pr(|X - \mu| \geq a\sigma) \geq 1 - \frac{1}{a^2}$$

and then observing that $1 - \Pr(|X - \mu| \geq a\sigma) = \Pr(|X - \mu| < a\sigma)$. Hence we have

$$\Pr(|X - \mu| < a\sigma) \geq 1 - \frac{1}{a^2} \tag{3-16}$$

EXAMPLE 3-15

A random variable X has mean $\mu = 5$ and variance $\sigma^2 = 4$, but the probability distribution is unknown. What can we say about

1 $\Pr(|X - 5| \geq 6)$
2 $\Pr(|X - 5| < 6)$
3 $\Pr(-1 < X < 11)$

Solution

For *1* we can use either (3-14) or (3-15). Since $\sigma = 2$, $a = 3$ and $b = 6$. Both forms yield $\Pr(|X - 5| \geq 6) \leq \frac{1}{9}$. From formula (3-16) we have immediately $\Pr(|X - 5| < 6) \geq \frac{8}{9}$. Finally, for *3* we observe that $\Pr(-1 < X < 11) = \Pr(-1 - 5 < X - 5 < 11 - 5) = \Pr(-6 < X - 5 < 6) = \Pr(|X - 5| < 6) \geq \frac{8}{9}$.

Chebyshev's inequality is not very useful in applied problems. It has, however, proved to be an extremely useful tool in constructing proofs for certain probability theorems.

EXERCISES

3-48 Suppose we are told that a random variable X has mean $\mu = 2$ and variance $\sigma^2 = 1$.
(a) What do we know about $\Pr(|X - 2| \geq 4)$?
(b) With the result in (a), what can be said about $\Pr(|X - 2| < 4)$?
(c) What can be said about $\Pr(-2 < X < 6)$?
(d) What value of t ensures that $\Pr(|X - 2| \geq t) \leq .04$?

3-49 A random variable X has mean $\mu = 50$ and variance $\sigma^2 = 100$. What can we say about
(a) $\Pr(|X - 50| \geq 15)$
(b) $\Pr(X \geq 65) + \Pr(X \leq 35)$
(c) $\Pr(|X - 50| < 20)$
(d) $\Pr(30 < X < 70)$
(e) Values of t that make $\Pr(|X - 50| \geq t) \leq .01$

3-50 Let X be a random variable with probability function

$$f(x) = \tfrac{1}{2} \qquad x = -2, 2$$

Obviously $\Pr(|X| \geq 2) = 1$ and $\Pr(|X| < 2) = 0$. What information does Chebyshev's inequality yield concerning the latter two probabilities?

3-51 Let X be a random variable with probability distribution

x	-1	0	1
$f(x)$	p_1	p_2	p_1

where $p_1 + p_2 + p_1 = 1$. It is easy to see that $\Pr(|X| \geq 1) = 2p_1$. What information does Chebyshev's inequality yield about $\Pr(|X| \geq 1)$?

3-52 For the random variable X whose probability distribution is

x	-2	-1	0	1	2
$f(x)$	$\tfrac{1}{100}$	$\tfrac{2}{100}$	$\tfrac{94}{100}$	$\tfrac{2}{100}$	$\tfrac{1}{100}$

compare the exact value of $\Pr(|X| \geq 2)$ with the bound for this probability given by Chebyshev's inequality.

3-1 A quantity whose value is determined by the outcome of a random experiment is called a random variable. If X is a random variable, then

$$F(x) = \Pr(X \leq x)$$

is called the cumulative distribution or distribution function.

3-2 If X is a random variable with probability function $f(x)$, and $Y = u(X)$ is any single-valued function of X that can be solved for X to yield one value $X = w(Y)$, then

$$g(y) = \Pr(Y = y) = \Pr[X = w(y)] = f[w(y)] \tag{3-1}$$

If the solution yields more than one value of X, say $x_1, x_2, \ldots, x_k$, then

$$g(y) = \Pr(Y = y) = \Pr(X = x_1) + \Pr(X = x_2) \\ + \cdots + \Pr(X = x_k) \tag{3-2}$$

3-3 The expected value of X is

$$E(X) = \sum_{i=1}^{k} x_i f(x_i) = \sum_{i=1}^{k} x_i \Pr(X = x_i) \tag{3-4}$$

if X has a finite number of outcomes, and

$$E(X) = \sum_{i=1}^{\infty} x_i f(x_i) = \sum_{i=1}^{\infty} x_i \Pr(X = x_i) \tag{3-5}$$

if X has a countable infinity of outcomes.

3-4 If X has probability function $f(x)$ and $Y = u(X)$, then

$$E(Y) = \sum_{i=1}^{k} u(x_i) f(x_i) \tag{3-6}$$

If X has probability function $f(x)$ and $E(X) = \mu$, then the variance of X is

$$\text{Var}(X) = E[(X - \mu)^2] = \sum_{i=1}^{k} (x_i - \mu)^2 f(x_i) \tag{3-7}$$

The computing formula for the variance is

$$\text{Var}(X) = E(X^2) - \mu^2 \tag{3-9}$$

If Y_1, Y_2, and Y_3 are random variables, then

$$E(Y_1 + Y_2 + Y_3) = E(Y_1) + E(Y_2) + E(Y_3) \tag{3-10}$$

For a random variable X,

$$E(a) = a \qquad (3\text{-}11)$$
$$E[au(X)] = aE[u(X)] \qquad (3\text{-}12)$$
$$E[au(X) + b] = aE[u(X)] + b \qquad (3\text{-}13)$$

where a and b are constants.

3-5 For a random variable X with $E(X) = \mu$ and $\mathrm{Var}(X) = \sigma^2 > 0$, with any constant $a > 0$,

$$\Pr(|X - \mu| \geq a\sigma) \leq \frac{1}{a^2} \qquad (3\text{-}14)$$

Further, if $a\sigma = b$,

$$\Pr(|X - \mu| \geq b) \leq \frac{\sigma^2}{b^2} \qquad (3\text{-}15)$$

CHAPTER FOUR

Some Important Probability Distributions

4-1 INTRODUCTION

In this chapter we shall investigate some probability distributions which arise so frequently, both in probability theory and applications, that they are worthy of special consideration. For most of these distributions we shall show how the probability functions and useful properties of the distributions functions are derived. We shall also describe the extensive tables that now exist for these distributions and that make the evaluation of probabilities a very simple chore.

In view of the comments of Sec. 1-5 we can regard any probability distribution as a model. It would not be inconsistent with common usage to say that the model is either (1) a formula or table that gives probabilities for all values of the random variable (that is, the probability function or the probability distribution) or (2) the assumptions that are necessary to derive the probabilities. For most of the situations to be discussed in this chapter we shall start with a set of assumptions that characterize a large class of experiments and actually derive the formula that is used to compute probabilities. As the examples will illustrate, before the model is used in a specific situation, some effort should be made to justify its appropriateness.

4-2 THE BINOMIAL DISTRIBUTION

There are several reasons for beginning our study of probability distributions with the binomial. First, it probably is used in more applications than any of the others to be discussed in this chapter. Second, the formula for computing probabilities is very easy to derive. Third, it is a straightforward task to investigate the reasonableness of the assumptions required to derive the probability function. Finally, the necessary computations are simple since tables of the binomial distribution are very extensive and easy to read.

Consider a series of events or experiments that have the following properties:

(a) *The result of each experiment can be classified into one of two categories, say, success and failure.*

(b) *The probability p of a success is the same for each experiment.*   (4-1)

(c) *Each experiment is independent of all the others.*

(d) *The series of experiments is performed a fixed number of times, say, n.*

These conditions appear to be satisfied when a symmetric die is rolled 10 times, regarding a 1 or 2 as a success and using $\frac{1}{3}$ for the value of p. Now suppose we would like to know the probability of obtaining exactly four successes. One way to achieve the result is $SSSSFFFFFF$, where S and F denote success and failure. That is, we could be successful the first four times and fail the last six. The probability that this happens is

$$(\tfrac{1}{3})(\tfrac{1}{3})(\tfrac{1}{3})(\tfrac{1}{3})(\tfrac{2}{3})(\tfrac{2}{3})(\tfrac{2}{3})(\tfrac{2}{3})(\tfrac{2}{3})(\tfrac{2}{3}) = (\tfrac{1}{3})^4(\tfrac{2}{3})^6$$

a result obtained by recognizing that the ten events are independent and by using the multiplication theorem for independent events, formula

(2-19). In other words

$$\Pr(SSSSFFFFFF) = \Pr(S)\Pr(S)\Pr(S)\Pr(S)\Pr(F)\Pr(F)\Pr(F)\Pr(F)$$
$$\Pr(F)\Pr(F)$$
$$= [\Pr(S)]^4[\Pr(F)]^6$$

Obviously there are lots of other orders of successes and failures, each order being an event, which will yield exactly four successes in ten throws of a die (for example, $SSSFSFFFFF$). The probability of obtaining any one of the orders is $(\frac{1}{3})^4(\frac{2}{3})^6$ and the total number of possible orders is $k = 10!/4!6! = \binom{10}{4}$, the number of ways 10 things taken 10 at a time can be permuted if four are alike and six are alike. Now if we let A_1, A_2, . . . , A_k be the events associated with the k orders, we recognize that these k events are mutually exclusive since the occurrence of one specified order excludes the occurrence of any other order. Hence, the probability of exactly four successes in ten throws is $\Pr(A_1 \cup A_2 \cup \cdots \cup A_k)$, which, by formula (2-4), is $\Pr(A_1) + \Pr(A_2) + \cdots + \Pr(A_k)$, where each of the latter probabilities is $(\frac{1}{3})^4(\frac{2}{3})^6$. Consequently,

$$\Pr(4 \text{ successes in } 10 \text{ trials}) = \binom{10}{4}(\tfrac{1}{3})^4(\tfrac{2}{3})^6$$

In the die example, the same die was thrown each time. Exactly the same experiment was repeated ten times. Each throw—and in general, each repetition of the experiment—is sometimes referred to as a "trial" of the experiment. The same terminology is used if ten dice (all symmetric) are thrown simultaneously. That is, each die tossed is called a trial of the experiment even though ten different (but equivalent) experiments have been performed. Thus each trial is either a separate experiment or a repetition of the same experiment. The kind of trials we are concerned with in this section satisfies conditions (4-1).

We now generalize the discussion about ten throws of a die to yield a formula for the probability of exactly x successes in n repetitions or trials. One way to achieve this number of successes is

$$\underbrace{S \cdots S}_{x \text{ times}}\underbrace{F \cdots F}_{n-x \text{ times}}$$

that is, be successful x times in a row, then fail the remaining $n - x$ times. According to the assumptions, each experiment is independent of all the others, and by the multiplication theorem, the probability of achieving the given order is

$$\underbrace{p \cdots p}_{x \text{ times}}\underbrace{q \cdots q}_{n-x \text{ times}} = p^x q^{n-x}$$

90

where $q = 1 - p$. There are many other orders that yield x successes and $n - x$ failures, the total number being $k = n!/x!(n - x)! = \binom{n}{x}$, the number of ways n things taken n at a time can be permuted if x are alike and $n - x$ are alike. The probability associated with each order is $p^x q^{n-x}$. Now let $A_1, A_2, \ldots, A_k$ be the events associated with the k orders; these events are mutually exclusive since only one order can occur at a time. Further, by the addition theorem

$$\begin{aligned}\Pr(A_1 \cup A_2 \cup \cdots \cup A_k) &= \Pr(A_1) + \cdots + \Pr(A_k) \\ &= p^x q^{n-x} + \cdots + p^x q^{n-x} \\ &= \binom{n}{x} p^x q^{n-x}\end{aligned}$$

Hence the probability function for the random variable X, the number of successes in n trials satisfying (4-1), is

$$b(x;n,p) = \binom{n}{x} p^x q^{n-x} \qquad x = 0, 1, 2, \ldots, n \tag{4-2}$$

The name "binomial" arises from the fact that $b(x;n,p)$ is a term in the expansion of the binomial $(q + p)^n$ (see Appendix A2).

In most interesting binomial problems we need a sum of terms rather than individual terms. Appendix B1 gives values of

$$\begin{aligned}B(r;n,p) &= \sum_{x=0}^{r} b(x;n,p) \\ &= b(0;n,p) + b(1;n,p) + \cdots + b(r;n,p) \\ &= \Pr(r \text{ or fewer successes in } n \text{ trials}) \\ &= \Pr(X \leq r)\end{aligned} \tag{4-3}$$

for some values of p and n. Several extensive tables of the binomial distribution have been published. The largest of these are the Harvard table [1] and the Ordnance Corps table [10]. The Harvard table gives right-hand sums [$\Pr(X \geq r)$ instead of $\Pr(X \leq r)$] to five decimal places for $n = 1(1)50(2)100(10)200(20)500(50)1000$ and $p = .01(.01).50, \frac{1}{16}, \frac{1}{12}, \frac{1}{8}, \frac{1}{6}, \frac{3}{16}, \frac{1}{3}, \frac{3}{8}, \frac{5}{12}, \frac{7}{16}$. This notation means that the table entries are given for $n = 1, 2, 3, \ldots, 49, 50, 52, 54, \ldots, 98, 100, 110, \ldots, 190, 200, 220, \ldots, 480, 500, 550, \ldots, 950, 1,000$, and $p = .01, .02, \ldots, .49, .50$ plus the nine fractional values. The Ordnance Corps table also gives right-hand sums, but to seven decimal places, for $n = 1(1)150$ and $p = .01(.01)(.50)$. Several smaller tables have been published including those prepared by the National Bureau of Standards [4], Owen [5], and Romig [9]. In addition, tables for small values of p have been prepared by Robertson [8], giving entries for $p = .001(.001).050$ for roughly the same values of n contained in the

Harvard table, and by Weintraub [11], giving entries for $p = .00001$, .0001(.0001).001(.001).10 and $n = 1(1)100$. Now we turn our attention to some examples.

EXAMPLE 4-1

If an ordinary die is thrown four times, what is the probability that exactly two 6's occur?

Solution

First we check conditions (4-1).

(a) Each throw results in a 6 or not a 6.
(b) For each throw $p = \frac{1}{6}$ remains constant as the probability of a success.
(c) Successive throws are independent.
(d) The die is thrown four times.

Hence the binomial distribution seems to be appropriate for this situation with $n = 4$, $x = 2$, and

$$b(2;4,\tfrac{1}{6}) = \binom{4}{2}\left(\frac{1}{6}\right)^2\left(\frac{5}{6}\right)^2 = \frac{150}{1,296}$$

EXAMPLE 4-2

In Example 4-1, what is the probability of two or fewer 6's?

Solution

By two or fewer 6's we mean zero, one, or two 6's. We get

$$\text{Pr(no 6's)} = \binom{4}{0}\left(\frac{1}{6}\right)^0\left(\frac{5}{6}\right)^4 = \frac{625}{1,296}$$

$$\text{Pr(one 6)} = \binom{4}{1}\left(\frac{1}{6}\right)^1\left(\frac{5}{6}\right)^3 = \frac{500}{1,296}$$

$$\text{Pr(two 6's)} = \binom{4}{2}\left(\frac{1}{6}\right)^2\left(\frac{5}{6}\right)^2 = \frac{150}{1,296}$$

Then,

$$\text{Pr(two or fewer 6's)} = \text{Pr}(X \leq 2) = \text{Pr}(X = 0)$$
$$+ \text{Pr}(X = 1) + \text{Pr}(X = 2)$$
$$= \frac{625}{1,296} + \frac{500}{1,296} + \frac{150}{1,296}$$
$$= \frac{1,275}{1,296}$$

92

EXAMPLE 4-3

In a 10-question true-false examination, what is the probability of getting 70 percent or better correct by guessing? Exactly 7 out of 10 correct?

Solution

Check conditions (4-1).

(a) Assuming each question is answered, it is right or wrong.
(b) For each question the probability of a correct guess is $p = \frac{1}{2}$.
(c) Successive questions are independent if each gives no information about the correctness of any of the others. A good examination should be so constructed.
(d) There are ten questions.

Hence the binomial distribution with $n = 10$, $p = \frac{1}{2}$ seems to be the appropriate one for computing probabilities. We want the probability of 7, 8, 9, or 10 correct. This is

$$\Pr(X \geqq 7) = 1 - \Pr(X \leqq 6)$$

$$= 1 - \sum_{x=0}^{6} b(x;10,\tfrac{1}{2})$$

$$= 1 - .82812 \quad \textit{from Appendix B1}$$
$$= .17188$$

The probability of getting exactly seven correct is

$$\Pr(X = 7) = \Pr(X \leqq 7) - \Pr(X \leqq 6)$$

$$= \sum_{x=0}^{7} b(x;10,\tfrac{1}{2}) - \sum_{x=0}^{6} b(x;10,\tfrac{1}{2})$$

$$= .94531 - .82812 \quad \textit{from Appendix B1}$$
$$= .11719$$

EXAMPLE 4-4

In a 20-question, 5-answer multiple-choice examination, what is the probability of getting 6 or more correct by guessing?

Solution

We observe that

(a) Assuming all questions are answered, each is right or wrong.
(b) For each question the probability of a correct guess is $p = \frac{1}{5}$.

(c) When each answer is a guess, presumably the answer to any question does not influence the answer to any other question. In many multiple-choice examinations there would be real doubt as to whether the independence condition is satisfied.

(d) The examination contains 20 questions.

Thus the random variable X = the number of correct answers appears to have a binomial distribution with $n = 20$, $p = \frac{1}{5} = .20$. We want

$$\begin{aligned}
\Pr(X \geq 6) &= \Pr(6 \text{ correct}) + \cdots + \Pr(20 \text{ correct}) \\
&= 1 - \Pr(X \leq 5) \\
&= 1 - \sum_{x=0}^{5} b(x;20,.20) \\
&= 1 - .80421 \qquad \textit{from Appendix B1} \\
&= .19579
\end{aligned}$$

EXAMPLE 4-5

It is estimated that 90 percent of a potato crop is good, the remainder having rotten centers that cannot be detected unless the potatoes are cut open. What is the probability of getting 20 or less good ones in a sack of 25 potatoes?

Solution

It is not unreasonable to assume

(a) The potato can be classified as good or bad.

(b) The probability of getting a good potato remains approximately $p = .90$ from trial to trial, since the crop consists of a very large number of potatoes.

(c) Potatoes are independently good or bad.

(d) In addition, the number of trials is a fixed number.

Thus the random variable X = the number of good potatoes appears to have a binomial distribution with $n = 25$, $p = .90$. We need

$$\Pr(X \leq 20) = \sum_{x=0}^{20} b(x;25,.90)$$

Since the tables give p's only up to .50 the latter sum cannot be read directly from the tables. However, if we let Y = the number of bad potatoes, then the above assumptions are the same with $p = .10$, and Y has a binomial distribution with $n = 25$, $p = .10$. Since getting 20 or fewer good ones is the same as getting 5 or more bad ones, it is apparent that

$$\Pr(X \leq 20) = \Pr(Y \geq 5) = \sum_{y=5}^{25} b(y;25,.10)$$

$$= 1 - \sum_{y=0}^{4} b(y;25,.10)$$

$$= 1 - .90201 \qquad \textit{from Appendix B1}$$
$$= .09799$$

Thus, when $p > \frac{1}{2}$ interchange the roles of p and q, success and failure.

In Examples 4-3 to 4-5 there is some room to doubt that the assumptions (4-1) are satisfied. This is not an uncommon state of affairs in probability problems. One often encounters situations in which the conditions needed to derive the probability distribution are not fulfilled. When this happens, the probability distribution may still give probabilities that are sufficiently close to actual probabilities for practical purposes. Some further experience with the particular random variable under consideration is then necessary to verify the adequacy of the model.

In Example 4-5 we encountered a p larger than $\frac{1}{2}$ and found that the roles of both p and q and success and failure had to be interchanged to enter the tables. We used a specific example to convince ourselves that if X has a binomial distribution with given values of n and p, then $Y = n - X$ has a binomial distribution with the same n and probability of success $q = 1 - p$. Then we observed that

$$\Pr(X \leq r) = \Pr(Y \geq n - r)$$

$$= \sum_{y=n-r}^{n} b(y;n,q)$$

$$= 1 - \sum_{y=0}^{n-r-1} b(y;n,q)$$

Expressed another way, this is

$$B(r;n,p) = 1 - B(n - r - 1; n,q) \tag{4-4}$$

Although we used a common-sense argument to get (4-4), the result can be obtained in a formal algebraic manner. In the right-hand side of

$$\Pr(X \leq r) = \sum_{x=0}^{r} \frac{n!}{x!(n - x)!} p^x q^{n-x}$$

let $y = n - x$ so that $x = n - y$. As x takes on the values 0, 1, 2, . . . , r, y assumes the values $n, n - 1, \ldots , n - r$ or $n - r$,

$n - r + 1, \ldots, n$ when listed in increasing order. Thus

$$\Pr(X \leq r) = \sum_{y=n-r}^{n} \frac{n!}{(n-y)!y!} p^{n-y} q^y$$

$$= \sum_{y=n-r}^{n} \binom{n}{y} q^y p^{n-y}$$

$$= \sum_{y=n-r}^{n} b(y;n,q)$$

as before. Perhaps it is easier to go through the procedure of interchanging p and q and success and failure each time than to remember and use (4-4).

Although in the next chapter we shall discover an easier method of finding the mean of a binomial random variable, it can be found directly from the definition. To do this we must evaluate

$$E(X) = \sum_{x=0}^{n} x \binom{n}{x} p^x q^{n-x} \tag{4-5}$$

Let us begin by considering the special case $n = 3$. Then

$$E(X) = 0 \binom{3}{0} p^0 q^3 + 1 \binom{3}{1} p^1 q^2 + 2 \binom{3}{2} p^2 q^1 + 3 \binom{3}{3} p^3 q^0$$

$$= 0 + 3pq^2 + 6p^2q + 3p^3$$
$$= 3p(q^2 + 2pq + p^2) = 3p(q + p)^2$$
$$= 3p(1) = 3p$$

This computation not only suggests that $E(X) = np$ but also indicates the method used to derive that result. Since the first term in (4-5) is 0, we might just as well start the sum at $x = 1$. Hence

$$E(X) = \sum_{x=1}^{n} x \frac{n!}{x!(n-x)!} p^x q^{n-x}$$

$$= \sum_{x=1}^{n} \frac{n!}{(x-1)!(n-x)!} p^x q^{n-x}$$

and, as in the case with $n = 3$, np can be factored out of every term in the sum, yielding

$$E(X) = np \sum_{x=1}^{n} \frac{(n-1)!}{(x-1)![(n-1)-(x-1)]} p^{x-1} q^{n-1-(x-1)}$$

Upon letting $y = x - 1$ the latter sum becomes

$$E(X) = np \sum_{y=0}^{n-1} \binom{n-1}{y} p^y q^{n-1-y}$$

But

$$\sum_{y=0}^{n-1} \binom{n-1}{y} p^y q^{n-1-y} = \Pr(Y = 0) + \Pr(Y = 1)$$

$$+ \cdots + \Pr(Y = n - 1)$$

$$= 1$$

where Y is a binomial random variable generated by $n - 1$ trials with probability of successes p, and the sum represents the sum of the probabilities in the distribution. Alternatively, using the algebraic result (A2-4) found in Appendix A2 we get

$$\sum_{y=0}^{n-1} \binom{n-1}{y} p^y q^{n-1-y} = (q + p)^{n-1} = (1)^{n-1} = 1$$

Consequently, we have the result

$$E(X) = np \tag{4-6}$$

The easier method of the next chapter will also be used to find the variance of a binomial random variable. That result is

$$\text{Var}(X) = npq \tag{4-7}$$

This result can also be found by first evaluating $E(X^2)$ and then using formula (3-9). The actual calculation is outlined in Exercise 4-10.

EXERCISES

Use Appendix B1 to evaluate probabilities in the following problems. Where appropriate comment upon the reasonableness of the binomial assumptions (4-1).

4-1 A well-known baseball player has a lifetime batting average of .300. If he comes to bat five times in his next game, what is the probability that he will get more than two hits? Three hits?

4-2 A four-answer multiple-choice examination has 100 questions. Assuming that a student only guesses and answers every question, what is the probability that he gets 30 or more correct? Suppose that the instructor decides that no grade (number correct) will be passing unless the probability of getting or exceeding that grade by guessing is less than .01. What is the minimum passing grade? Find the mean, variance, and standard deviation of the number of correct answers.

4-3   A missile manufacturer claims that his missiles are successful 90 percent of the time. The Air Force checks the stock by firing 10 missiles and obtains five successes. What is the probability of obtaining 5 or fewer successes if $p = .90$? What conclusion is one apt to draw?

4-4   Suppose that 25 voters are chosen independently of one another and asked if they favor a certain proposal. If 40 percent of the voters favor the proposal, what is the probability that a majority of the 25 voters chosen will favor the proposal?

4-5   Suppose that you believe that you can hit a bull's-eye with a dart one time in ten. If this is the case, compute the probability of scoring 15 or more bull's-eyes in 100 throws. Find the mean, variance, and standard deviation of the number of bull's-eyes in 100 throws.

4-6   It is known that 25 percent of all rabbits inoculated with a serum containing a certain disease germ will contract the disease. If 20 rabbits are inoculated, what is the probability that at least 3 get positive reactions?

4-7   Weather-bureau records in a certain locality show that 40 percent of the days in April are cloudy. Find the probability that, during the first 20 days of next April, at most 5 will be cloudy.

4-8   In order to select its beer tasters, a brewery gives an applicant a tasting examination. The applicant is presented with four glasses, one of which contains ale and three of which contain beer, and is asked to identify the one containing ale. If the procedure is repeated 10 times and the brewery requires 7 or more correct answers for a satisfactory score, what is the probability that an applicant will pass the test if he cannot discriminate and only guesses each time? Suppose that the applicant does possess discriminatory ability and has a certain probability p of making a correct choice on each trial. What is the smallest value of p of those given in Appendix B1 that will guarantee the applicant at least a probability of .85 of passing the test?

4-9   The standard cure for tuberculosis is successful 30 percent of the time. A new cure is tried on a group of 50 patients and is successful in 29 cases. What conclusion is one apt to draw and why?

4-10 If X has the probability function given by (4-2), find $E(X^2)$. To do this first observe that

$$X^2 = X(X - 1) + X$$

so that

$$E(X^2) = E[X(X - 1) + X] \\ = E[X(X - 1)] + E(X)$$

Then

$$E[X(X-1)] = \sum_{x=0}^{n} x(x-1) \binom{n}{x} p^x q^{n-x}$$

$$= \sum_{x=2}^{n} x(x-1) \binom{n}{x} p^x q^{n-x}$$

Cancel as was done in the evaluation of $E(X)$ and then let $y = x - 2$. Deduce that $E[X(X-1)] = n(n-1)p^2$ and use this result to show that $\sigma^2 = npq$.

4-3   THE NEGATIVE BINOMIAL DISTRIBUTION

Consider a series of events or experiments having the following properties:

(a)   *The result of each experiment can be classified into one of two categories, say, success and failure.*

(b)   *The probability p of a success is the same for each experiment.*   (4-8)

(c)   *Each experiment is independent of all the others.*

(d)   *The series of experiments is performed a variable number of times until a fixed number of successes, say, c, is achieved.*

Only (d) differs from assumptions (4-1), which were used to obtain an expression for the probability of x success in n trials. Now we shall derive a formula for the probability that exactly n repetitions of the experiment are required to achieve c successes.

As an example, suppose that we continue to roll a die and regard a 1 or 2 as a success, so that $p = \frac{1}{3}$, until $c = 4$ successes have occurred, and we would like to know the probability that $n = 10$ trials are required. One way to achieve the desired result is *SSSFFFFFFS*, that is, succeed the first three times, fail the next six, and then succeed. Obviously, the last trial must be a success or four successes would have been achieved sooner. By the multiplication theorem, the probability of the above sequence is $(\frac{1}{3})^4(\frac{2}{3})^6$. Once more, many other sequences, all mutually exclusive events with probability of occurrence $(\frac{1}{3})^4(\frac{2}{3})^6$, yield the desired result. Hence, again we need to count the number of ways the S's and F's can be permuted. Since the last letter must be an S, only the remaining nine letters can be rearranged, the number of rearrangements being

$$\frac{9!}{3!\,6!} = \binom{9}{3}$$

Consequently,

$$\text{Pr}(10 \text{ trials are required to achieve 4 successes}) = \binom{9}{3}(\tfrac{1}{3})^4(\tfrac{2}{3})^6$$

The preceding argument is easily generalized. One way to achieve c successes in exactly n trials is

$$\underbrace{S \cdots S}_{c\,-\,1 \text{ times}} \underbrace{F \cdots F}_{n\,-\,c \text{ times}} S$$

That is, be successful $c - 1$ times, fail the next $n - c$ times, and succeed the last time. By the multiplication theorem, the probability that this happens is

$$\underbrace{p \cdots p}_{c\,-\,1 \text{ times}} \underbrace{q \cdots q}_{n\,-\,c \text{ times}} p = p^c q^{n-c}$$

Next we count all the mutually exclusive events corresponding to the different orders that yield the cth success on the nth trial. Since the last letter in every sequence must be an S, this leaves $c - 1$ of the S's and $n - c$ of the F's for forming permutations. The number that can be constructed is

$$\frac{(n-1)!}{(c-1)!(n-c)!} = \binom{n-1}{c-1}$$

the total number of ways to permute $n - 1$ things taken $n - 1$ at a time if $c - 1$ are alike and $n - c$ are alike. Since every order has the same probability of occurrence, the addition theorem yields the formula

$$b^*(n;c,p) = \binom{n-1}{c-1} p^c q^{n-c} \qquad n = c, c+1, c+2, \ldots \qquad (4\text{-}9)$$

for the probability that the cth success occurs on the nth trial. The random variable N cannot be less than c since at least c repetitions are required to produce c successes. The name negative binomial arises from the fact that $b^*(n;c,p)$ is a term in the expansion of $p^c(1 - q)^{-c}$.

Tables of the negative binomial have been prepared by Williamson and Bretherton [12]. These give $B^*(r;c,p) = \text{Pr}(N \le r)$ to six decimal places. However, negative binomial probabilities can be evaluated from an ordinary binomial table. Thus, if one has a good binomial table, it is not necessary to have the Williamson and Bretherton table also. It is known that

$$B^*(r;c,p) = \sum_{n=c}^{r} \binom{n-1}{c-1} p^c q^{n-c} = \sum_{x=c}^{r} \binom{r}{x} p^x q^{r-x}$$

$$= 1 - \sum_{x=0}^{c-1} \binom{r}{x} p^x q^{r-x}$$

$$= 1 - B(c-1; r,p) \tag{4-10}$$

In other words

$$\Pr(N \leq r) = \Pr(X \geq c)$$

where N has a negative binomial distribution with known constants c and p and X has a binomial distribution with known constants r and p. Thus a left-hand sum of negative binomial probabilities is equal to a right-hand sum of ordinary binomial probabilities.

The result (4-10) is not terribly surprising once it is pointed out. For the sake of argument let us again consider the die example where a 1 or 2 was called a success. Suppose we set as a goal the obtaining of 4 successes but agree to roll no more than 10 times. Our goal can be achieved in two ways. First, if the 4th success is obtained on either the 4th, 5th, 6th, 7th, 8th, 9th, or 10th roll the goal is accomplished. The probability that this series of events happens is $\Pr(N \leq 10)$ where $c = 4$, $p = \frac{1}{3}$. The second way to achieve the goal in 10 tries is to plan to roll exactly 10 times and count 4, 5, 6, 7, 8, 9, or 10 successes. The probability of the latter series of events is $\Pr(X \geq 4)$ where $n = 10$, $p = \frac{1}{3}$. In both cases we are considering the event "the goal is achieved in 10 trials" so that the two probabilities must be the same. If we replace 4 by c, 10 by r, and let p be any number between 0 and 1, the argument is perfectly general. Algebraic proofs of (4-10) have appeared in various places in the literature, in particular in a paper by Patil [6].

EXAMPLE 4-6

A student takes a five-answer multiple-choice examination orally. He continues to answer questions until he gets five correct answers. What is the probability that he gets them on or before the twenty-fifth question if he guesses at each answer?

Solution

The reasonableness of assumptions (a) to (c) of (4-8) was discussed in Example 4-4. Also, (d) is satisfied, since the number of questions required to achieve five successes is a random variable. Thus we use the negative binomial probability function with $c = 5$, $r = 25$, $p = \frac{1}{5} = .20$ to compute probabilities.   We seek

$$Pr(N \leq 25) = \sum_{n=5}^{25} b^*(n;5,.20)$$

$$= \sum_{x=5}^{25} b(x;25,.20)$$

$$= 1 - \sum_{x=0}^{4} b(x;25,.20)$$

$$= 1 - .42067 \qquad \text{from Appendix B1}$$
$$= .57933$$

EXAMPLE 4-7

Consider again the potato crop of Example 4-5.   Suppose that a cook needs 20 good potatoes for a meal, so he selects potatoes at random, cuts them open, and throws away the bad.   What is the probability that he must cut open more than 25 potatoes?

Solution

The assumptions (a) to (c) of (4-8) have already been discussed in Example 4-5.   Since the number of successes is a fixed quantity, the negative binomial with $c = 20$, $p = .9$ is appropriate for the calculation of probabilities.   We need

$$Pr(N \geq 26) = \sum_{n=26}^{\infty} b^*(n;20,.9)$$

$$= 1 - \sum_{n=20}^{25} b^*(n;20,.9)$$

$$= 1 - \sum_{x=20}^{25} b(x;25,.9)$$

$$= \sum_{x=0}^{19} b(x;25,.9)$$

$$= \sum_{x=6}^{25} b(x;25,.1)$$

$$= 1 - \sum_{x=0}^{5} b(x;25,.1)$$

$$= 1 - .96660 \qquad \text{from Appendix B1}$$
$$= .03340$$

This result could also have been obtained as follows:

$$
\begin{aligned}
\Pr(N \geq 26) &= 1 - \Pr(N \leq 25) \\
&= 1 - B^*(25;20,.9) \\
&= 1 - [1 - B(19;25,.9)] \qquad by\ (4\text{-}10) \\
&= B(19;25,.9) \\
&= 1 - B(25 - 19 - 1; 25,.1) \qquad by\ (4\text{-}4) \\
&= 1 - .96660 \qquad from\ Appendix\ B1 \\
&= .03340
\end{aligned}
$$

EXAMPLE 4-8

To determine who pays for coffee, three people each toss a coin and the odd man pays. If the coins all show heads or all show tails, they are tossed again. What is the probability that a decision is reached in five repetitions or sooner?

Solution

A decision is reached on any trial if the result is two heads and a tail or two tails and a head. To compute the probability of these events, the binomial is appropriate since

(a) Each coin will show either a head or a tail.
(b) The probability of resulting in a head is $\frac{1}{2}$ for each coin.
(c) The three coins are tossed independently of one another.
(d) Exactly three coins are tossed in each repetition.

Thus $n = 3$, $p = \frac{1}{2}$, and

$$
\Pr(1\ \text{head},\ 2\ \text{tails}) = \binom{3}{1}\left(\frac{1}{2}\right)\left(\frac{1}{2}\right)^2 = \frac{3}{8}
$$

$$
\Pr(2\ \text{heads},\ 1\ \text{tail}) = \binom{3}{2}\left(\frac{1}{2}\right)^2\left(\frac{1}{2}\right) = \frac{3}{8}
$$

and the probability of a decision is $\frac{3}{8} + \frac{3}{8} = .75$.

Next each set of tosses is characterized by these facts:

(a) A decision is reached or not reached with each set.
(b) The probability of a decision is .75 for all sets.
(c) The result of each set is independent of the result for any other set.
(d) A variable number of sets is required to produce 1 decision.

Thus the negative binomial with $c = 1$, $p = .75$ is appropriate to determine the probability needed to answer the original question. We have

$$
\begin{aligned}
\Pr(N \leq 5) &= B^*(5;1,.75) \\
&= 1 - B(0;5,.75) \qquad \text{by (4-10)} \\
&= 1 - [1 - B(4;5,.25)] \qquad \text{by (4-4)} \\
&= B(4;5,.25) \\
&= .99902 \qquad \text{by Appendix B1}
\end{aligned}
$$

Since the sum of all the probabilities in a probability distribution is 1, we have

$$
\sum_{n=c}^{\infty} \binom{n-1}{c-1} p^c q^{n-c} = 1 \tag{4-11}
$$

The result (4-11) can be used to find the mean and variance of N. By definition we have

$$
E(N) = \sum_{n=c}^{\infty} \frac{n(n-1)!}{(c-1)!(n-c)!} p^c q^{n-c}
$$

Multiplying both numerator and denominator by cp we get

$$
E(N) = \frac{c}{p} \sum_{n=c}^{\infty} \frac{n!}{c!(n-c)!} p^{c+1} q^{n-c} = \frac{c}{p} \sum_{n=c}^{\infty} \binom{n}{c} p^{c+1} q^{n-c}
$$

In the latter sum let $n = y - 1$, $y = n + 1$. Then

$$
E(N) = \frac{c}{p} \sum_{y=c+1}^{\infty} \binom{y-1}{c+1-1} p^{c+1} q^{y-(c+1)}
$$

But this last sum is exactly the sum appearing on the left-hand side of (4-11) except that $c + 1$ has replaced c (and y has replaced n). Hence, this sum has the value 1 and

$$
E(N) = \frac{c}{p} \tag{4-12}
$$

The above procedure can also be used to show

$$
\mathrm{Var}(N) = \frac{cq}{p^2} \tag{4-13}
$$

The actual steps are outlined in Exercise 4-17.

In summary we note that the binomial and the negative binomial are derived under almost the same set of conditions. In the first case the number of trials n is fixed and the number of successes X is a random variable. In the second case the number of trials N is a random variable. As previously noted, X can assume only one of the $n + 1$ values 0, 1,

2, . . . , n whereas the set of values available for N is all positive integers c or larger.

In establishing Eq. (4-10) we used a probability argument. Reference was made to the fact that algebraic proofs can be used to derive this result. The algebraic derivation is, however, much more complicated. This example demonstrates that complicated mathematical proofs can often be considerably simplified by using probability arguments. Let us give one further example. If we divide both sides of Eq. (4-11) by p^c we have immediately

$$\sum_{n=c}^{\infty} \binom{n-1}{c-1} q^{n-c} = p^{-c} = (1-q)^{-c} \qquad 0 < q < 1$$

an identity that is established in Appendix A2 with considerably more difficulty [compare with Eq. (A2-12)]. Of course, we first had to derive the formula for the probability function of the negative binomial distribution, but once that result is available the above identity follows immediately. The unfortunate drawback to probability proofs is that we usually stumble on them by accident when working another problem. Frequently proofs based upon probability arguments depend upon how clever we are in selecting the correct starting point to begin the proof.

EXERCISES

Use Appendix B1 to evaluate probabilities in the following problems. Where appropriate comment on the reasonableness of the negative binomial assumptions (4-8).

4-11 A well-known baseball player has a lifetime batting average of .300. He needs 32 more hits to up his lifetime total to 3,000. What is the probability that 100 or fewer times at bat are required to achieve his goal? Find the expected number of times at bat required to get 32 hits. What is the variance of N?

4-12 A student takes a four-answer multiple choice examination orally. He continues to answer questions until he gets 10 correct answers. What is the probability that more than 25 questions are required if he guesses at each answer? Find the mean and variance of the number of trials required.

4-13 A missile manufacturer claims that his missiles are successful 90 percent of the time. The Air Force checks the stock by firing until 4 successes are obtained, and 11 trials are required. What is the probability that 11 or more trials are required if $p = .90$? What conclusion is one apt to draw?

4-14 Suppose it is estimated that 40 percent of a certain large group are strong supporters of a project and willing to volunteer their services if asked. The remaining 60 percent will decline to volunteer. In order to get five people for a committee, members of the group are contacted one at a time until the committee is complete. Twenty-one contacts are required. If the 40 percent figure is correct, what is the probability that 21 or more contacts are required? Are you inclined to believe the 40 percent figure? If the 40 percent figure is correct, find the expected number of contacts required.

4-15 A scientist needs three diseased rabbits for an experiment. He has 20 rabbits available and inoculates them one at a time with a serum containing a disease germ, quitting if and when he gets 3 positive reactions. If the probability is .25 that a rabbit can contract the disease from the serum, what is the probability that the scientist is able to get 3 diseased rabbits from 20? What is the expected number of inoculations to get 3 positive reactions?

4-16 To determine who buys coffee, four people toss one coin each. This is repeated until someone has a result different from the other three. What is the expected number of repetitions required to reach a decision?

4-17 If N has the probability function given by (4-9), find $E(N^2)$. To do this first observe that $N^2 = N(N + 1) - N$ so that $E(N^2) = E[N(N + 1) - N] = E[N(N + 1)] - E(N)$. Then

$$E[N(N + 1)] = \sum_{n=c}^{\infty} n(n + 1) \binom{n - 1}{c - 1} p^c q^{n-c}$$

Show that this can be rewritten as

$$E[N(N + 1)] = \frac{c(c + 1)}{p^2} \sum_{n=c}^{\infty} \binom{n + 1}{c + 1} p^{c+2} q^{n-c}$$

Now let $n = y - 2$, $y = n + 2$, and observe that the new sum can be written like (4-11) with $c + 2$ replacing c. Hence deduce that $E[N(N + 1)] = c(c + 1)/p^2$ and use the result to show that $\sigma^2 = cq/p^2$.

4-4 THE UNIFORM DISTRIBUTION

Suppose an experiment can terminate in k mutually exclusive ways, all equally likely. A single throw of a true die is an example of this kind of experiment. Any one of the sides is as likely to show as any other. A more practical situation is created by writing k numbers, one each on k

slips of paper of equal size, placing the numbers in a hat or box, mixing the slips well, and then drawing one.

For a six-sided symmetric die, the probability associated with each outcome is $\frac{1}{6}$. If X is the number of spots showing, then the probability function of the random variable X is

$$f(x;6) = \frac{1}{6} \qquad x = 1, 2, 3, 4, 5, 6$$

For any experiment where the outcomes are 1, 2, . . . , k occurring with equal probabilities the probability function is

$$f(x;k) = \frac{1}{k} \qquad x = 1, 2, \ldots, k \qquad (4\text{-}14)$$

We shall generalize (4-14) slightly by letting the random variable X assume any one of k values with equal probability. This yields

$$f(x;k) = \frac{1}{k} \qquad x = x_1, x_2, \ldots, x_k \qquad (4\text{-}15)$$

A random variable X having the probability function (4-15) is said to have a uniform distribution.

Obviously, the calculation of probabilities presents no problem in the case of the uniform distribution. Hence it is not necessary to have a table corresponding to Appendix B1 for the binomial.

Extensive tables have been prepared that make it possible to generate random variables having the uniform distribution. One such table has been published by the Rand Corporation [7]. A small part of that table appears in Appendix B2. We could construct such a table by using the hat procedure, mentioned at the beginning of the section, with numbers 0, 1, 2, . . . , 8, 9. (The table was, of course, prepared in more efficient manner using high-speed computers.) After each draw the number would be recorded and replaced before the next draw. Thus, the probability function for the one-digit numbers appearing in the table is

$$f(x;10) = \frac{1}{10} \qquad x = 0, 1, 2, \ldots, 8, 9 \qquad (4\text{-}16)$$

so that each of the 10 integers is supposed to occur about 10 percent of the time. The table can also be regarded as being composed of a series of two, three, or more digit numbers. Suppose we put our finger over a two-digit number and ask, "What is the probability that the number is 27?" Since the probability that the first digit is 2 is $\frac{1}{10}$ and the probability that the second digit is 7 is $\frac{1}{10}$, the answer is $\frac{1}{100}$. Similarly, the probability of selecting any given two-digit number is $\frac{1}{100}$. Thus, the probability function for two-digit numbers is

$$f(x;100) = \frac{1}{100} \qquad x = 0, 1, 2, \ldots, 99 \qquad (4\text{-}17)$$

107

Similarly, for three-digit numbers we get

$$f(x;1000) = \frac{1}{1000} \qquad x = 0, 1, 2, \ldots, 999 \qquad (4\text{-}18)$$

and so on for four or more digits.

The Rand table can also be used to generate a uniform distribution where k is not a multiple of 10. Suppose we put our finger over a two-digit number. If it is more than 27 we disregard it. If it is less than or equal to 27, we ask, "What is the probability that the number is 15?" To get the answer let A_1 be the event "the number is less than or equal to 27" and A_2 be the event "the number is 15." We observe that $\Pr(A_1A_2) = \frac{1}{100}$, $\Pr(A_1) = \frac{28}{100}$, and by formula (2-10), $\Pr(A_2|A_1) = (\frac{1}{100})/(\frac{28}{100}) = \frac{1}{28}$, a result that is intuitively obvious. The answer is the same if 15 is replaced by any number 27 or less. Hence, if we disregard all the numbers larger than 27, the remaining numbers have the probability function $f(x;28) = \frac{1}{28}$, $x = 0, 1, 2, \ldots, 27$. Similarly, if we disregard zero and all digits larger than k, the remaining numbers in the table have the probability function (4-14).

The main use of the Rand table arises in sampling when it is required to "draw a random sample from a finite population." By this is meant that k objects are under consideration (the population) from which we wish to select a subset of n of the objects, $n < k$ (the sample), in such a way that every sample of size n has the same probability of being selected. To see how this is done suppose a finite population consists of seven individuals or objects (with which we may or may not associate numbers $x_1, \ldots, x_7$). Let the individuals or objects be designated by $1, 2, \ldots, 7$ from which we want to select a random sample of three. Start at an arbitrary place in the table and read down a column until three numbers between 1 and 7 inclusive are obtained. Suppose we get $3, 5,$ and 6. Then our random sample consists of the objects or individuals designated by $3, 5, 6$. We have yet to demonstrate that the probability of drawing any sample is the same as the probability of drawing any other sample if we use this procedure. In selecting the first number for the sample, the possible choices have the uniform distribution with probability function $f(x;7) = \frac{1}{7}$, $x = 1, 2, \ldots, 7$. The probability of getting 3, 5, or 6 on the first selection is $\frac{3}{7}$. Suppose we drew 5 first. Then the next selection is to be made from the uniform distribution with probability function $f(x;6) = \frac{1}{6}$, $x = 1, 2, 3, 4, 6, 7$. The probability of getting 3 or 6 is $\frac{2}{6}$. Suppose we draw 6 second. Then the last selection is made from numbers having the uniform distribution with probability function $f(x;5) = \frac{1}{5}$, $x = 1, 2, 3, 4, 7$, and the probability of getting a 3 is $\frac{1}{5}$. Consequently, the probability of drawing the sample consisting of objects 3, 5, 6 is $(\frac{3}{7})(\frac{2}{6})(\frac{1}{5}) = 3!4!/7! = 1 \bigg/ \binom{7}{3}$.

The argument is the same for each of the $\binom{7}{3}$ possible samples, so that the

108

probability of drawing each is $1 \Big/ \binom{7}{3}$. Consequently, the procedure has produced a random sample.

The above discussion is easily generalized to a population of k items and a sample of size n. On each selection the distribution of the numbers assigned to the items is uniform. The probability that one of a specified set of n items is drawn the first time is n/k; the probability that one of the remaining $n - 1$ items is selected the second time is $(n - 1)/(k - 1)$; . . . ; the probability that the one remaining at the time of the last draw is then selected is $1/[k - (n - 1)]$. Consequently, the probability of drawing any specified set of n items is

$$\frac{n}{k} \frac{n-1}{k-1} \cdot \cdot \cdot \frac{1}{k-n+1} = \frac{n!(k-n)!}{k!} = \frac{1}{\binom{k}{n}}$$

No matter which set of n is selected, the probability of drawing the sample is $1 \Big/ \binom{k}{n}$ for any one of the $\binom{k}{n}$ possible samples. We observe that the distribution of samples is uniform with probability function

$$f(x;m) = \frac{1}{\binom{k}{n}} \qquad x = S_1, S_2, \ldots, S_m$$

where $m = \binom{k}{n}$ and $S_1, S_2, \ldots, S_m$ are the m possible samples. Here the variable X is n dimensional (the kind to be considered in the next chapter), since each sample contains n items.

The Rand publication includes suggestions for the use of the table. A starting position in the table is determined by selecting a seven-digit number in some arbitrary manner. We might, for example, take the last seven numbers in the first row of the table. This yields 9274945. We use the first five digits to locate a row (rows are numbered from 00000 to 19999). Since we have 92749, which is out of the range, we reduce the first digit to a 1 (to a zero if the first digit is even) and use row 12749. The last two numbers, 45 in our example, are used to locate the column (there are 50 columns in each line). Whenever the latter number exceeds 50, we subtract 50 from it to determine the column number. Hence if the last two digits were 77, we would begin in column 27. Since Appendix B2 contains only 200 rows, a five-digit number will suffice for locating the starting point. After using a five-digit number for this purpose (seven-digit with the original table), it is advisable to draw a line through it so that the same starting point is not used in the future. From the starting point we can proceed to read off any size numbers we desire by going down a column, up a column, left to right as we would read a book, right to left as if we were reading backwards, across diagonals, and many other ways. To be uniform we shall read down the column until the bottom of the page is reached. Then we shall proceed to the top of the page and into the next column on the right made up of the same size numbers.

EXAMPLE 4-9

Draw a random sample of size 5 from items designated by 1, 2, . . . , 353 by using Appendix B2 and the five-digit number in the upper left-hand corner of the second page to determine the starting point.

Solution

The five-digit number in the upper left-hand corner of the second page of Appendix B2 is 32179. Thus we shall start in row 121, column 29. The first three-digit number is 216. Below 216 we find 777, 761, 769 (all too large and are disregarded). Next we read 200, 655, 676, 099, 594, 127, 199.

Our sample consists of the items numbered 216, 200, 99, 127, 199, the first five numbers between 1 and 353 inclusive, disregarding any repeats (here there were none).

EXAMPLE 4-10

Suppose a box contains 106 oranges. We want to select 6 at random, cut them open to determine if they are good or spoiled, and, on the basis of the 6, decide in some way whether or not to purchase the box. Use Appendix B2 to draw the sample and the five-digit number 56058 to determine the starting point.

Solution

The five-digit number 56058 determines our starting point. Hence we shall start in row 160, column 8 with 584. To number the oranges we could lay them out in a row so that we know which orange goes with which number. Reading down the column from the starting point we get 338, 292, 004. Hence orange number 4 is in the sample. Continuing down the column, 091 is the only other number less than or equal to 106 obtained before the bottom of the page is reached. Then going to the top of the page, column 11, the first three-digit number is 824. Proceeding downward looking for numbers less than or equal to 106, we find 047 and 073 before reaching the bottom of the page. Then we go to the top of the page, column 14 and read 742. Proceeding downward again we find 018 and 014. Hence we take oranges numbered 4, 91, 47, 73, 18, 14 for our random sample of six. In actual practice one might attempt the hat procedure with the oranges. That is, select oranges one at a time from various places in the box and hope that a random sample is obtained.

In Exercise 3-45 we found the mean and variance of the random variable X having the probability function (4-14). Those results were

$$E(X) = \frac{k+1}{2}$$

$$\text{(4-19)}$$

$$\text{Var}(X) = \frac{k^2 - 1}{12}$$

For the more general case given by (4-15), the best we can do without knowing the specific values of X is write

$$E(X) = \frac{1}{k} \sum_{i=1}^{k} x_i \qquad \text{(4-20)}$$

$$\text{Var}(X) = \frac{1}{k} \sum_{i=1}^{k} x_i^2 - [E(X)]^2 \qquad \text{(4-21)}$$

EXAMPLE 4-11

If X has the probability function given by (4-14) with $k = 7$, find the mean, variance, and standard deviation of X. If X has the probability function given by (4-15) with $x_1 = 1$, $x_2 = 5$, $x_3 = 6$, $x_4 = 9$, $x_5 = 12$. $x_6 = 16$, $x_7 = 21$, find the mean and variance of X.

Solution

For the first situation we substitute $k = 7$ into (4-19). We find $E(X) = 4$, $\text{Var}(X) = 4$, $\sigma = 2$. For the second situation we find from (4-20) and (4-21) that

$$E(X) = \frac{1 + 5 + 6 + 9 + 12 + 16 + 21}{7} = \frac{70}{7} = 10$$

$$\text{Var}(X) = \frac{1^2 + 5^2 + 6^2 + 9^2 + 12^2 + 16^2 + 21^2}{7} - (10)^2$$

$$= \frac{984}{7} - 100 = \frac{284}{7}$$

EXERCISES

4-18   Suppose that your class contains 80 students numbered 1 to 80. Select a random sample of size 10 following the procedure used in the examples and using the first five-digit number found in row 00030 to determine the starting point in the tables.

4-19   A committee contains 10 people. It is desired to select 3 of these at random to form a subcommittee to prepare a report. To do this the committee members are each assigned a number 1 to 10 and the table of random numbers is used. The starting point is determined by the first five digits in row 00199 and the procedure outlined in the examples is used. Which people are on the subcommittee?

4-20 If X has the probability function given by (4-14) with $k = 25$, find the mean and variance of X.

4-21 If X has the probability function given by (4-15) with $x_1 = 1.6$, $x_2 = 2.3$, $x_3 = 3.5$, $x_4 = 3.1$, $x_5 = 3.5$, find the mean and variance of X.

4-5 THE HYPERGEOMETRIC DISTRIBUTION

Let us suppose that a sack of fruit contains six apples and four oranges. We may be interested in the probability of getting two apples and three oranges if we were to draw five pieces of fruit from the sack one at a time without looking. (We would hope that the five pieces so drawn constitute a random sample.) Knowing how to count combinations makes the calculation of such a probability a fairly easy problem. The total number of ways the drawing can be accomplished is $\binom{10}{5}$, the number of combinations of 10 things taken 5 at a time. If the sample is selected randomly, then each possible sample has probability $1 \Big/ \binom{10}{5}$ of being selected. In other words, the sample space for our experiment consists of $\binom{10}{5}$ points with each being assigned probability $1 \Big/ \binom{10}{5}$. Now all we have to do is count the number of sample points which correspond to the event of interest. Since 2 apples can be selected from 6 in $\binom{6}{2}$ ways and, with each of these selections, 3 oranges can be picked from 4 in $\binom{4}{3}$ ways, the total number of ways to draw 2 apples and 3 oranges is $\binom{6}{2}\binom{4}{3}$. Since we have agreed that all $\binom{10}{5}$ sample points should be assigned equal weights, the probability of drawing two apples and three oranges is

$$\frac{\binom{6}{2}\binom{4}{3}}{\binom{10}{5}} = \frac{(6!/2!4!)(4!/3!1!)}{10!/5!5!} = \frac{5}{21}$$

112

Before attempting to generalize the preceding discussion, we note that the following conditions characterize the situation:

(a) *The result of each draw can be classified into one of two categories, say, success (apples) and failure (oranges).*
(b) *The probability of a success changes on each draw.*   (4-22)
(c) *Successive draws are dependent.*
(d) *The drawing is repeated a fixed number of times.*

The fact that drawings are made *without replacement* changes (b) and (c) from the binomial situation. Obviously, if an apple is drawn the first time, the probability of drawing an apple the second time is reduced from an original $6/10$ to $5/9$. Since the outcome of any draw after the first is affected by what has happened on preceding draws, the drawings are dependent events. If the apple or orange is replaced after every draw and the fruit is mixed before the next draw, then the probability of a success remains constant, drawings are independent, and the binomial model is the appropriate one to use.

The conditions satisfied in the fruit example can be generalized as follows:

(a) *The drawing is made from N items.*
(b) *A random sample of size n is selected (drawing without replacement).*   (4-23)
(c) *k of the N items have some characteristic (for example, they may be defective).*

We then seek the probability of obtaining x defectives in n draws. The total number of ways to draw n items is $\binom{N}{n}$. Since the drawing is done at random, the sample space we use consists of $\binom{N}{n}$ points, one for each possible sample, each assigned probability $1 \Big/ \binom{N}{n}$. To obtain exactly x defectives implies that $n - x$ nondefective items must be drawn from a total of $N - k$ available. The total number of ways to get exactly x defectives and $n - x$ nondefectives is $\binom{k}{x}\binom{N-k}{n-x}$. Consequently, since all the $\binom{N}{n}$ points in the sample space are assigned equal weights, definition (1-5) yields for the probability that x of the n items are defective

$$p(N,n,k,x) = \frac{\binom{k}{x}\binom{N-k}{n-x}}{\binom{N}{n}} \tag{4-24}$$

The three numbers N, n, and k are constants and X is the random variable. A random variable X having probability function (4-24) is said to be a *hypergeometric random variable* and (4-24) is called the *hypergeometric probability function*. In the apple-orange illustration we can make the identification $N = 10$, $n = 5$, $k = 6$, $x = 2$. In this case the possible choices for X are 1, 2, 3, 4, 5 or $n - (N - k) \leq X \leq n$ (we have to draw at least one apple). The alternative identification $N = 10$, $n = 5$, $k = 4$, $x = 3$ would work just as well if an orange is regarded as a defective. Now we see that the choices for X are 0, 1, 2, 3, 4 or $0 \leq X \leq k$. Thus X can be no smaller than the larger of the two numbers 0 and $n - (N - k)$, and it can be no larger than the smaller of k and n. Hence, X can take on all integer values such that

$$\text{Maximum } [0, n - (N - k)] \leq X \leq \text{minimum } [k,n] \qquad (4\text{-}25)$$

Appendix B3 gives values of $p(N,n,k,x)$ and

$$P(N,n,k,r) = \sum_{x = \max [0, n - (N - k)]}^{r} p(N,n,k,x) = \Pr(X \leq r) \qquad (4\text{-}26)$$

for $N = 10$. An extensive table has been published by Lieberman and Owen [2], where both $p(N,n,k,x)$ and $P(N,n,k,r)$ are given to six decimal places for $N = 1(1)50(10)100$. In addition, some values are given for $N = 1000(100)2000$. A table for $N = 1(1)20$ is found in Owen's book of tables [5]. In order to enter Appendix B3 (and the Lieberman and Owen table) it may be necessary to use the relationships

$$p(N,n,k,x) = p(N,k,n,x) \quad \text{and} \quad P(N,n,k,x) = P(N,k,n,x) \qquad (4\text{-}27)$$

That is, n and k can be interchanged, a result easily verified by writing down (4-27) in factorial form (see Exercise 4-29). (When using the Lieberman and Owen table with $N > 25$, additional symmetries found on page 4 of that publication may be required.)

EXAMPLE 4-12

What is the probability of drawing two or fewer apples from a sack containing six apples and four oranges if five pieces of fruit are selected at random?

Solution

The preceding discussion has indicated that the hypergeometric model is appropriate. We have already identified $N = 10$, $n = 5$, $k = 6$. Now we want

$$\Pr(X \leq 2) = \Pr(1 \leq X \leq 2) = P(10,5,6,2)$$
$$= P(10,6,5,2)$$
$$= .261905 \quad \textit{from Appendix B3}$$

Without the table we compute

$$\frac{\binom{6}{1}\binom{4}{4}}{\binom{10}{5}} + \frac{\binom{6}{2}\binom{4}{3}}{\binom{10}{5}} = \frac{1}{42} + \frac{5}{21} = \frac{11}{42} = .261095$$

EXAMPLE 4-13

Ten vegetable cans, all the same size, have lost their labels. It is known that five contain tomatoes and five contain corn. If five are selected at random, what is the probability that all contain tomatoes? What is the probability that three or more contain tomatoes?

Solution

We check conditions (4-23).

(a)   There are $N = 10$ items to draw from.
(b)   The number of draws is $n = 5$, made at random.
(c) $k = 5$ of the 10 cans are tomatoes.

Hence the hypergeometric model is appropriate and the probability that all 5 are tomatoes is

$$\mathrm{Pr}(X = 5) = p(10,5,5,5) = \frac{\binom{5}{5}\binom{5}{0}}{\binom{10}{5}} = .003968 \qquad \textit{from Appendix B3}$$

The probability of obtaining three or more cans of tomatoes is 1 minus the probability of drawing two or fewer. Thus we want

$$
\begin{aligned}
\mathrm{Pr}(X \geqq 3) &= 1 - \mathrm{Pr}(X \leqq 2) \\
&= 1 - P(10,5,5,2) \\
&= 1 - .500000 \qquad \textit{from Appendix B3} \\
&= .500000
\end{aligned}
$$

We observed that the binomial model is not appropriate when conditions (4-23) are satisfied because the probability of a success is not constant and successive draws are dependent. If N is large and n/N is small, then the probability of a success changes very little from draw to draw and events are practically independent. Consequently, one would expect that the binomial would give good approximations to hypergeometric probabilities in this situation. For example, if 3,000 out of 10,000 people in a community favor a proposal, then given any specific order of results for the first 10 randomly selected people, the probability that the eleventh person selected favors the proposal is near .3 (actually between $2{,}990/9{,}990 = .2997$ if all 10 favored and $3{,}000/9{,}990 = .3003$ if all 10 opposed). It has been verified by use of hypergeometric and binomial tables that the approximation

$$P(N,n,k,r) \cong \sum_{x=0}^{r} b\left(x;n, \frac{k}{N}\right) \qquad (4\text{-}28)$$

is reasonably good if $n/N \le .1$, $N > 50$ (for $N \le 50$ we would use the Lieberman and Owen tables), $k \ge n$. If $k < n$, then interchange the rolls of n, k in (4-28) getting

$$P(N,n,k,r) \cong \sum_{x=0}^{r} b\left(x;k, \frac{n}{N}\right) \qquad (4\text{-}29)$$

and require that $k/N \le .1$. The Lieberman and Owen publication also contains a good discussion on approximations to the hypergeometric distribution.

EXAMPLE 4-14

Suppose an organization contains 90 people of whom 45 are Republicans. A random sample of 5 members is selected to form a committee. What is the probability that 3 or less members of the committee are Republicans?

Solution

We check conditions (4-23).

(*a*)   There are $N = 90$ items to draw from.
(*b*)   The number of draws is $n = 5$, made at random.
(*c*) $k = 45$ of the 90 people are Republicans.

Hence the hypergeometric model is appropriate and we want $P(90,5,45,3)$. This entry is not found in Appendix B3. The most efficient thing to do is to get the Lieberman and Owen tables and read immediately

$$P(90,5,45,3) = P(90,45,5,3) = .819643$$

Alternatively, we can try approximation (4-28) since $n/N = \tfrac{5}{90} < .1$. We get

$$P(90,5,45,3) \cong \sum_{x=0}^{3} b(x;5,.50) = .81250$$

Thus the correct probability differs from the approximated value by about .007.

EXAMPLE 4-15

A company receives a moderately priced piece of equipment in lots of 100. A random sample of 10 pieces is selected and each piece tested. If the sample contains 0 or 1 defective pieces the lot is accepted. If the sample contains 2 or more defective pieces, all 100 items are tested before a decision is made. Assuming that the lot contains 5 defective pieces, what is the probability that all 100 will be inspected?

Solution

We check conditions (4-23).

(a)   There are $N = 100$ items to draw from.
(b)   The number of draws is $n = 10$, made at random.
(c) $k = 5$ of the 100 items are defective.

Hence the hypergeometric model is appropriate and we want

$$\Pr(X \geqq 2) = 1 - \Pr(X \leqq 1) = 1 - P(100,10,5,1)$$

where X is the number of defectives in the sample. Again, this can be evaluated from the Lieberman-Owen table, which yields $P(100,10,5,1) = .923143$. Thus the exact probability is $1 - .923143 = .076857$. Alternatively, since $k < n$ and $k/N = 5/100 < .1$, we can use approximation (4-29). We get $P(100,10,5,1) \cong \sum_{x=0}^{1} b(x;5,.10) = .91854$ so that the approximate answer is $1 - .91854 = .08146$.

To summarize concerning approximations, if the exact probability is immediately available in tables, it is senseless to use an approximation. However, when the range of existing tables is exceeded and the conditions for the use of an approximation are satisfied, then the approximate probability would be obtained and would be satisfactory for most applications.

The mean and variance of X can be found directly from (4-24) using the same procedures that produced these constants for the binomial, negative binomial, and uniform distributions. These calculations are complicated slightly by the fact that the limits on X are given by (4-25). The actual derivations are outlined in Exercise 4-30 in which it is to be shown that

$$E(X) = \frac{nk}{N} = np \tag{4-30}$$

$$\mathrm{Var}(X) = \frac{nk(N-k)(N-n)}{N^2(N-1)} = npq\,\frac{N-n}{N-1} \tag{4-31}$$

where $p = k/N$ and $q = 1 - p$.

In each case discuss the appropriateness of the hypergeometric model. Use Appendix B3 when possible.

4-22 The names of five men and five women are written on slips of paper and placed in a hat. After thorough mixing, four names are drawn. What is the probability that two are men and two are women? What is the probability that two or less are men?

4-23 A 5-card hand is dealt from a well-shuffled deck of 52 cards. What is the probability that it contains 3 aces? What is the probability that 3 or more cards in the hand are aces? What is the expected number of aces in a 5-card hand?

4-24 From a crate of 100 oranges, some of which are frozen, 10 are selected at random. Suppose that it is necessary to cut open an orange to determine whether or not it is frozen. In order to find the probability that 7 or more oranges out of the 10 are frozen if the box contains 20 frozen oranges, we could use the hypergeometric tables. What entry would we seek? (This turns out to be .000392.) Approximate the probability by using the binomial distribution. What conclusion is one apt to draw if a sample of 10 contains 7 frozen oranges?

4-25 In Exercise 4-24 find the mean and variance of X, the number of frozen oranges in the sample, if the box contains 20 frozen oranges.

4-26 A box of 10 screws contains 3 silver-colored screws and 7 black screws. If a random sample of 4 screws is selected, what is the probability that all 3 silver screws are in the sample? What is the probability that the number of silver screws in the sample is 3 or less? 2 or less?

4-27 The game of sheepshead is played with an ordinary deck of cards after all cards with denominations 2, 3, 4, 5, and 6 have been removed. In other words, the sheepshead deck contains 32 cards. Of the kings in a deck of cards, 3 have two eyes, and 1 has one eye. An 8-card hand is dealt from a well-shuffled sheepshead deck. To find the probability that the hand contains 0 or 1 two-eyed kings by using tables, we would seek which entry? (The actual probability is .853226.) Approximate the result using the binomial. The needed entry is not found in Appendix B1 but the calculation is easy.

4-28 An urn contains six red balls and four black balls from which a random sample of five balls is drawn. Let X be the number of red balls in the sample. Exhibit the probability distribution of X in table form.

4-29 Verify that $p(N,n,k,x) = p(N,k,n,x)$.

118

4-30   Use the fact that

$$\sum_{x=a}^{b} \frac{\dbinom{k}{x}\dbinom{N-k}{n-x}}{\dbinom{N}{n}} = 1$$

where $a = $ maximum $[0, n - (N - k)]$, $b = $ minimum $[k,n]$ (in other words, the sum of the probabilities in a probability distribution must be 1) to show that the mean and variance of a hypergeometric random variable are given by (4-30) and (4-31).

Hint: By definition

$$E(X) = \sum_{x=a}^{b} x \frac{\dbinom{k}{x}\dbinom{N-k}{n-x}}{\dbinom{N}{n}}$$

Show that this can be rewritten as

$$E(X) = \frac{nk}{N} \sum_{x=a}^{b} \frac{\dbinom{k-1}{x-1}\dbinom{N-k}{n-x}}{\dbinom{N-1}{n-1}}$$

Now let $x - 1 = y$, $x = y + 1$ and get (if $a > 0$)

$$E(X) = \frac{nk}{N} \sum_{y=a'}^{b'} \frac{\dbinom{k-1}{y}\dbinom{N-1-(k-1)}{n-1-y}}{\dbinom{N-1}{n-1}}$$

where $a' = $ maximum $[0, n - 1 - (N - 1 - k + 1)]$, $b' = $ minimum $[k - 1, n - 1]$. Hence, the last sum is the sum of all the probabilities in a hypergeometric distribution with constants $N - 1, k - 1, n - 1$, a sum which is 1. If $a = 0$, then

$$E(X) = \sum_{x=0}^{b} x \frac{\dbinom{k}{x}\dbinom{N-k}{n-x}}{\dbinom{N}{n}} = \sum_{x=1}^{b} x \frac{\dbinom{k}{x}\dbinom{N-k}{n-x}}{\dbinom{N}{n}}$$

$$= \frac{nk}{N} \sum_{y=0}^{b'} \frac{\dbinom{k-1}{y}\dbinom{N-1-(k-1)}{n-1-y}}{\dbinom{N-1}{n-1}} = \frac{nk}{N}$$

as before.   To find the variance we need

$$E(X^2) = E[X(X - 1) + X] = E[X(X - 1)] + E(X)$$

Show that

$$E[X(X - 1)] = \frac{n(n - 1)k(k - 1)}{N(N - 1)} \sum_{x=a}^{b} \frac{\dbinom{k - 2}{x - 2}\dbinom{N - k}{n - x}}{\dbinom{N - 2}{n - 2}}$$

Then let $x - 2 = y$ and proceed as in the evaluation of $E(X)$ to show that the latter sum is 1.   Finally, put the previous results together to show that (4-31) is correct.

4-6 THE POISSON DISTRIBUTION

A distribution often used to compute probabilities for random variables distributed over time and space is the Poisson distribution.   Experience has shown that the Poisson is an excellent model to use for computing probabilities associated with the number of calls coming into a telephone switchboard during a fixed period of time.   We may know that the board can handle up to 20 calls per minute without being overtaxed.   If the probability of 21 or more calls is sufficiently small, the board will provide efficient service.   If it is not, then perhaps more lines or operators are needed to provide a satisfactory standard of service.   The Poisson model is used to compute the required probability.   Some other random variables for which the Poisson has been used to evaluate probabilities are:

1 The number of automobile deaths per month in a large city
2 The number of bacteria in a given culture
3 The number of red blood cells in a specimen of blood
4 The number of meteorites located on an acre of desert land
5 The number of typing errors per page
6 The number of defects in a manufactured article
7 The number of atoms disintegrating per second from radioactive material
8 The number of buzz-bomb hits on a square mile of London in 1944
9 The number of calls an individual receives per day
10 The number of deaths from horse kicks per year for each army corps in the Prussian Army over a period of 20 years

In each of the above situations the following assumptions seem reasonable:

120

(a)   *Events that occur in one time interval (or region of space)
are independent of those occurring in any other nonoverlapping time interval (or region of space).*

(b)   *For a small time interval (or region of space) the probability that an event occurs is proportional to the length of*   (4-32)
the time interval (or region of space).

(c)   *The probability that two or more events occur in a very small
time interval (or region of space) is so small that it can be
neglected.*

By using the assumptions (4-32), it is possible to show that the probability that exactly x successes occur in a given time interval (or given
region of space) is

$$p(x;\mu) = \frac{e^{-\mu}\mu^x}{x!} \qquad x = 0, 1, 2, \ldots \tag{4-33}$$

where $e = 2.71828 \cdot \cdot \cdot$ and μ (as we shall show) is the average number
of successes occurring in the given time interval (or given region of space).
Appendix B4 contains cumulative sums

$$P(r;\mu) = \sum_{x=0}^{r} p(x;\mu) = \Pr(X \leq r) \tag{4-34}$$

for some values of μ. A fairly extensive table has been published by
Molina [3]. His table gives $\Pr(X \geq r)$ to at least six decimal places for
$\mu = .001(.001).010(.01).30(.1)15(1)100$.

It is not as easy as with some of the preceding models to convince one-
self that the assumptions (4-32) of the Poisson are reasonable. Sometimes it is obvious that the independence condition is not satisfied. For
example, we might be tempted to use the Poisson to compute the probability distribution of the number of cornborers found in a hill of corn. A
little reflection reveals that events are not independent, since insects are
usually hatched in batches. As we have previously mentioned, however,
models sometimes give fairly accurate probabilities even though all the
assumptions are not satisfied. To pass final judgement on the appropriateness of the model, one has to rely on accumulated experimental
evidence.

EXAMPLE 4-16 ───

If a person receives five calls on the average during a day, what is the
probability that he will receive fewer than five calls tomorrow? Exactly
five calls?

Solution

According to previous discussion, experience has shown that the Poisson model is appropriate for this situation. The average $\mu = 5$ is given. Thus we need

$$\Pr(X \leq 4) = P(4;5) = \sum_{x=0}^{4} p(x;5) = .44049 \qquad \text{by Appendix B4}$$

to answer the first question. The probability of receiving exactly 5 calls is

$$\begin{aligned} \Pr(X \leq 5) - \Pr(X \leq 4) &= P(5;5) - P(4;5) \\ &= .61596 - .44049 \qquad \text{by Appendix B4} \\ &= .17547 \end{aligned}$$

That is, the probability of receiving exactly five calls is equal to the probability of receiving five or fewer minus the probability of receiving four or fewer.

EXAMPLE 4-17

A secretary claims that she averages one error per page. A sample page is selected at random from some of her work, and five errors are counted. What is the probability of her making five or more errors on a page if her claim is correct?

Solution

Perhaps the errors made are not independent of one another. If one or two errors occurring relatively near each other disturb the secretary, she may be inclined to make more errors relatively soon. However, letting X be the number of errors on a page and assuming that the Poisson is appropriate, the probability is

$$\begin{aligned} \Pr(X \geq 5) = \sum_{x=5}^{\infty} p(x;1) &= 1 - \sum_{x=0}^{4} p(x;1) \\ &= 1 - P(4;1) \\ &= 1 - .99634 \qquad \text{from Appendix B4} \\ &= .00366 \end{aligned}$$

In view of the small probability we might be inclined to conclude one of the following:

1 The Poisson model is correct and a near miracle has occurred.
2 The model is correct but the wrong average value μ has been used.
3 The model is incorrect.

Probably 2 is the most plausible.

Besides being a probability model in its own right, the Poisson is sometimes used to approximate binomial probabilities when n is large and p is close to 0 or 1. Of course, it would make no sense to approximate a binomial sum that can be read directly from the tables. The individual terms of the binomial are replaced with the corresponding terms from the Poisson with $\mu = np$.

EXAMPLE 4-18 ━━━━━━━━━━━━━━━━━━━━━━━━━━━━━━━━━━━━━━━

A life insurance company has found that the probability is .00001 that a person in the 40 to 50 age bracket dies during a year period from a certain rare disease. If the company has 100,000 policyholders in this group, what is the probability that it must pay off more than four claims during a year because of death from this cause?

Solution

The binomial model is not unreasonable since

(a) A person either dies from the disease during the year or he does not.
(b) The records give $p = .00001$, which we shall assume is constant for each person.
(c) Presumably, whether a person dies from the disease in no way affects what happens to another person in the group.
(d) The number of trials is $n = 100,000$.

Thus, the probability of more than four claims is

$$\sum_{x=5}^{100,000} b(x;100,000, .00001) = 1 - \sum_{x=0}^{4} b(x;100,000, .00001)$$

Since $\mu = np = 100,000(.00001) = 1$, we use the approximation

$$\sum_{x=0}^{4} b(x;100,000, .00001) \cong \sum_{x=0}^{4} p(x;1)$$
$$= .99634 \qquad \textit{by Appendix B4}$$

Hence $\Pr(X \geq 5) \cong 1 - .99634 = .00366$.

━━━

The mean $E(X)$ can be found by making use of the fact

$$\sum_{x=0}^{\infty} \frac{e^{-\mu}\mu^{x}}{x!} = 1 \qquad\qquad\qquad (4\text{-}35)$$

123

That is, the sum of the probabilities in a probability distribution must equal 1. By the definition of expected value we have

$$E(X) = \sum_{x=0}^{\infty} x \frac{e^{-\mu}\mu^x}{x!} = \sum_{x=1}^{\infty} x \frac{e^{-\mu}\mu^x}{x!}$$

$$= \sum_{x=1}^{\infty} \frac{e^{-\mu}\mu^x}{(x-1)!}$$

Now, let $y = x - 1$, $x = y + 1$ in the latter sum and we get

$$E(X) = \sum_{y=0}^{\infty} \frac{e^{-\mu}\mu^{y+1}}{y!} = \mu \sum_{y=0}^{\infty} \frac{e^{-\mu}\mu^{y}}{y!}$$

which, because of (4-35), reduces to

$$E(X) = \mu \tag{4-36}$$

Hence we are justified in using the symbol μ for the constant appearing in the Poisson formula. By similar calculations we can show

$$\text{Var}(X) = \mu \tag{4-37}$$

The actual steps are outlined in Exercise 4-36.

EXERCISES

Comment upon the reasonableness of the Poisson model where necessary.

4-31 A city has, on the average, five traffic deaths per month. What is the probability that this average is exceeded in any given month?

4-32 If a typist makes two errors per page on the average, what is the probability of her typing a page with no errors? With one error?

4-33 A taxicab company has, on the average, 10 flat tires per week. During the past week they had 20. Assuming that the Poisson model is appropriate, what is the probability of having 20 or more flats during a week? Would you suspect foul play?

4-34 An intercontinential ballistic missile has 10,000 parts. The probability that each part does not fail during a flight is .99995, and parts work independently of one another. What is the probability of a successful flight?

4-35 An automobile insurance company has found that the probability of paying off on any given policy during a year is .001. What is the probability that the company has to pay 15 or more claims next year if it holds 10,000 policies?

4-36   Use the fact that

$$\sum_{x=0}^{\infty} \frac{e^{-\mu}\mu^x}{x!} = 1$$

to find the variance of a Poisson random variable. *Hint:* Find

$$E(X^2) = E[X(X - 1)] + E(X)$$

where

$$E[X(X - 1)] = \sum_{x=0}^{\infty} x(x - 1)\frac{e^{-\mu}\mu^x}{x!} = \sum_{x=2}^{\infty} \frac{e^{-\mu}\mu^x}{(x - 2)!}$$

By letting $y = x - 2$ show that the latter sum has value μ^2.

REFERENCES

1   HARVARD UNIVERSITY COMPUTATION LABORATORY: "Tables of the Cumulative Binomial Probability Distribution," Harvard University Press, Cambridge, Mass., 1955.

2   LIEBERMAN, G. J., and D. B. OWEN: "Tables of the Hypergeometric Probability Distribution," Stanford University Press, Stanford, Calif., 1961.

3   MOLINA, E. C.: "Poisson's Exponential Binomial Limit," D. Van Nostrand Company, Inc., Princeton, N.J., 1949.

4   NATIONAL BUREAU OF STANDARDS: Tables of the Binomial Probability Distribution, *Applied Mathematics*, Series 6, U.S. Government Printing Office, Washington, D.C., 1950.

5   OWEN, D. B.: "Handbook of Statistical Tables," Addison-Wesley Publishing Company, Inc., Reading, Mass., 1962.

6   PATIL, G. P.: On the Evaluation of the Negative Binomial Distribution with Examples, *Technometrics*, Vol. 2, No. 4, pp. 501–505, 1960.

7   RAND CORPORATION, "A Million Random Digits with 100,000 Normal Deviates," The Free Press of Glencoe, New York, 1955.

8   ROBERTSON, WILLIAM H.: "Tables of the Binomial Distribution Function for Small Values of *p*," Sandia Corporation, Albuquerque, New Mexico, 1960. (Available from the Office of Technical Services, Department of Commerce, Washington, D.C.)

9   ROMIG, H. G.: "50–100 Binomial Tables," John Wiley & Sons, Inc. New York, 1953.

10 U.S. ARMY ORDNANCE CORPS: Tables of Cumulative Binomial Probabilities, *Ordnance Corps Pamphlet ORDP*20-1, September, 1952.

11 WEINTRAUB, SOL: "Tables of the Cumulative Binomial Probability Distribution for Small Values of *p*," The Free Press of Glencoe, New York, 1963.

12 WILLIAMSON, ERIC, and MICHAEL BRETHERTON: "Tables of the Negative Binomial Probability Distribution," John Wiley & Sons, Inc., New York, 1963.

4-2   The probability function of X, the number of successes in n trials, is

$$b(x;n,p) = \binom{n}{x} p^x q^{n-x} \qquad x = 0, 1, \ldots, n \qquad (4\text{-}2)$$

Conditions (4-1) must be satisfied.

$$B(r;n,p) = \Pr(X \leqq r) = \sum_{x=0}^{r} b(x;n,p) \qquad (4\text{-}3)$$

To enter the binomial table with $p > \frac{1}{2}$ use

$$B(r;n,p) = 1 - B(n - r - 1; n, q) \qquad (4\text{-}4)$$

For a binomial random variable

$$E(X) = np \qquad (4\text{-}6)$$

$$\operatorname{Var}(X) = npq \qquad (4\text{-}7)$$

4-3   The probability function of N, the number of trials required for c successes, is

$$b^*(n;c,p) = \binom{n-1}{c-1} p^c q^{n-c} \qquad n = c, c + 1, c + 2, \ldots \qquad (4\text{-}9)$$

Conditions (4-8) must be satisfied.   Then

$$B^*(r;c,p) = \Pr(N \leqq r) = \sum_{n=c}^{r} b^*(n;c,p)$$

$$= 1 - B(c - 1; r, p) \qquad (4\text{-}10)$$

For a negative binomial random variable,

$$E(N) = \frac{c}{p} \qquad (4\text{-}12)$$

$$\operatorname{Var}(N) = \frac{cq}{p^2} \qquad (4\text{-}13)$$

4-4   The probability function of X, the uniform random variable, is

$$f(x;k) = \frac{1}{k} \qquad x = 1, 2, \ldots, k \qquad (4\text{-}14)$$

or, slightly more generally,

$$f(x;k) = \frac{1}{k} \qquad x = x_1, x_2, \ldots, x_k \qquad (4\text{-}15)$$

For form (4-14),

$$E(X) = \frac{k+1}{2}$$

$$\text{Var}(X) = \frac{k^2 - 1}{12}$$

(4-19)

while for form (4-15),

$$E(X) = \frac{1}{k} \sum_{i=1}^{k} x_i$$

(4-20)

$$\text{Var}(X) = \frac{1}{k} \sum_{i=1}^{k} x_i^2 - [E(X)]^2$$

(4-21)

4-5 The probability function of X, the hypergeometric random variable, is

$$p(N,n,k,x) = \frac{\binom{k}{x}\binom{N-k}{n-x}}{\binom{N}{n}}$$

(4-24)

Conditions (4-23) must be satisfied. Then

$$P(N,n,k,r) = \text{Pr}(X \leqq r)$$

(4-26)

To approximate sums of hypergeometric probabilities we may use

$$P(N,n,k,r) \cong \sum_{x=0}^{r} b\left(x;n,\frac{k}{N}\right)$$

(4-28)

if $k \geqq n$, $n/N \leqq .1$, $N > 50$, or

$$P(N,n,k,r) \cong \sum_{x=0}^{r} b\left(x;k,\frac{n}{N}\right)$$

(4-29)

if $k < n$, $k/N \leqq .1$, $N > 50$.

For the hypergeometric random variable

$$E(X) = \frac{nk}{N} = np$$

(4-30)

$$\text{Var}(X) = \frac{nk(N-k)(N-n)}{N^2(N-1)} = npq\,\frac{N-n}{N-1}$$

(4-31)

where $p = k/N$, $q = 1 - p$.

4-6 The probability function of X, the Poisson random variable, is

$$p(x;\mu) = \frac{e^{-\mu}\mu^x}{x!} \qquad x = 0, 1, 2, \ldots \tag{4-33}$$

Conditions (4-32) should seem reasonable. Then

$$P(r;\mu) = \Pr(X \leqq r) = \sum_{x=0}^{r} p(x;\mu) \tag{4-34}$$

For the Poisson random variable

$$E(X) = \mu \tag{4-36}$$

$$\mathrm{Var}(X) = \mu \tag{4-37}$$

CHAPTER FIVE

Random Variables and Distributions in Two and More Dimensions

5-1 JOINT DISTRIBUTIONS

In Chap. 3 we defined a random variable as a quantity whose value is determined by the outcome of a random experiment. Frequently the performance of the experiment determines more than one characteristic of interest. As a simple illustration, consider a card drawn at random from an ordinary deck. For many card games we would be interested in both the denomination and the suit. If we let X be a random variable which can assume the values 1, 2, . . . , 13 corresponding respectively to the denominations Ace, 2, 3, . . . , 10, J, Q, K and let Y be a random variable which can assume the values 1, 2, 3, 4 corresponding respectively to the suits clubs, diamonds, hearts, and spades, then the outcome of random drawing can be described by the two-dimensional random variable (X, Y). Since the drawing is done at random, the probability of selecting

129

Table 5-1 Joint probability distribution for the card example

y \ x	1	2	3	4	5	6	7	8	9	10	11	12	13	Row totals
1	$\frac{1}{52}$	$\frac{1}{52}$	$\frac{1}{52}$	$\frac{1}{52}$	$\frac{1}{52}$	$\frac{1}{52}$	$\frac{1}{52}$	$\frac{1}{52}$	$\frac{1}{52}$	$\frac{1}{52}$	$\frac{1}{52}$	$\frac{1}{52}$	$\frac{1}{52}$	$\frac{13}{52}$
2	$\frac{1}{52}$	$\frac{1}{52}$	$\frac{1}{52}$	$\frac{1}{52}$	$\frac{1}{52}$	$\frac{1}{52}$	$\frac{1}{52}$	$\frac{1}{52}$	$\frac{1}{52}$	$\frac{1}{52}$	$\frac{1}{52}$	$\frac{1}{52}$	$\frac{1}{52}$	$\frac{13}{52}$
3	$\frac{1}{52}$	$\frac{1}{52}$	$\frac{1}{52}$	$\frac{1}{52}$	$\frac{1}{52}$	$\frac{1}{52}$	$\frac{1}{52}$	$\frac{1}{52}$	$\frac{1}{52}$	$\frac{1}{52}$	$\frac{1}{52}$	$\frac{1}{52}$	$\frac{1}{52}$	$\frac{13}{52}$
4	$\frac{1}{52}$	$\frac{1}{52}$	$\frac{1}{52}$	$\frac{1}{52}$	$\frac{1}{52}$	$\frac{1}{52}$	$\frac{1}{52}$	$\frac{1}{52}$	$\frac{1}{52}$	$\frac{1}{52}$	$\frac{1}{52}$	$\frac{1}{52}$	$\frac{1}{52}$	$\frac{13}{52}$
Column totals	$\frac{4}{52}$	$\frac{4}{52}$	$\frac{4}{52}$	$\frac{4}{52}$	$\frac{4}{52}$	$\frac{4}{52}$	$\frac{4}{52}$	$\frac{4}{52}$	$\frac{4}{52}$	$\frac{4}{52}$	$\frac{4}{52}$	$\frac{4}{52}$	$\frac{4}{52}$	1

any card is $\frac{1}{52}$ and $\Pr(X = x, Y = y) = f(x,y) = \frac{1}{52}$, $x = 1, 2, \ldots, 13$, $y = 1, 2, 3, 4$. The function $f(x,y)$ that yields probabilities for all X and Y is called the *joint probability function* of the random variables X and Y. These probabilities can be given in the form of a table (see Table 5-1), which we shall refer to as a *joint probability distribution*.

In the card example two random variables were generated by the same card. Many times the two random variables will be associated with two repetitions of the same or similar type experiments. For example, suppose we toss a four-sided symmetric die twice and let $X_1 =$ number of spots on the down side on the first toss, $X_2 =$ number of spots on the down side on the second toss. Then, we easily obtain

$$\Pr(X_1 = x_1, X_2 = x_2) = f(x_1,x_2) = \frac{1}{16} \qquad \begin{matrix} x_1 = 1, 2, 3, 4 \\ x_2 = 1, 2, 3, 4 \end{matrix}$$

for the joint probability function and the joint probability distribution appearing in Table 5-2.

Table 5-2 Joint probability distribution for four-sided die example

x_2 \ x_1	1	2	3	4	Row totals
1	$\frac{1}{16}$	$\frac{1}{16}$	$\frac{1}{16}$	$\frac{1}{16}$	$\frac{4}{16}$
2	$\frac{1}{16}$	$\frac{1}{16}$	$\frac{1}{16}$	$\frac{1}{16}$	$\frac{4}{16}$
3	$\frac{1}{16}$	$\frac{1}{16}$	$\frac{1}{16}$	$\frac{1}{16}$	$\frac{4}{16}$
4	$\frac{1}{16}$	$\frac{1}{16}$	$\frac{1}{16}$	$\frac{1}{16}$	$\frac{4}{16}$
Column totals	$\frac{4}{16}$	$\frac{4}{16}$	$\frac{4}{16}$	$\frac{4}{16}$	

Let us consider one further probability distribution for which all probabilities are not the same. Suppose that two ordinary six-sided symmetric dice are tossed. For each die let the result 1, 2, 3 be designated as category C_1, the result 4 or 5 be designated as category C_2, and the result 6 as category C_3. Let

X_1 = number of occurrences of category C_1
X_2 = number of occurrences of category C_2

Then, the number of occurrences of category C_3 is $2 - X_1 - X_2$ and both X_1 and X_2 can assume the values 0, 1, and 2 subject to the restriction that $X_1 + X_2 \leq 2$. To illustrate the calculation of probabilities, suppose we find $\Pr(X_1 = 1, X_2 = 0)$. The event $(X_1 = 1, X_2 = 0)$ is achieved if either of the mutually exclusive events C_1C_3 or C_3C_1 happens. That is, die 1 produces category 1 and die 2 produces category 3 or vice versa. Hence,

$$\Pr(X_1 = 1, X_2 = 0) = (\tfrac{3}{6})(\tfrac{1}{6}) + (\tfrac{1}{6})(\tfrac{3}{6}) = \tfrac{6}{36}$$

Similarly, we can find $\Pr(X_1 = 2, X_2 = 0)$ by observing that the event $(X_1 = 2, X_2 = 0)$ is only produced by C_1C_1. Hence,

$$\Pr(X_1 = 2, X_2 = 0) = (\tfrac{3}{6})(\tfrac{3}{6}) = \tfrac{9}{36}$$

Further calculations of the same type allow us to complete Table 5-3. The joint probability function is

$$f(x_1,x_2) = \frac{2!}{x_1!x_2!(2 - x_1 - x_2)!} \left(\frac{3}{6}\right)^{x_1} \left(\frac{2}{6}\right)^{x_2} \left(\frac{1}{6}\right)^{2-x_1-x_2} \qquad \begin{array}{l} x_1 = 0, 1, 2 \\[4pt] x_2 = 0, 1, 2 \\[4pt] 0 \leq x_1 + x_2 \leq 2 \end{array}$$

a special case of the multinomial distribution, which we shall consider later.

Table 5-3   Joint probability distribution of X_1 and X_2

x_2	x_1 0	1	2	Row totals
0	$\tfrac{1}{36}$	$\tfrac{6}{36}$	$\tfrac{9}{36}$	$\tfrac{16}{36}$
1	$\tfrac{4}{36}$	$\tfrac{12}{36}$	0	$\tfrac{16}{36}$
2	$\tfrac{4}{36}$	0	0	$\tfrac{4}{36}$
Column totals	$\tfrac{9}{36}$	$\tfrac{18}{36}$	$\tfrac{9}{36}$	1

From the joint probability distribution of two random variables we can get the one-dimensional probability distribution of each of the random variables. For example, using Table 5-3 we see that the column totals are

$$\Pr(X_1 = 0) = \Pr(X_1 = 0, X_2 = 0) + \Pr(X_1 = 0, X_2 = 1)$$
$$+ \Pr(X_1 = 0, X_2 = 2)$$

$$= \sum_{x_2=0}^{2} f(0,x_2) = \tfrac{9}{36} = \tfrac{1}{4}$$

$$\Pr(X_1 = 1) = \Pr(X_1 = 1, X_2 = 0) + \Pr(X_1 = 1, X_2 = 1)$$
$$+ \Pr(X_1 = 1, X_2 = 2)$$

$$= \sum_{x_2=0}^{1} f(1,x_2) = \tfrac{18}{36} = \tfrac{2}{4}$$

$$\Pr(X_1 = 2) = \Pr(X_1 = 2, X_2 = 0) + \Pr(X_1 = 2, X_2 = 1)$$
$$+ \Pr(X_1 = 2, X_2 = 2)$$

$$= \sum_{x_2=0}^{0} f(2,x_2) = \tfrac{9}{36} = \tfrac{1}{4}$$

or

$$f(x_1,x_2) = \sum_{x_2=0}^{2-x_1} f(x_1,x_2) \qquad x_1 = 0, 1, 2$$

The upper limit on the sum changes due to the restriction $x_1 + x_2 \leqq 2$, or $x_2 \leqq 2 - x_1$, which makes some of the probabilities in Table 5-3 equal to 0. Similarly the row totals are $\Pr(X_2 = 0)$, $\Pr(X_2 = 1)$ and $\Pr(X_2 = 2)$. The probability distributions of X_1 and of X_2 are given in Table 5-4 (after expressing the probabilities with a smaller common denominator). These are called marginal probability distributions. In

Table 5-4 Marginal probability distributions for the X_1 and X_2 of Table 5-3

x_1	0	1	2
$f_1(x_1)$	$\tfrac{1}{4}$	$\tfrac{2}{4}$	$\tfrac{1}{4}$
x_2	0	1	2
$f_2(x_2)$	$\tfrac{4}{9}$	$\tfrac{4}{9}$	$\tfrac{1}{9}$

general, the marginal probabilities are found by evaluating

$$f_1(x_1) = \sum_{x_2} f(x_1, x_2) \tag{5-1}$$

$$f_2(x_2) = \sum_{x_1} f(x_1, x_2) \tag{5-2}$$

where the sums are taken over all values of the other random variable. Thus we see that when the joint probability distribution is available (that is, a table such as Table 5-3), then the column totals are the probabilities that appear in the marginal probability distribution of X_1 and the row totals are the probabilities that appear in the marginal probability distribution of X_2.

Next suppose that for the random variables of Table 5-3 we are given that $X_2 = 0$. With this information and our knowledge of conditional probability we can find the *conditional probability distribution* of X_1 given $X_2 = 0$. We have

$$\Pr(X_1 = 0 | X_2 = 0) = \frac{\Pr(X_1 = 0, X_2 = 0)}{\Pr(X_2 = 0)} = \frac{1/36}{16/36} = 1/16$$

$$\Pr(X_1 = 1 | X_2 = 0) = \frac{\Pr(X_1 = 1, X_2 = 0)}{\Pr(X_2 = 0)} = \frac{6/36}{16/36} = 6/16$$

$$\Pr(X_1 = 2 | X_2 = 0) = \frac{\Pr(X_1 = 2, X_2 = 0)}{\Pr(X_2 = 0)} = \frac{9/36}{16/36} = 9/16$$

In general, we shall use the notation

$$g(x_1 | x_2) = \Pr(X_1 = x_1 | X_2 = x_2) = \frac{\Pr(X_1 = x_1, X_2 = x_2)}{\Pr(X_2 = x_2)} = \frac{f(x_1, x_2)}{f_2(x_2)} \tag{5-3}$$

and

$$h(x_2 | x_1) = \Pr(X_2 = x_2 | X_1 = x_1) = \frac{\Pr(X_1 = x_1, X_2 = x_2)}{\Pr(X = x_1)} = \frac{f(x_1, x_2)}{f_1(x_1)} \tag{5-4}$$

for the conditional probability functions of X_1 given X_2 and X_2 given X_1, respectively. [In (5-3) and (5-4) we must, of course, require $f_2(x_2) > 0$, $f_1(x_1) > 0$.] Thus, using our above calculations, the probability distribution of X_1 given $X_2 = 0$ can be written

$$g(x_1 | 0) = 1/16 \qquad x_1 = 0$$
$$= 6/16 \qquad x_1 = 1$$
$$= 9/16 \qquad x_1 = 2$$

133

Similarly, by using (5-4) and the probabilities in Table 5-3 we get for the probability distribution of X_2 given $X_1 = 0$

$$
\begin{aligned}
h(x_2|0) &= \tfrac{1}{9} & x_2 &= 0 \\
&= \tfrac{4}{9} & x_2 &= 1 \\
&= \tfrac{4}{9} & x_2 &= 2
\end{aligned}
$$

Let us return to the random variable of Table 5-1. We note that marginal probability functions are

$$
\begin{aligned}
f_1(x) &= \tfrac{1}{13} & x &= 1, 2, \ldots, 13 \\
f_2(y) &= \tfrac{1}{4} & y &= 1, 2, 3, 4
\end{aligned}
$$

It is easy to see that for all values of X and Y we have

$$
f(x,y) = f_1(x)f_2(y) \tag{5-5}
$$

or

$$
\Pr(X = x, Y = y) = \Pr(X = x)\Pr(Y = y)
$$

Thus, according to the discussion of Sec. 2-5 and formula (2-18), the events $X = x$ and $Y = y$ are independent. Whenever Eq. (5-5) holds for all possible choices of X and Y, then the random variables X and Y are said to be *independent*. Random variables that are not independent are said to be *dependent*. If condition (5-5) fails for one choice of X and Y, then the random variables are dependent. For X_1 and X_2 of Table 5-3, we have

$$
f_1(0) = \tfrac{9}{36} = \tfrac{1}{4} \qquad f_2(0) = \tfrac{16}{36} = \tfrac{4}{9}
$$

and

$$
f_1(0)f_2(0) = (\tfrac{1}{4})(\tfrac{4}{9}) = (\tfrac{1}{9}) \neq \tfrac{1}{36} = f(0,0)
$$

Thus X_1 and X_2 are dependent random variables.

We observe that if the random variables X and Y are independent and the marginal probability functions are known, then it is a very simple task to construct the joint probability function. All we have to do is form the product on the right-hand side of Eq. (5-5). Similar remarks can be made for the situation in which a conditional probability function is known together with the marginal probability function of the given random variable. For example, if we know $g(x_1|x_2)$ and $f_2(x_2)$, then from (5-3) we see that the joint probability function is

$$
f(x_1,x_2) = g(x_1|x_2)f_2(x_2) \tag{5-6}
$$

EXAMPLE 5-1

Suppose that a six-sided symmetric die and an ordinary coin are each thrown twice. Let X be the number of 6's the die produces and Y be the number of heads resulting from the throw of the coin. Find the joint probability function of X and Y and write out the joint probability distribution and the marginal totals.

Solution

We know from our discussion in Sec. 4-2 that X has a binomial distribution with $n = 2$, $p = \frac{1}{6}$ so that

$$f_1(x) = \binom{2}{x}\left(\frac{1}{6}\right)^x\left(\frac{5}{6}\right)^{2-x} \qquad x = 0, 1, 2$$

Similarly Y has a binomial distribution with $n = 2$, $p = \frac{1}{2}$ and

$$f_2(y) = \binom{2}{y}\left(\frac{1}{2}\right)^y\left(\frac{1}{2}\right)^{2-y} \qquad y = 0, 1, 2$$

Hence

$$f(x,y) = \binom{2}{x}\left(\frac{1}{6}\right)^x\left(\frac{5}{6}\right)^{2-x}\binom{2}{y}\left(\frac{1}{2}\right)^y\left(\frac{1}{2}\right)^{2-y} \qquad \begin{matrix} x = 0, 1, 2 \\ y = 0, 1, 2 \end{matrix}$$

since X and Y are independent. To form the joint probability distribution with independent random variables, first find the marginal probabilities and fill in the row and column totals; then fill in the cells of the table by taking products. We get

$$f_1(0) = \frac{25}{36} \qquad f_1(1) = \frac{10}{36} \qquad f_1(2) = \frac{1}{36}$$

$$f_2(0) = \frac{1}{4} \qquad f_2(1) = \frac{2}{4} \qquad f_2(2) = \frac{1}{4}$$

With these values we easily construct Table 5-5.

Table 5-5 *Joint probability distribution for Example 5-1*

y	x 0	1	2	Row totals
0	$^{25}\!/_{144}$	$^{10}\!/_{144}$	$^{1}\!/_{144}$	$\frac{1}{4}$
1	$^{50}\!/_{144}$	$^{20}\!/_{144}$	$^{2}\!/_{144}$	$\frac{2}{4}$
2	$^{25}\!/_{144}$	$^{10}\!/_{144}$	$^{1}\!/_{144}$	$\frac{1}{4}$
Column totals	$^{25}\!/_{36}$	$^{10}\!/_{36}$	$^{1}\!/_{36}$	1

EXAMPLE 5-2

Suppose that a coin is tossed twice, and let Y be the number of heads. After the value of Y, say, y, is determined, a die is rolled y times with the occurrence of a 1 or 2 being regarded as a success. Let X be the number of successes obtained with the die. Find the joint probability function of X and Y and write out the joint probability distribution and the marginal totals.

Solution

We already observed in Example 5-1 that the probability function of Y is

$$f_2(y) = \binom{2}{y}\left(\frac{1}{2}\right)^y \left(\frac{1}{2}\right)^{2-y} \qquad y = 0, 1, 2$$

For each given value of Y, say, y, X has a binomial distribution with $n = y$, $p = \frac{1}{3}$ so that

$$g(x|y) = \binom{y}{x}\left(\frac{1}{3}\right)^x \left(\frac{2}{3}\right)^{y-x} \qquad 0 \le x \le y \le 2$$

According to (5-6) the joint probability function is

$$f(x,y) = \binom{2}{y}\left(\frac{1}{2}\right)^y \left(\frac{1}{2}\right)^{2-y} \binom{y}{x}\left(\frac{1}{3}\right)^x \left(\frac{2}{3}\right)^{y-x} \qquad 0 \le x \le y \le 2$$

In constructing Table 5-6 we can fill in all cells with zeros when $X > Y$. Of course, the row totals will be the same as in Table 5-5. Hence, for the cells with nonzero probability we need only $g(x|y)$ for each case. We easily compute

$$g(0|0) = 1$$

$$g(0|1) = \frac{2}{3} \qquad g(1|1) = \frac{1}{3}$$

$$g(0|2) = \frac{4}{9} \qquad g(1|2) = \frac{4}{9} \qquad g(2|2) = \frac{1}{9}$$

Table 5-6 *Joint probability distribution for Example 5-2*

	x			
y	**0**	**1**	**2**	**Row totals**
0	$\frac{1}{4}$	0	0	$\frac{1}{4}$
1	$\frac{4}{12}$	$\frac{2}{12}$	0	$\frac{2}{4}$
2	$\frac{4}{36}$	$\frac{4}{36}$	$\frac{1}{36}$	$\frac{1}{4}$
Column totals	$\frac{25}{36}$	$\frac{10}{36}$	$\frac{1}{36}$	1

and fill in the missing cells. We observe that X has a binomial distribution with $n = 2$ and $p = (\frac{1}{2})(\frac{1}{3}) = \frac{1}{6}$. This is a special case of the result sought in Exercise 5-9.

EXAMPLE 5-3

From a sack of fruit containing three oranges, two apples, and three bananas a random sample of four pieces of fruit is selected. If X is the number of oranges and Y is the number of apples in the sample find the joint probability distribution of X and Y and write down the joint probability function.

Solution

It is easy to enumerate the possible pairs of values for X and Y, say, (x,y). These are $(0,1)$, $(0,2)$, $(1,0)$, $(1,1)$, $(1,2)$, $(2,0)$, $(2,1)$, $(2,2)$, $(3,0)$, $(3,1)$. It is, of course, impossible to have the pairs $x = 0$, $y = 0$ and $x = 3$, $y = 2$ since this would require drawing four bananas with only three in the sack to achieve the first event, and having a sample of five to achieve the second. Now we shall illustrate the calculation of probabilities. Suppose we want $\Pr(X = 1, Y = 2)$. The drawing is done at random and there are $\binom{8}{4}$ points, so each point is assigned probability $1 \Big/ \binom{8}{4}$. Now, all we have to do is count the number of sample points that correspond to the event $X = 1$, $Y = 2$ (implying one banana is drawn). Since one orange can be selected from three in $\binom{3}{1}$ ways, two apples can be selected from two in $\binom{2}{2}$ ways, and one banana selected from three in $\binom{3}{1}$ ways, the total number of sample points corresponding to the event $X = 1$, $Y = 2$ is $\binom{3}{1}\binom{2}{2}\binom{3}{1}$ and

$$\Pr(X = 1, Y = 2) = \binom{3}{1}\binom{2}{2}\binom{3}{1} \Big/ \binom{8}{4} = \frac{9}{70}$$

Similarly,

$$\Pr(X = 0, Y = 1) = \binom{3}{0}\binom{2}{1}\binom{3}{3} \Big/ \binom{8}{4} = \frac{2}{70}$$

etc. Eight more calculations of this type allow us to complete the entries in Table 5-7. All ten calculations can be summarized by the joint probability function

$$f(x,y) = \frac{\binom{3}{x}\binom{2}{y}\binom{3}{4 - x - y}}{\binom{8}{4}}$$

Table 5-7 Joint probability distribution for Example 5-3

y \ x	0	1	2	3	Row totals
0	0	$\tfrac{3}{70}$	$\tfrac{9}{70}$	$\tfrac{3}{70}$	$\tfrac{15}{70}$
1	$\tfrac{2}{70}$	$\tfrac{18}{70}$	$\tfrac{18}{70}$	$\tfrac{2}{70}$	$\tfrac{40}{70}$
2	$\tfrac{3}{70}$	$\tfrac{9}{70}$	$\tfrac{3}{70}$	0	$\tfrac{15}{70}$
Column totals	$\tfrac{5}{70}$	$\tfrac{30}{70}$	$\tfrac{30}{70}$	$\tfrac{5}{70}$	1

for the 10 points listed previously. To make the combination symbols meaningful, we must have $0 \leqq x \leqq 3$, $0 \leqq y \leqq 2$, $0 \leqq 4 - x - y \leqq 3$ all satisfied. The latter pair of inequalities can also be written as $1 \leqq x + y \leqq 4$.

The marginal probability distributions in Table 5-7 were obtained by computing the joint probabilities first and then adding entries across rows and down columns. It is easy to recognize that both X and Y have hypergeometric distributions. For example, consider the sack as being composed of three oranges and five non-oranges. Then if four pieces are selected at random, the probability function of X is, according to the discussion of Sec. 4-5,

$$f_1(x) = \frac{\binom{3}{x}\binom{5}{4-x}}{\binom{8}{4}} \qquad x = 0, 1, 2, 3$$

Similarly, the probability function of Y is

$$f_2(y) = \frac{\binom{2}{y}\binom{6}{4-y}}{\binom{8}{4}} \qquad y = 0, 1, 2$$

That $f_1(x)$ and $f_2(y)$ yield the column totals and the row totals can be quickly verified.

In all the previous examples, only a finite number of points in the joint distribution of X and Y had nonzero probabilities. Obviously, as with one-dimensional random variables, this need not be the case. Suppose X_1 and X_2 are independent random variables having Poisson distributions with means μ_1 and μ_2 respectively. Then the joint probability function

Table 5-8   Joint probability distribution of two independent Poisson random variables

x_2 \ x_1	0	1	2	$\cdots$	Row totals
0	$e^{-\mu_1}e^{-\mu_2}$	$e^{-\mu_1}\mu_1 e^{-\mu_2}$	$\dfrac{e^{-\mu_1}\mu_1^2}{2!}e^{-\mu_2}$ $\cdots$		$e^{-\mu_2}$
1	$e^{-\mu_1}e^{-\mu_2}\mu_2$	$e^{-\mu_1}\mu_1 e^{-\mu_2}\mu_2$	$\dfrac{e^{-\mu_1}\mu_1^2}{2!}e^{-\mu_2}\mu_2$ $\cdots$		$e^{-\mu_2}\mu_2$
2	$e^{-\mu_1}\dfrac{e^{-\mu_2}\mu_2^2}{2!}$	$e^{-\mu_1}\mu_1\dfrac{e^{-\mu_2}\mu_2^2}{2!}$	$\dfrac{e^{-\mu_1}\mu_1^2}{2!}\dfrac{e^{-\mu_2}\mu_2^2}{2!}$ $\cdots$		$\dfrac{e^{-\mu_2}\mu_2^2}{2!}$
$\cdots$	$\cdots\cdots\cdots$	$\cdots\cdots\cdots$	$\cdots\cdots\cdots$		$\cdots\cdots$
Column totals	$e^{-\mu_1}$	$e^{-\mu_1}\mu_1$	$\dfrac{e^{-\mu_1}\mu_1^2}{2!}$	$\cdots$	1

of X_1 and X_2 is

$$f(x_1,x_2) = \frac{e^{-\mu_1}\mu_1^{x_1}}{x_1!}\,\frac{e^{-\mu_2}\mu_2^{x_2}}{x_2!} \qquad \begin{array}{l} x_1 = 0, 1, 2, \ldots \\ x_2 = 0, 1, 2, \ldots \end{array}$$

The joint probability distribution can be constructed as previously and appears in Table 5-8.   If further we know that $\mu_1 = .1$ and $\mu_2 = .2$, we can use Appendix B4 to get specific values for the probabilities.   We find that the marginal probability distributions of X_1 and X_2 are as given in Table 5-9.   From these entries we get the joint probabilities by multiplication (Table 5-10).

Table 5-9   Probability distributions of Poisson random variables with $\mu_1 = .1$, $\mu_2 = .2$

x_1	0	1	2	3	4	
$f_1(x_1)$	.90484	.09048	.00453	.00015	.00000	$\cdots$
x_2	0	1	2	3	4	5
$f_2(x_2)$	.81873	.16375	.01637	.00109	.00006	.00000 $\cdots$

Table 5-10 *Joint distribution of two independent Poisson random variables with $\mu_1 = .1$, $\mu_2 = .2$*

x_2 \\ x_1	0	1	2	3	4	$\cdots$	Row totals
0	.74082	.07408	.00371	.00012	.00000	$\cdots$	.81873
1	.14817	.01482	.00074	.00002	.00000	$\cdots$	.16375
2	.01481	.00148	.00007	.00000	.00000	$\cdots$	.01637
3	.00099	.00010	.00000				.00109
4	.00005						.00005
$\cdots$							$\cdots\cdots\cdots$
Column totals	.90484	.09048	.00453	.00015	.00000		

EXERCISES

5-1 The random variables X and Y have the following joint probability distribution:

y \\ x	1	2	3	4
1	$\frac{1}{30}$	$\frac{1}{30}$	$\frac{2}{30}$	$\frac{1}{30}$
2	$\frac{2}{30}$	$\frac{2}{30}$	$\frac{4}{30}$	$\frac{2}{30}$
3	$\frac{3}{30}$	$\frac{3}{30}$	$\frac{6}{30}$	$\frac{3}{30}$

Find the marginal probability distributions and determine whether X and Y are independent or dependent.

5-2 Let X_1 and X_2 be independent random variables with the following probability distributions:

x_1	0	1	2	
$f_1(x_1)$	$\frac{1}{8}$	$\frac{5}{8}$	$\frac{2}{8}$	
x_2	0	1	2	3
$f_2(x_2)$	$\frac{1}{9}$	$\frac{2}{9}$	$\frac{4}{9}$	$\frac{2}{9}$

Construct the joint probability distribution of X_1 and X_2.

5-3 Let X_1 and X_2 be independent random variables with the following probability distributions:

x_1	1	2	3	4	5
$f_1(x_1)$	.1	.2	.4	.2	.1
x_2	1	2	3	4	
$f_2(x_2)$	.3	.3	.2	.2	

Construct the joint probability distribution of X_1 and X_2.

5-4 Let X_1 and X_2 be independent random variables both having the following probability distribution:

x	0	1	2	3
$f(x)$	$\frac{1}{8}$	$\frac{3}{8}$	$\frac{3}{8}$	$\frac{1}{8}$

Construct the joint probability distribution of X_1 and X_2.

5-5 Let X_1 have a binomial distribution with $n = n_1$ and $p = p_1$, and let X_2 have a binomial distribution with $n = n_2$ and $p = p_2$. If X_1 and X_2 are independent, write down the joint probability function of X_1 and X_2.

5-6 Verify that the marginal probability functions for X_1 and X_2 of Table 5-4 are

$$f_1(x_1) = \binom{2}{x_1}\left(\frac{1}{2}\right)^{x_1}\left(\frac{1}{2}\right)^{2-x_1} \qquad x_1 = 0, 1, 2$$

$$f_2(x_2) = \binom{2}{x_2}\left(\frac{1}{3}\right)^{x_2}\left(\frac{2}{3}\right)^{2-x_2} \qquad x_2 = 0, 1, 2$$

Next verify that the probability function of X_1 given $X_2 = 0$ is

$$g(x_1|0) = \binom{2}{x_1}\left(\frac{3}{4}\right)^{x_1}\left(\frac{1}{4}\right)^{2-x_1} \qquad x_1 = 0, 1, 2$$

5-7 From a box containing three black screws, two red screws, and five white screws, a random sample of five screws is selected. If X is the number of black screws in the sample and Y is the number of red screws in the sample, find the joint probability function of X and Y and write out the joint probability distribution complete with row and column totals. Verify that X and Y are dependent.

5-8　A table of random numbers is used to select one of the integers 1, 2, 3, 4. Let X denote the number. After X is determined, say, $X = x$, then a second number is selected at random from the integers 1, . . . , x. Let Y denote the second number chosen. Find the joint probability function of X and Y and write out the joint probability distribution complete with row and column tables. Verify that X and Y are dependent.

5-9　Let Y have a binomial distribution with $p = p_1$ so that the probability function of Y is

$$f_2(y) = \binom{n}{y} p_1^y (1 - p_1)^{n-y} \qquad y = 0, 1, \ldots, n$$

For each given value of Y, say, $Y = y$, X has a binomial distribution with constants y and p_2 so that

$$g(x|y) = \binom{y}{x} p_2^x (1 - p_2)^{y-x} \qquad x = 0, 1, \ldots, y$$

Write down the joint probability function of X and Y and from it deduce that the marginal probability function of X is binomial with constants n and $p = p_1 p_2$. *Hint:* For each x, sum on y from x to n. After replacing combination symbols by factorials, factor the term $\binom{n}{x} p_2^x$ out of the sum. Now let $u = y - x$ in the sum and the sum becomes

$$p_1^x \sum_{u=0}^{n-x} \binom{n-x}{u} [p_1(1 - p_2)]^u [1 - p_1]^{n-x-u}$$

Now, using formula (A2-3), the latter expression can be written

$$p_1^x [p_1(1 - p_2) + (1 - p_1)]^{n-x}$$

This term, together with the term factored out previously, yields the desired result.

5-10　During each of the next 100 days an individual plans to toss a coin in order to determine whether he will have tea or coffee during his morning break. If a head shows, he will drink coffee; otherwise, he will drink tea. He never uses sugar with tea but if he drinks coffee he tosses the coin again using sugar if a head shows. Find the probability that he uses sugar 20 or less times. *Hint:* (1) Let Y = number of times coffee is used and X = number of times sugar is used. Then the conditions used in Exercise 5-9 are satisfied. (2) Alternatively, sugar is used only if the event HH happens. Hence we can regard the problem as representing 100 independent repetitions of an experiment with the probability of a success (using sugar) being $\frac{1}{4}$ on each repetition. This suggests that the result of Exercise 5-9 can be established by a fairly easy probability argument. We have, in fact, done this for a specific case.

5-11 Let Y have a Poisson distribution so that the probability function of Y is

$$f_2(y) = \frac{e^{-\mu}\mu^y}{y!} \qquad y = 0, 1, 2, \ldots$$

For each given value of Y, say, $Y = y$, X has a binomial distribution with constants y and p so that

$$g(x|y) = \binom{y}{x} p^x (1 - p)^{y-x} \qquad x = 0, 1, 2, \ldots, y$$

Write down the joint probability function of X and Y and from it deduce that the marginal probability function of X is Poisson with mean $p\mu$. *Hint:* For each x, sum on y from x to infinity. After replacing combination symbols by factorials, factor the term $e^{-\mu}p^x/x!$ out of the sum. Now let $u = y - x$ and the sum becomes

$$\mu^x \sum_{u=0}^{\infty} \frac{[\mu(1 - p)]^u}{u!}$$

Now, multiply and divide each term in the sum by $e^{-\mu(1-p)}$. Then, using (4-35), the term involving the sum on u can be rewritten $\mu^x e^{\mu(1-p)}$. This term, together with the term factored out previously, yields the desired result.

5-12 A family receives on the average 12 telephone calls a day. Suppose the number of calls received in a day has a Poisson distribution. Further, suppose it is reasonable to assume that no matter how many calls are received in a day, the number of calls for adult members of the family has a binomial distribution with $p = \frac{1}{3}$. What is the probability that adults in the family will receive 4 or more calls tomorrow? *Hint:* Use the result derived in Exercise 5-11.

5-13 For the random variables whose joint probability distribution is given in Table 5-7, find the conditional probability distribution of X given $Y = 1$. Find the conditional probability distribution of Y given $X = 2$.

5-14 In Example 5-3 we found the joint probability function of X and Y and the marginal probability functions of X and Y. Use these to find the conditional probability function of X given Y and the conditional probability function of Y given X. From the formulas verify the results of Exercise 5-13.

5-2 MORE ON DERIVED RANDOM VARIABLES

In this section we shall extend some of the notions of Sec. 3-2 to the situation in which we have a pair of random variables, say, X_1 and X_2, and a joint probability function $f(x_1,x_2)$. Two standard problems immediately suggest themselves. First, we may be given two functions of X_1 and X_2, say, $Y_1 = u_1(X_1,X_2)$ and $Y_2 = u_2(X_1,X_2)$, and seek the joint probability function (or probability distribution) of Y_1 and Y_2. Second, we may have only one function of X_1 and X_2, say, $Z = w(X_1,X_2)$, and require the probability function (or probability distribution) of Z.

When the joint probability distribution is available, both types of problems usually involve a fairly easy enumeration. To illustrate the second type of problem let us again consider the example in which a four-sided die is tossed twice. Table 5-2 gives the joint probability distribution. Let us find the probability distribution of $Z = \max (X_1,X_2)$, that is, the largest of the two random variables. We see that $Z = 1$ only if $X_1 = 1$, $X_2 = 1$, and $\Pr(Z = 1) = \frac{1}{16}$. Next, $Z = 2$ for three choices of (X_1,X_2), these being (1,2), (2,2), (2,1), and $\Pr(Z = 2) = \frac{3}{16}$. Five choices of (X_1,X_2), which are (1,3), (2,3), (3,3), (3,2), (3,1), yield $Z = 3$ and $\Pr(Z = 3) = \frac{5}{16}$. Finally seven choices of (X_1,X_2) produce $Z = 4$ and $\Pr(Z = 4) = \frac{7}{16}$. Collecting these probabilities yields Table 5-11. In summary, we first observed which values Z can assume and then evaluated

$$g(z) = \Pr[w(X_1,X_2) = z]$$

by enumerating the pairs (X_1,X_2) for which $Z = z$.

To illustrate the first type of problem using the four-sided die example,

Table 5-11 Probability distribution of Z

z	1	2	3	4
$g(z)$	$\frac{1}{16}$	$\frac{3}{15}$	$\frac{5}{16}$	$\frac{7}{16}$

Table 5-12 (Y_1,Y_2) *corresponding to* (X_1,X_2) *of Table 5-2*

x_2 \ x_1	1	2	3	4
1	(2,0)	(3,1)	(4,2)	(5,3)
2	(3,1)	(4,0)	(5,1)	(6,2)
3	(4,2)	(5,1)	(6,0)	(7,1)
4	(5,3)	(6,2)	(7,1)	(8,0)

let $Y_1 = X_1 + X_2$ and $Y_2 = |X_1 - X_2|$. Thus, to get Y_2 we take the difference between X_1 and X_2 and discard the sign. For each of the 16 values of (X_1,X_2) represented in Table 5-2, we can calculate Y_1 and Y_2. We get the 16 pairs (Y_1,Y_2) that appear in Table 5-12. Obviously each event listed in Table 5-12 has probability of occurrence $\frac{1}{16}$ since it is derived from an event that has probability of occurrence $\frac{1}{16}$. Finally, to get the joint probability distribution of Y_1 and Y_2 that appears in Table 5-13, we sum probabilities corresponding to the same event. Although we did not seek the marginal probability distributions of Y_1 and Y_2 we quickly obtain them as a byproduct of our computations. In summary, we first converted each (X_1,X_2) to (Y_1,Y_2) and observed that probabilities of occurrence are the same for corresponding pairs. Then we summed all probabilities corresponding to the same event.

Table 5-13 *Joint probability distribution of Y_1 and Y_2*

y_2 \ y_1	2	3	4	5	6	7	8	Row totals
0	$\frac{1}{16}$	0	$\frac{1}{16}$	0	$\frac{1}{16}$	0	$\frac{1}{16}$	$\frac{4}{16}$
1	0	$\frac{2}{16}$	0	$\frac{2}{16}$	0	$\frac{2}{16}$	0	$\frac{6}{16}$
2	0	0	$\frac{2}{16}$	0	$\frac{2}{16}$	0	0	$\frac{4}{16}$
3	0	0	0	$\frac{2}{16}$	0	0	0	$\frac{2}{16}$
Column totals	$\frac{1}{16}$	$\frac{2}{16}$	$\frac{3}{16}$	$\frac{4}{16}$	$\frac{3}{16}$	$\frac{2}{16}$	$\frac{1}{16}$	1

EXAMPLE 5-4

For the random variables X_1, X_2 whose joint probability distribution is given in Table 5-3, find the probability distribution of $Z = X_1 + X_2$. Find the probability distribution of $W = \min(X_1, X_2)$, that is, the smallest of X_1 and X_2.

Solution

We have

$$\Pr(Z = 0) = \Pr(X_1 = 0, X_2 = 0) = \tfrac{1}{36}$$
$$\Pr(Z = 1) = \Pr(X_1 = 1, X_2 = 0) + \Pr(X_1 = 0, X_2 = 1)$$
$$= \tfrac{6}{36} + \tfrac{4}{36} = \tfrac{10}{36}$$
$$\Pr(Z = 2) = \Pr(X_1 = 2, X_2 = 0) + \Pr(X_1 = 1, X_2 = 1)$$
$$+ \Pr(X_1 = 0, X_2 = 2)$$
$$= \tfrac{9}{36} + \tfrac{12}{36} + \tfrac{4}{36} = \tfrac{25}{36}$$

We cannot have $Z = 3$ or $Z = 4$ since $\Pr(X_1 = 2, X_2 = 1) = \Pr(X_1 = 1, X_2 = 2) = \Pr(X_1 = 2, X_2 = 2) = 0$. Hence the probability distribution of Z is

z	0	1	2
$g(z)$	$\tfrac{1}{36}$	$\tfrac{10}{36}$	$\tfrac{25}{36}$

Next,

$$\Pr(W = 0) = \Pr(X_1 = 0, X_2 = 0) + \Pr(X_1 = 0, X_2 = 1)$$
$$+ \Pr(X_1 = 0, X_2 = 2)$$
$$+ \Pr(X_1 = 1, X_2 = 0) + \Pr(X_1 = 2, X_2 = 0)$$
$$= \tfrac{1}{36} + \tfrac{4}{36} + \tfrac{4}{36} + \tfrac{6}{36} + \tfrac{9}{36} = \tfrac{24}{36}$$
$$\Pr(W = 1) = \Pr(X_1 = 1, X_2 = 1) + \Pr(X_1 = 1, X_2 = 2)$$
$$+ \Pr(X_1 = 2, X_2 = 1)$$
$$= \tfrac{12}{36} + 0 + 0 = \tfrac{12}{36}$$
$$\Pr(W = 2) = \Pr(X_1 = 2, X_2 = 2) = 0$$

Thus, the probability distribution of W is

$$g(w) = \tfrac{2}{3} \quad w = 0$$
$$= \tfrac{1}{3} \quad w = 1$$

EXAMPLE 5-5

For the random variables X_1 and X_2 whose joint probability distribution is given in Table 5-2, find the joint probability distribution of

$$Y_1 = \max(X_1, X_2) \qquad Y_2 = \min(X_1, X_2)$$

146

Table 5-14 *Joint distribution of Y_1 and Y_2 for Example 5-5*

y_2 \ y_1	1	2	3	4	Row totals
1	$\frac{1}{16}$	$\frac{2}{16}$	$\frac{2}{16}$	$\frac{2}{16}$	$\frac{7}{16}$
2	0	$\frac{1}{16}$	$\frac{2}{16}$	$\frac{2}{16}$	$\frac{5}{16}$
3	0	0	$\frac{1}{16}$	$\frac{2}{16}$	$\frac{3}{16}$
4	0	0	0	$\frac{1}{16}$	$\frac{1}{16}$
Column totals	$\frac{1}{16}$	$\frac{3}{16}$	$\frac{5}{16}$	$\frac{7}{16}$	1

Solution

The value of Y_1 necessarily must be at least as large as the value of Y_2. We get

$\Pr(Y_1 = 1,\ Y_2 = 1) = \Pr(X_1 = 1,\ X_2 = 1) = \frac{1}{16}$
$\Pr(Y_1 = 2,\ Y_2 = 1) = \Pr(X_1 = 2,\ X_2 = 1) + \Pr(X_1 = 1,\ X_2 = 2) = \frac{2}{16}$
$\Pr(Y_1 = 3,\ Y_2 = 1) = \Pr(X_1 = 3,\ X_2 = 1) + \Pr(X_1 = 1,\ X_2 = 3) = \frac{2}{16}$
$\Pr(Y_1 = 4,\ Y_2 = 1) = \Pr(X_1 = 4,\ X_2 = 1) + \Pr(X_1 = 1,\ X_2 = 4) = \frac{2}{16}$
$\Pr(Y_1 = 2,\ Y_2 = 2) = \Pr(X_1 = 2,\ X_2 = 2) = \frac{1}{16}$
$\Pr(Y_1 = 3,\ Y_2 = 2) = \Pr(X_1 = 3,\ X_2 = 2) + \Pr(X_1 = 2,\ X_2 = 3) = \frac{2}{16}$
$\Pr(Y_1 = 4,\ Y_2 = 2) = \Pr(X_1 = 4,\ X_2 = 2) + \Pr(X_1 = 2,\ X_2 = 4) = \frac{2}{16}$
$\Pr(Y_1 = 3,\ Y_2 = 3) = \Pr(X_1 = 3,\ X_2 = 3) = \frac{1}{16}$
$\Pr(Y_1 = 4,\ Y_2 = 3) = \Pr(X_1 = 4,\ X_2 = 3) + \Pr(X_1 = 3,\ X_2 = 4) = \frac{2}{16}$
$\Pr(Y_1 = 4,\ Y_2 = 4) = \Pr(X_1 = 4,\ X_2 = 4) = \frac{1}{16}$

The calculations are summarized in Table 5-14 and the marginal probability distributions have been added.

Next, suppose that instead of the joint probability distribution we have the joint probability function $f(x_1, x_2)$ and we seek the joint probability function of $Y_1 = u_1(X_1, X_2)$ and $Y_2 = u_2(X_1, X_2)$. To keep the discussion as simple as possible we shall restrict ourselves to cases in which there is a one-to-one correspondence between the pairs (X_1, X_2) and (Y_1, Y_2). As an illustration let

$$f(x_1, x_2) = \left(\frac{1}{3}\right)^{x_1}\left(\frac{2}{3}\right)^{1-x_1}\binom{2}{x_2}\left(\frac{1}{2}\right)^2 \qquad \begin{aligned} x_1 &= 0,\ 1 \\ x_2 &= 0,\ 1,\ 2 \end{aligned}$$

which yields the probability distribution in Table 5-15. Now let $Y_1 = X_1 + X_2$, $Y_2 = X_2$ so that $X_1 = Y_1 - Y_2$, $X_2 = Y_2$, and each pair

Table 5-15 *Joint probability distribution of X_1 and X_2*

x_2	x_1 0	1	Row totals
0	$\frac{2}{12}$	$\frac{1}{12}$	$\frac{1}{4}$
1	$\frac{4}{12}$	$\frac{2}{12}$	$\frac{2}{4}$
2	$\frac{2}{12}$	$\frac{1}{12}$	$\frac{1}{4}$
Column totals	$\frac{2}{3}$	$\frac{1}{3}$	1

(X_1, X_2) yields one (Y_1, Y_2) and vice versa. We have

$$g(y_1,y_2) = \Pr(Y_1 = y_1,\ Y_2 = y_2) = \Pr(X_1 + X_2 = y_1,\ X_2 = y_2)$$
$$= \Pr(X_1 = y_1 - y_2,\ X_2 = y_2)$$
$$= \left(\frac{1}{3}\right)^{y_1 - y_2} \left(\frac{2}{3}\right)^{1 - y_1 + y_2} \binom{2}{y_2} \left(\frac{1}{2}\right)^2$$

for the six pairs of values that correspond to the six pairs of (X_1, X_2). These are $(0,0)$, $(1,0)$, $(1,1)$, $(2,1)$, $(2,2)$, $(3,2)$. The joint probability distribution found in Table 5-16 is easily constructed. It is obvious that we still have the same six nonzero probabilities, but that most of them now correspond to different pairs than previously. To generalize the discussion let $Y_1 = u_1(X_1, X_2)$ and $Y_2 = u_2(X_1, X_2)$ be solved yielding

Table 5-16 *Joint probability distribution of Y_1 and Y_2*

y_2	y_1 0	1	2	3	Row totals
0	$\frac{2}{12}$	$\frac{1}{12}$	0	0	$\frac{3}{12}$
1	0	$\frac{4}{12}$	$\frac{2}{12}$	0	$\frac{6}{12}$
2	0	0	$\frac{2}{12}$	$\frac{1}{12}$	$\frac{3}{12}$
Column totals	$\frac{2}{12}$	$\frac{5}{12}$	$\frac{4}{12}$	$\frac{1}{12}$	1

$X_1 = w_1(Y_1, Y_2)$, $X_2 = w_2(Y_1, Y_2)$.   Then

$$g(y_1,y_2) = \Pr(Y_1 = y_1,\ Y_2 = y_2) = \Pr[u_1(X_1,X_2) = y_1,\ u_2(X_1,X_2) = y_2]$$
$$= \Pr[X_1 = w_1(y_1,y_2),\ X_2 = w_2(y_1,y_2)]$$
$$= f[w_1(y_1,y_2),\ w_2(y_1,y_2)] \qquad (5\text{-}7)$$

In other words, to get the joint probability function of Y_1 and Y_2, replace x_1 and x_2 in $f(x_1,x_2)$ by their solutions in terms of y_1 and y_2.   The new set of nonzero probabilities will be the same as the old but in general will be associated with different pairs.

Perhaps one of the most interesting and useful applications of derived random variables with joint probability functions arises when we seek the marginal probability function of $Y_1 = u_1(X_1, X_2)$, where X_1 and X_2 are independent random variables and $f_1(x_1)$ and $f_2(x_2)$ are known.   The function of X_1 and X_2 that probably arises more frequently than any other is $Y_1 = X_1 + X_2$.   We shall now consider several important examples.

Let X_1 and X_2 be independent random variables both having a Poisson distribution with means μ_1 and μ_2, respectively.   Hence, the joint probability function is

$$f(x_1,x_2) = \frac{e^{-\mu_1}\mu_1^{x_1}}{x_1!} \frac{e^{-\mu_2}\mu_2^{x_2}}{x_2!} \qquad \begin{array}{l} x_1 = 0, 1, 2, \ldots \\[4pt] x_2 = 0, 1, 2, \ldots \end{array} \qquad (5\text{-}8)$$

Suppose we want the probability function of $Y_1 = X_1 + X_2$.   If we let $Y_2 = X_2$, then $X_1 = Y_1 - Y_2$, $X_2 = Y_2$ and the joint probability function of Y_1 and Y_2 is

$$g(y_1,y_2) = \frac{e^{-(\mu_1+\mu_2)}\mu_1^{y_1-y_2}\mu_2^{y_2}}{(y_1 - y_2)!\,y_2!} \qquad \begin{array}{l} y_1 = 0, 1, 2, \ldots \\[4pt] y_2 = 0, 1, 2, \ldots \\[4pt] y_2 \leqq y_1 \end{array}$$

The marginal probability function

$$g_1(y_1) = \sum_{y_2=0}^{y_1} g(y_1,y_2)$$

is the result we desire.   This is

$$g_1(y_1) = \frac{e^{-(\mu_1+\mu_2)}}{y_1!} \sum_{y_2=0}^{y_1} \frac{y_1!}{y_2!(y_1 - y_2)!} \mu_2^{y_2}\mu_1^{y_1-y_2}$$
$$= \frac{e^{-(\mu_1+\mu_2)}}{y_1!}(\mu_1 + \mu_2)^{y_1} \qquad y_1 = 0, 1, 2, \ldots \qquad (5\text{-}9)$$

since the sum and $(\mu_1 + \mu_2)^{y_1}$ are equal by formula (A2-3). Thus Y_1 has a Poisson distribution with mean $\mu_1 + \mu_2$. We could have formed the probability distribution of Y by turning to Table 5-8 and enumerating probabilities as we have done in previous examples. We would get

$$\Pr(Y_1 = 0) = \Pr(X_1 = 0,\ X_2 = 0) = e^{-(\mu_1+\mu_2)}$$
$$\Pr(Y_1 = 1) = \Pr(X_1 = 1,\ X_2 = 0) + \Pr(X_1 = 0,\ X_2 = 1)$$
$$= e^{-(\mu_1+\mu_2)}(\mu_1 + \mu_2)$$
$$\Pr(Y_1 = 2) = \Pr(X_1 = 2,\ X_2 = 0) + \Pr(X_1 = 1,\ X_2 = 1)$$
$$+ \Pr(X_1 = 0,\ X_2 = 2)$$
$$= e^{-(\mu_1+\mu_2)}\left[\frac{\mu_1^2}{2!} + \mu_1\mu_2 + \frac{\mu_2^2}{2!}\right] = e^{-(\mu_1+\mu_2)}\frac{(\mu_1 + \mu_2)^2}{2!}$$

. . . .

results that suggest that Y_1 has a Poisson distribution with mean $\mu_1 + \mu_2$.

Next suppose we seek the probability function of $Y_1 = X_1 + X_2$ where X_1 and X_2 are independent random variables having binomial distributions with the same p but with $n = n_1$ and $n = n_2$, respectively. The joint probability function of X_1 and X_2 is

$$f(x_1,x_2) = \binom{n_1}{x_1}p^{x_1}(1 - p)^{n_1-x_1}\binom{n_2}{x_2}p^{x_2}(1 - p)^{n_2-x_2} \quad \begin{aligned} x_1 &= 0, 1, 2, \ldots, n_1 \\ x_2 &= 0, 1, 2, \ldots, n_2 \end{aligned}$$

If we let $Y_2 = X_2$, then the joint probability function of Y_1 and Y_2 is

$$g(y_1,y_2) = \binom{n_1}{y_1 - y_2}\binom{n_2}{y_2}p^{y_1}(1 - p)^{n_1+n_2-y_1} \quad \begin{aligned} y_1 &= 0, 1, 2, \ldots, n_1 + n_2 \\ y_2 &= 0, 1, 2, \ldots, n_2 \\ y_2 &\leq y_1 \leq y_2 + n_1 \end{aligned}$$

If we could evaluate

$$g_1(y_1) = \sum_{y_2} g(y_1,y_2)$$

summing over the correct values of y_2, we would have the marginal probability function of Y_1. Although this can be done, the algebra is slightly complicated and we shall use another argument to derive the probability function of Y_1. Observe that X_1 is the number of successes when a binomial experiment with probability of success p is repeated n_1 times. Similarly X_2 is the number of successes when a binomial experiment with probability of success p is repeated n_2 times. Hence $Y_1 = X_1 + X_2$ is the number of successes when a binomial experiment with probability of success p is repeated $n_1 + n_2$ times. But we know from Sec. 4-2 that the

150

probability function of such a random variable is

$$f(y_1) = \binom{n_1 + n_2}{y_1} p^{y_1}(1 - p)^{n_1+n_2-y_1} \qquad y_1 = 0, 1, 2, \ldots, n_1 + n_2$$

$$(5\text{-}10)$$

Thus Y_1 has a binomial distribution with constants p and $n_1 + n_2$.

EXAMPLE 5-6

If the joint probability function of X_1 and X_2 is

$$f(x_1,x_2) = \frac{2!}{x_1!x_2!(2 - x_1 - x_2)!} \left(\frac{3}{6}\right)^{x_1} \left(\frac{2}{6}\right)^{x_2} \left(\frac{1}{6}\right)^{2-x_1-x_2} \qquad \begin{array}{l} x_1 = 0, 1, 2 \\ x_2 = 0, 1, 2 \\ 0 \leq x_1 + x_2 \leq 2 \end{array}$$

find the joint probability function of $Y_1 = X_1 + X_2$ and $Y_2 = X_2$. Then, using this result, find the marginal probability function of Y_1.

Solution

Since $X_1 = Y_1 - Y_2$, $X_2 = Y_2$ we get

$$g(y_1,y_2)$$
$$= \frac{2!}{(y_1 - y_2)!y_2!(2 - y_1)!} \left(\frac{3}{6}\right)^{y_1-y_2} \left(\frac{2}{6}\right)^{y_2} \left(\frac{1}{6}\right)^{2-y_1} \qquad \begin{array}{l} y_1 = 0, 1, 2 \\ y_2 = 0, 1, 2 \\ 0 \leq y_2 \leq y_1 \leq 2 \end{array}$$

The six pairs of (X_1,X_2) for which we have nonzero probability (see Table 5-3) are (0,0), (1,0), (0,1), (2,0), (1,1), (0,2). When converted in terms of (Y_1,Y_2) these become (0,0), (1,0), (1,1), (2,0), (2,1), (2,2). If we prefer, we could list the latter six pairs instead of giving the values for which $g(y_1,y_2)$ is nonzero in terms of equations and inequalities.

The marginal probability function of Y_1 is

$$g_1(y_1) = \sum_{y_2=0}^{y_1} g(y_1,y_2)$$

$$= \frac{2!}{y_1!(2 - y_1)!} \left(\frac{1}{6}\right)^{2-y_1} \sum_{y_2=0}^{y_1} \frac{y_1!}{y_2!(y_1 - y_2)!} \left(\frac{2}{6}\right)^{y_2} \left(\frac{3}{6}\right)^{y_1-y_2}$$

$$= \binom{2}{y_1} \left(\frac{1}{6}\right)^{2-y_1} \left(\frac{3}{6} + \frac{2}{6}\right)^{y_1} \qquad \text{by formula (A2-3)}$$

$$= \binom{2}{y_1} \left(\frac{5}{6}\right)^{y_1} \left(\frac{1}{6}\right)^{2-y_1} \qquad y_1 = 0, 1, 2$$

so that Y_1 has a binomial distribution with $p = \frac{5}{6}$, $n = 2$.

151

EXAMPLE 5-7

One telephone receives on the average three calls per day. A second telephone receives on the average seven calls per day. If the number of calls received on the first telephone is independent of the number of calls received on the second, and for each telephone the number of calls follows a Poisson distribution, then find the probability that the sum of the calls received on the two telephones is eight or less for a given day.

Solution

Let X_1 = number of calls received on the first telephone and X_2 = number of calls received on the second telephone. Then X_1 has a Poisson distribution with mean 3 and X_2 has a Poisson distribution with mean 7. We proved that $Y_1 = X_1 + X_2$ has a Poisson distribution with mean $3 + 7 = 10$. Hence

$$Pr(Y_1 \leq 8) = \sum_{y_1 = 0}^{8} p(y_1; 10) = .33282 \qquad \text{from Appendix B4}$$

EXERCISES

5-15 If X and Y have the joint probability distribution given in Table 5-5, find the probability distribution of $Z = XY$.

5-16 If X and Y have the joint probability distribution given in Table 5-5, find the probability distribution of $Z = X + Y$.

5-17 If X and Y have the joint probability distribution given in Table 5-5, find the joint probability distribution of $Z_1 = XY$ and $Z_2 = X + Y$.

5-18 If X and Y have the joint probability distribution given in Table 5-5, find the probability distribution of $Z_1 = \max(X, Y)$. Find the probability distribution of $Z_2 = \min(X, Y)$.

5-19 The random variable X_1 has a binomial distribution with $n_1 = 3$, $p_1 = \frac{1}{2}$, and the random variable X_2 has a binomial distribution with $n_2 = 2$, $p_2 = \frac{1}{3}$. If X_1 and X_2 are independent, find the probability distribution of $Y = X_1 + X_2$.

5-20 For the random variables X_1 and X_2 of Exercise 5-19 find the joint probability distribution of $Y_1 = \max(X_1, X_2)$ and $Y_2 = \min(X_1, X_2)$. From the resulting table find the two marginal distributions.

5-21 The independent random variables X_1 and X_2 both have the probability function

$$f(x) = \frac{1}{4} \qquad x = 1, 2, 3, 4$$

Find the probability distribution of $Y = X_1 + X_2$.

5-22   If X_1 and X_2 are independent random variables both having the same probability function

$$f(x) = \left(\frac{1}{4}\right)^x \left(\frac{3}{4}\right)^{1-x} \qquad x = 0, 1$$

(the binomial distribution with $p = \frac{1}{4}$, $n = 1$), find the joint probability function of $Y_1 = X_1 + X_2$ and $Y_2 = X_2$.   Be sure to list those values of (Y_1, Y_2) for which the probabilities are nonzero. Then write out the joint probability distribution.   Show that the marginal probability distribution of Y_1 is binomial with $p = \frac{1}{4}$, $n = 2$.

5-23   The joint probability function of X_1 and X_2 is

$$f(x_1, x_2) = \frac{n!}{x_1! x_2! (n - x_1 - x_2)!} p_1^{x_1} p_2^{x_2} (1 - p_1 - p_2)^{n - x_1 - x_2}$$

$$x_1 = 0, 1, 2, \ldots, n \qquad \qquad 0 \leqq x_1 + x_2 \leqq n$$
$$x_2 = 0, 1, 2, \ldots, n$$

where p_1 and p_2 are both between zero and 1 and $p_1 + p_2 < 1$. Find the joint probability function of $Y_1 = X_1 + X_2$ and $Y_2 = X_2$. Show that the marginal probability function of Y_1 is a binomial with constants n and $p_1 + p_2$.   Observe that for each value of Y_1, Y_2 can take on the values $y_2 = 0, 1, 2, \ldots, y_1$.   Then use formula (A2-3).

5-24   One typist makes on the average one error per page.   A second typist makes on the average three errors per page.   If both type a complete page, what is the probability that the total number of errors on both pages is seven or more?   Assume in each case that the numbers of errors per page has a Poisson distribution.

5-25   A student takes a five-answer multiple-choice examination consisting of 25 questions and guesses at the answer to every question.   A few weeks later he takes a second five-answer multiple-choice examination consisting of 25 questions and again guesses at the answer to every question.   What is the probability that his total number of correct answers on both examinations is 10 or more? *Hint:* Justify the use of (5-10).

5-26   Let X_1 be a random variable having a negative binomial distribution with constants c_1 and p.   Let X_2 be a second random variable, independent of X_1, having a negative binomial distribution with constants c_2 and p.   Use an argument similar to the one used to derive (5-10) to prove that $Y_1 = X_1 + X_2$ has a negative binomial distribution with constants $c_1 + c_2$ and p.

Let $Z = w(X_1, X_2)$ be a function of two random variables X_1 and X_2 having a joint probability distribution. To find $E(Z)$, the logical way to proceed is: (1) First construct the probability distribution of Z by the methods of Sec. 5-2. (2) Evaluate $E(Z)$ by using the definition of expected value for the one-dimensional case (Sec. 3-3). Although the procedure is straightforward and leads to no difficulties, it is usually more convenient to evaluate $E[(w(X_1, X_2)]$ directly from the joint distribution. This is possible because of the following theorem:

Let (X_1, X_2) be a two-dimensional random variable with nonzero probabilities $\Pr(X_1 = x_1, X_2 = x_2)$ being given a joint probability distribution or a joint probability function $f(x_1, x_2)$. If $w(X_1, X_2)$ is a function of X_1 and X_2, then the expected value of $w(X_1, X_2)$ is

$$E(Z) = E[w(X_1, X_2)] = \sum_{x_2} \sum_{x_1} w(x_1, x_2) \Pr(X_1 = x_1, X_2 = x_2) \qquad (5\text{-}11)$$

where the sums are taken over all values of (x_1, x_2) for which there are nonzero probabilities.

Before we prove the theorem, let us consider some examples.

EXAMPLE 5-8

For the random variables X_1, X_2 whose joint probability distribution is given in Table 5-3, evaluate $E(X_1 + X_2)$.

Solution

Using (5-11) we get

$$
\begin{aligned}
E(X_1 + X_2) &= (0+0)(\tfrac{1}{36}) + (1+0)(\tfrac{6}{36}) + (2+0)(\tfrac{9}{36}) \\
&\quad + (0+1)(\tfrac{4}{36}) + (1+1)(\tfrac{12}{36}) \\
&\quad + (0+2)(\tfrac{4}{36}) \\
&= 0(\tfrac{1}{36}) + 1[(\tfrac{6}{36}) + (\tfrac{4}{36})] + 2[(\tfrac{9}{36}) + (\tfrac{12}{36}) + (\tfrac{4}{36})] \\
&= 0(\tfrac{1}{36}) + 1(\tfrac{10}{36}) + 2(\tfrac{25}{36}) \\
&= \tfrac{60}{36} = \tfrac{5}{3} = 1.67
\end{aligned}
$$

In Example 5-4 we found the probability distribution of $Z = X_1 + X_2$. Thus, alternatively, we can calculate

$$E(Z) = 0(\tfrac{1}{36}) + 1(\tfrac{10}{36}) + 2(\tfrac{25}{36}) = \tfrac{60}{36}$$

getting the same answer as before.

EXAMPLE 5-9

For the random variables X_1 and X_2 whose joint probability distribution is given by Table 5-3, find $E[(X_1 - 1)(X_2 - \frac{2}{3})]$.

Solution

Using (5-11) we get

$$
\begin{aligned}
E[(X_1 - 1)(X_2 - \tfrac{2}{3})] &= (0-1)(0-\tfrac{2}{3})(\tfrac{1}{36}) + (1-1)(0-\tfrac{2}{3})(\tfrac{6}{36}) \\
&\quad + (2-1)(0-\tfrac{2}{3})(\tfrac{9}{36}) \\
&\quad + (0-1)(1-\tfrac{2}{3})(\tfrac{4}{36}) \\
&\quad + (1-1)(1-\tfrac{2}{3})(\tfrac{12}{36}) \\
&\quad + (0-1)(2-\tfrac{2}{3})(\tfrac{4}{36}) \\
&= (\tfrac{2}{3})(\tfrac{1}{36}) + (-\tfrac{2}{3})(\tfrac{9}{36}) + (-\tfrac{1}{3})(\tfrac{4}{36}) \\
&\quad + (-\tfrac{4}{3})(\tfrac{4}{36}) \\
&= (\tfrac{2}{3})[(\tfrac{1}{36}) - (\tfrac{9}{36}) - (\tfrac{2}{36}) - (\tfrac{8}{36})] \\
&= (\tfrac{2}{3})(-\tfrac{18}{36}) = -\tfrac{1}{3}
\end{aligned}
$$

To prove the theorem, we shall use Example 5-8 as a guide. We had

$$
\begin{aligned}
E(X_1 + X_2) &= 0[\Pr(X_1 = 0,\, X_2 = 0)] + 1[\Pr(X_1 = 1,\, X_2 = 0) \\
&\quad + \Pr(X_1 = 0,\, X_2 = 1)] \\
&\quad + 2[\Pr(X_1 = 2,\, X_2 = 0) + \Pr(X_1 = 1,\, X_2 = 1) \\
&\quad + \Pr(X_1 = 0,\, X_2 = 2)] \\
&= 0[\Pr(Z = 0)] + 1[\Pr(Z = 1)] + 2[\Pr(Z = 2)] \\
&= \sum_z z\Pr(Z = z) = E(Z)
\end{aligned}
$$

where the sum is over all values of Z, here $Z = 0$, $Z = 1$, and $Z = 2$. In general, let $Z = w(X_1, X_2)$ produce values of Z equal to z_1, z_2, z_3, ... Then from the joint probability distribution of X_1 and X_2 we can collect terms as follows:

$$
\begin{aligned}
E[w(X_1, X_2)] &= z_1[\text{sum of } \Pr(X_1 = x_1,\, X_2 = x_2) \text{ for which } w(x_1, x_2) = z_1] \\
&\quad + z_2[\text{sum of } \Pr(X_1 = x_1,\, X_2 = x_2) \text{ for which } w(x_1, x_2) = z_2] \\
&\quad + z_3[\text{sum of } \Pr(X_1 = x_1,\, X_2 = x_2) \text{ for which } w(x_1, x_2) = z_3] \\
&\quad + \cdots\cdots\cdots\cdots\cdots\cdots\cdots\cdots\cdots\cdots \\
&= z_1[\Pr(Z = z_1)] + z_2[\Pr(Z = z_2)] + z_3[\Pr(Z = z_3)] + \cdots \\
&= \sum_i z_i[\Pr(Z = z_i)] = E(Z)
\end{aligned}
$$

where the sum is taken over all values of Z.

Turning again to Table 5-3 we find

$$
\begin{aligned}
E(X_1) &= 0(\tfrac{9}{36}) + 1(\tfrac{18}{36}) + 2(\tfrac{9}{36}) = 1 \\
E(X_2) &= 0(\tfrac{16}{36}) + 1(\tfrac{16}{36}) + 2(\tfrac{4}{36}) = \tfrac{24}{36} = \tfrac{2}{3}
\end{aligned}
$$

In Example 5-8 we found $E(X_1 + X_2) = \frac{5}{3}$. Since

$$E(X_1) + E(X_2) = 1 + (\tfrac{2}{3}) = \tfrac{5}{3}$$

this suggests and illustrates the result

$$E(X_1 + X_2) = E(X_1) + E(X_2) \tag{5-12}$$

which holds for any two random variables X_1 and X_2. To prove (5-12) we can use (5-11) and write

$$E(X_1 + X_2) = \sum_{x_2} \sum_{x_1} (x_1 + x_2) \Pr(X_1 = x_1,\, X_2 = x_2)$$

$$= \sum_{x_2} \sum_{x_1} x_1 \Pr(X_1 = x_1, X_2 = x_2) + \sum_{x_2} \sum_{x_1} x_2 \Pr(X_1 = x_1, X_2 = x_2)$$

Thinking in terms of the probability distribution we shall evaluate the first term by summing over columns first. Hence for each x_1 we find

$$\sum_{x_2} \Pr(X_1 = x_1,\, X_2 = x_2) = \Pr(X_1 = x_1)$$

and the first term is

$$\sum_{x_1} \sum_{x_2} x_1 \Pr(X_1 = x_1,\, X_2 = x_2) = \sum_{x_1} x_1 \sum_{x_2} \Pr(X_1 = x_1,\, X_2 = x_2)$$

$$= \sum_{x_1} x_1 \Pr(X_1 = x_1) = E(X_1)$$

Similarly, we shall evaluate the second term by summing over rows first and get

$$\sum_{x_2} \sum_{x_1} x_2 \Pr(X_1 = x_1,\, X_2 = x_2) = \sum_{x_2} x_2 \sum_{x_1} \Pr(X_1 = x_1,\, X_2 = x_2)$$

$$= \sum_{x_2} x_2 \Pr(X_2 = x_2) = E(X_2)$$

Hence the result (5-12).

If $u_1(X_1)$ and $u_2(X_2)$ are any two functions of X_1 and X_2, then (5-12) is readily generalized to

$$E[u_1(X_1) + u_2(X_2)] = E[u_1(X_1)] + E[u_2(X_2)] \tag{5-13}$$

If, in (5-12), X_1 is replaced by $u(X_1)$ and X_2 is replaced by $u(X_2)$, then we immediately have (5-13). Thus, since $u_1(X_1)$ and $u_2(X_2)$ are random variables, (5-13) is actually a repetition of (5-12).

Another result, similar to (3-10) and proved in the same way, is

$$E[u_1(X_1,X_2) + u_2(X_1,X_2) + u_3(X_1,X_2)]$$
$$= E[u_1(X_1,X_2)] + E[u_2(X_1,X_2)] + E[u_3(X_1,X_2)] \tag{5-14}$$

which can, of course, be readily extended to more than three functions of X_1 and X_2.

An interesting and useful result concerning independent random variables is the following:

Let X_1 and X_2 be independent random variables and let $u_1(X_1)$ be a function of X_1, $u_2(X_2)$ be a function of X_2. Then

$$E[u_1(X_1)u_2(X_2)] = E[u_1(X_1)]E[u_2(X_2)] \tag{5-15}$$

To demonstrate this result, let us evaluate $E(XY^2)$ for the random variables whose joint distribution is given in Table 5-5. We can write

$$
\begin{aligned}
E(XY^2) &= 0(0^2)(^{25}\!/_{144}) + 0(1^2)(^{50}\!/_{144}) + 0(2^2)(^{25}\!/_{144}) + 1(0^2)(^{10}\!/_{144}) \\
&\quad + 1(1^2)(^{20}\!/_{144}) \\
&\quad + 1(2^2)(^{10}\!/_{144}) + 2(0^2)(^{1}\!/_{144}) + 2(1^2)(^{2}\!/_{144}) + 2(2^2)(^{1}\!/_{144}) \\
&= 0(^{25}\!/_{36})[(0^2)(^1\!/_4) + (1^2)(^2\!/_4) + 2^2(^1\!/_4)] + 1(^{10}\!/_{36})[0^2(^1\!/_4) \\
&\quad + 1^2(^2\!/_4) + 2^2(^1\!/_4)] + 2(^1\!/_{36})[0^2(^1\!/_4) + 1^2(^2\!/_4) + 2^2(^1\!/_4)] \\
&= [0(^{25}\!/_{36}) + 1(^{10}\!/_{36}) + 2(^1\!/_{36})][0^2(^1\!/_4) + 1^2(^2\!/_4) + 2^2(^1\!/_4)] \\
&= E(X)E(Y^2)
\end{aligned}
$$

To prove (5-15) we follow the above scheme. By (5-11) we have

$$E[u_1(X_1)u_2(X_2)] = \sum_{x_1} \sum_{x_2} u_1(x_1)u_2(x_2)\Pr(X_1 = x_1, X_2 = x_2)$$

Since X_1 and X_2 are independent we have by (5-5)

$$\Pr(X_1 = x_1, X_2 = x_2) = \Pr(X_1 = x_1)\Pr(X_2 = x_2)$$

for all values of X_1 and X_2. Thus

$$
\begin{aligned}
E[u_1(X_1)u_2(X_2)] &= \sum_{x_1} \sum_{x_2} u_1(x_1)u_2(x_2)\Pr(X_1 = x_1)\Pr(X_2 = x_2) \\
&= \sum_{x_1} u_1(x_1)\Pr(X_1 = x_1) \sum_{x_2} u_2(X_2)\Pr(X_2 = x_2) \\
&= E[u_1(X_1)]E[u_2(X_2)]
\end{aligned}
$$

One important special case of (5-15) is

$$E(X_1X_2) = E(X_1)E(X_2) \tag{5-16}$$

Equation (5-16) does not usually hold with dependent random variables. To illustrate this, consider the random variables of Table 5-3. We have

$$E(X_1) = 1 \qquad E(X_2) = ^2\!/_3$$
$$E(X_1X_2) = ^1\!/_3 \qquad E(X_1X_2) \neq E(X_1)E(X_2)$$

Example 5-10 demonstrates that (5-16) may hold with dependent random variables.

Table 5-17 Joint probability distribution of X_1 and X_2

x_2	x_1			Row totals
	-1	0	1	
0	⅓	0	⅓	⅔
1	0	⅓	0	⅓
Column totals	⅓	⅓	⅓	1

EXAMPLE 5-10

Let X_1 and X_2 have the joint probability distribution given in Table 5-17. Show that X_1 and X_2 are dependent, yet (5-16) holds.

Solution

Since $\Pr(X_1 = -1, X_2 = 0) = \frac{1}{3}$ and $\Pr(X_1 = -1)\Pr(X_2 = 0) = \frac{2}{9} \neq \frac{1}{3}$, X_1 and X_2 are dependent. Next

$E(X_1) = -1(\frac{1}{3}) + 0(\frac{1}{3}) + 1(\frac{1}{3}) = 0$
$E(X_2) = 0(\frac{2}{3}) + 1(\frac{1}{3}) = \frac{1}{3}$
$E(X_1 X_2) = -1(0)(\frac{1}{3}) + 0(1)(\frac{1}{3}) + 1(0)(\frac{1}{3}) = 0$

Hence (5-16) is satisfied.

In Sec. 3-4 we showed that for any random variable X

$$E(X - \mu) = 0 \qquad (5\text{-}17)$$

We defined the variance of X as

$$\text{Var}(X) = E[(X - \mu)^2] \qquad (5\text{-}18)$$

With two random variables X_1 and X_2 we have not only two means but also two variances, $\text{Var}(X_1)$ and $\text{Var}(X_2)$, which can be computed using (5-18), and the marginal probability distributions of X_1 and X_2, respectively. Another important constant associated with the joint distribution of X_1 and X_2 is

$$\text{Cov}(X_1, X_2) = E[(X_1 - \mu_1)(X_2 - \mu_2)] \qquad (5\text{-}19)$$

called the *covariance* of X_1 and X_2. Here $\mu_1 = E(X_1)$, $\mu_2 = E(X_2)$, the mean for X_1 and X_2, respectively. We have already calculated a covariance in Example 5-9.

158

The covariance arises naturally when we seek the variance of the sum of two random variables. Since (5-12) yields

$$E(X_1 + X_2) = E(X_1) + E(X_2) = \mu_1 + \mu_2$$

we get by using the definition of variance (5-18)

$$
\begin{aligned}
\text{Var}(X_1 + X_2) &= E\{[X_1 + X_2 - (\mu_1 + \mu_2)]^2\} \\
&= E\{[(X_1 - \mu_1) + (X_2 - \mu_2)]^2\} \\
&= E[(X_1 - \mu_1)^2 + 2(X_1 - \mu_1)(X_2 - \mu_2) + (X_2 - \mu_2)^2]
\end{aligned}
$$

Using Eq. (5-14) we get

$$\text{Var}(X_1 + X_2) = E[(X_1 - \mu_1)^2] + 2E[(X_1 - \mu_1)(X_2 - \mu_2)] + E[(X_2 - \mu_2)^2]$$

Now, rearranging the order of terms on the right-hand side of the last equation, we can write

$$\text{Var}(X_1 + X_2) = \text{Var}(X_1) + \text{Var}(X_2) + 2\text{Cov}(X_1, X_2) \qquad (5\text{-}20)$$

If X_1 and X_2 are independent, then by (5-15)

$$
\begin{aligned}
E[(X_1 - \mu_1)(X_2 - \mu_2)] &= E(X_1 - \mu_1)E(X_2 - \mu_2) \\
&= 0 \cdot 0 = 0
\end{aligned}
$$

by (5-17). Thus for independent random variables (5-20) can be simplified to

$$\text{Var}(X_1 + X_2) = \text{Var}(X_1) + \text{Var}(X_2) \qquad (5\text{-}21)$$

As with the variance, we can express covariance in another form more adaptable to computation. We can write

$$
\begin{aligned}
E[(X_1 - \mu_1)(X_2 - \mu_2)] &= E[X_1 X_2 - \mu_2 X_1 - \mu_1 X_2 + \mu_1 \mu_2] \\
&= E(X_1 X_2) - \mu_2 E(X_1) - \mu_1 E(X_2) + \mu_1 \mu_2
\end{aligned}
$$

Observe that except for sign the last three terms are equal and their sum is $-\mu_1\mu_2$. Hence the computational form for covariance is

$$\text{Cov}(X_1, X_2) = E(X_1 X_2) - \mu_1 \mu_2 \qquad (5\text{-}22)$$

EXAMPLE 5-11

Use formula (5-22) to compute $\text{Cov}(X_1, X_2)$ for the random variables whose joint probability distribution is given by Table 5-3. [We already computed this covariance by using formula (5-19) in Example 5-9.] Then find $\text{Var}(X_1 + X_2)$.

Solution

$\mu_1 = E(X_1) = 0(\frac{9}{36}) + 1(\frac{18}{36}) + 2(\frac{9}{36}) = 1$
$\mu_2 = E(X_2) = 0(\frac{16}{36}) + 1(\frac{16}{36}) + 2(\frac{4}{36}) = \frac{2}{3}$
$E(X_1 X_2) = 1(\frac{12}{36}) = \frac{1}{3}$ (The other 8 terms are all 0)
$\text{Cov}(X_1, X_2) = \frac{1}{3} - (1)(\frac{2}{3}) = -\frac{1}{3}$ as before

Next,

$$E(X_1^2) = 0(\%_{36}) + 1(^{18}\!\%_{36}) + 4(\%_{36}) = \%_2$$
$$\mathrm{Var}(X_1) = \%_2 - (1)^2 = \frac{1}{2}$$
$$E(X_2^2) = 0(^{16}\!\%_{36}) + 1(^{16}\!\%_{36}) + 4(\frac{4}{36}) = \%_9$$
$$\mathrm{Var}(X_2) = \%_9 - (\frac{2}{3})^2 = \frac{4}{9}$$

$$\mathrm{Var}(X_1 + X_2) = \frac{1}{2} + \frac{4}{9} + 2\left(-\frac{1}{3}\right) = \frac{9 + 8 - 12}{18} = \frac{5}{18}$$

Occasionally we shall need expected values for conditional distributions. Since one of the random variables is fixed, a conditional probability distribution is a one-dimensional distribution. Except for slight differences in notation, results concerning expected values are the same as those given in Chap. 3. Thus the expected value of X_2 given $X_1 = x_1$ is

$$\mu_{X_2|x_1} = E(X_2|x_1) = \sum_{x_2} x_2 \mathrm{Pr}(X_2 = x_2|X_1 = x_1) \tag{5-23}$$

which corresponds to (3-4) and (3-5). Formula (3-6) also holds but is now written (for a given value of X_1)

$$E[u(X_2)|x_1] = \sum_{x_2} u(x_2) \mathrm{Pr}(X_2 = x_2|X_1 = x_1) \tag{5-24}$$

In particular, the variance of X_2 given $X_1 = x_1$ is

$$\mathrm{Var}(X_2|x_1) = E[(X_2 - \mu_{X_2|x_1})^2|x_1] = \sum_{x_2} (x_2 - \mu_{X_2|x_1})^2 \mathrm{Pr}(X_2 = x_2|X_1 = x_1)$$
$$\tag{5-25}$$
$$= E(X_2^2|x_1) - \mu_{X_2|x_1}^2 \tag{5-26}$$

where (5-26) is the computational form corresponding to (3-9). Similarly, if X_2 is given,

$$\mu_{X_1|x_2} = E(X_1|x_2) = \sum_{x_1} x_1 \mathrm{Pr}(X_1 = x_1|X_2 = x_2) \tag{5-27}$$

$$E[u(X_1)|x_2] = \sum_{x_1} u(x_1) \mathrm{Pr}(X_1 = x_1|X_2 = x_2) \tag{5-28}$$

$$\mathrm{Var}(X_1|x_2) = \sum_{x_1} (x_1 - \mu_{X_1|x_2})^2 \mathrm{Pr}(X_1 = x_1|X_2 = x_2) \tag{5-29}$$

$$= E(X_1^2|x_2) - \mu_{X_1|x_2}^2 \tag{5-30}$$

EXAMPLE 5-12 ────────────────────────────────

In Exercise 5-13 we found

$$\mathrm{Pr}(X = 0|Y = 1) = \%_{40} \qquad \mathrm{Pr}(X = 1|Y = 1) = ^{18}\!\%_{40}$$
$$\mathrm{Pr}(X = 2|Y = 1) = ^{18}\!\%_{40}$$

and

$$\Pr(X = 3 | Y = 1) = \tfrac{2}{40}$$

Find $E(X|1)$ and $\text{Var}(X|1)$.

Solution

$E(X|1) = 0(\tfrac{2}{40}) + 1(\tfrac{18}{40}) + 2(\tfrac{18}{40}) + 3(\tfrac{2}{40}) = \tfrac{3}{2}$

$E(X^2|1) = 0^2(\tfrac{2}{40}) + 1^2(\tfrac{18}{40}) + 2^2(\tfrac{18}{40}) + 3^2(\tfrac{2}{40}) = \tfrac{27}{10}$

$\text{Var}(X|1) = \tfrac{27}{10} - (\tfrac{3}{2})^2 = \tfrac{9}{20}$

EXERCISES

5-27   For the random variables X and Y whose joint probability distribution is given in Table 5-1, find $E(X + Y)$. In Sec. 5-1 we observed that X and Y are independent. Use this fact in finding $\text{Var}(X + Y)$. The formulas derived in Exercise 3-45 will shorten the arithmetic.

5-28   For the independent random variables whose joint probability distribution is given in Table 5-2, find $E(X_1 + X_2)$ and $\text{Var}(X_1 + X_2)$.

5-29   For the independent random variables whose joint probability distribution is given in Table 5-5, find $E(X + Y)$ and $\text{Var}(X + Y)$.

5-30   For the random variables whose joint probability distribution is given in Table 5-6, find $E(X + Y)$ and $\text{Var}(X + Y)$.

5-31   For the random variables whose joint probability distribution is given in Table 5-7, find $E(X + Y)$ and $\text{Var}(X + Y)$.

5-32   Let X_1 have a binomial distribution with constants n_1 and p_1. Let X_2 have a binomial distribution with constants n_2 and p_2. Find $E(X_1 + X_2)$. If we also know that X_1 and X_2 are independent, find $\text{Var}(X_1 + X_2)$.

5-33   Let X_1 and X_2 have Poisson distributions with constants μ_1 and μ_2, respectively. Find $E(X_1 + X_2)$. If we also know that X_1 and X_2 are independent, find $\text{Var}(X_1 + X_2)$.

5-34   For the random variables whose joint probability distribution was found in Exercise 5-8, find $E(X + Y)$ and $\text{Var}(X + Y)$.

5-35   In Exercise 5-13 we found two conditional probability distributions. Use the appropriate one to find $E(Y|2)$ and $\text{Var}(Y|2)$.

5-36   In Exercise 5-14 we found two conditional probability functions. Use these and formulas (4-30) and (4-31) to show

$$E(X|y) = \frac{4 - y}{2} \qquad \text{Var}(X|y) = \frac{(4 - y)(2 + y)}{20}$$

both results holding for $y = 0, 1, 2$, and

$$E(Y|x) = \frac{2(4 - x)}{5} \qquad \text{Var}(Y|x) = \frac{3(4 - x)(1 + x)}{50}$$

both results holding for $x = 0, 1, 2, 3$.

5-37 Let X_1 and X_2 be random variables having the joint probability function

$$f(x_1,x_2) = \frac{x_1 + x_2}{18} \qquad (x_1,x_2) = (0,0), (0,1), (0,2), (1,0), (1,1), (1,2),$$
$$(2,0), (2,1), (22,)$$

Show that

$$g(x_1|x_2) = \frac{x_1 + x_2}{3(x_2 + 1)} \qquad \begin{array}{l} x_1 = 0, 1, 2 \\ x_2 = 0, 1, 2 \end{array}$$

Then show that

$$E(X_1|x_2) = \frac{5 + 3x_2}{3(x_2 + 1)} \qquad x_2 = 0, 1, 2$$

and verify the above results by using the joint probability distribution to perform the necessary calculations.

5-38 For the random variables X and Y of Exercise 5-9 find $\text{Cov}(X,Y)$. *Hint:* Use (5-19) and write

$$E[(X - np_1p_2)(Y - np_1)]$$
$$= E[(X - Yp_2 + Yp_2 - np_1p_2)(Y - np_1)]$$
$$= E[(X - Yp_2)(Y - np_1)] + E[p_2(Y - np_1)^2]$$

Then

$$E[(X - Yp_2)(Y - np_1)] = \sum_{y=0}^{n} \sum_{x=0}^{y} (x - yp_2)(y - np_1)g(x|y)f_2(y)$$

$$= \sum_{y=0}^{n} (y - np_1)f_2(y) \sum_{x=0}^{y} (x - yp_2)\binom{y}{x} p_2^x (1 - p_2)^{y-x}$$

Observe that

$$\sum_{x=0}^{y} (x - yp_2)\binom{y}{x} p_2^x (1 - p_2)^{y-x} = 0$$

being a special case of formula (5-17).   Hence

$$\text{Cov}(X,Y) = E[p_2(Y - np_1)^2] = p_2 E[(Y - np_1)^2]$$

The answer follows by use of formula (4-7).

5-39 For the random variables X and Y of Exercise 5-11 find $\text{Cov}(X,Y)$. *Hint:* Follow the scheme outlined in Exercise 5-38.

5-40 Generalize (5-12) to show that $E(a_1X_1 + a_2X_2) = a_1E(X_1) + a_2E(X_2)$ where a_1 and a_2 are constants.

5-41 Generalize (5-20) to show that $\text{Var}(a_1X_1 + a_2X_2) = a_1^2\text{Var}(X_1) + a_2^2\text{Var}(X_2) + 2a_1a_2\text{Cov}(X_1,X_2)$ where a_1 and a_2 are constants.

5-42   In Exercises 5-38 and 5-39 we have made use of the following theorem: Let $w(X_1, X_2) = w$ be a function of two random variables X_1 and X_2. Now let $E(w|x_2) = \phi(x_2)$. Then $E[\phi(X_2)] = E(w)$. Thus, we see that if it is convenient to do so, we can find the expected value of a function of two random variables by breaking up the calculation into two parts, each of which requires the evaluation of the expected value of a single random variable. Prove the theorem. *Hint:* Observe that

$$\phi(x_2) = E(w|x_2) = \sum_{x_1} w(x_1, x_2) \Pr(X_1 = x_1 | X_2 = x_2)$$

and

$$E[\phi(X_2)] = \sum_{x_2} \phi(x_2) \Pr(X_2 = x_2)$$

Substitute the value of $\phi(x_2)$ in the second equation, replace $\Pr(X_1 = x_1 | X_2 = x_2)$ by its equivalent as given by formula (5-3), and compare the result with (5-11).

5-4   MORE THAN TWO DIMENSIONS

Most of the concepts discussed in the first three sections of this chapter can be generalized to the case of n random variables. The joint probability function will be denoted by

$$f(x_1, x_2, \ldots, x_n) = \Pr(X_1 = x_1, X_2 = x_2, \ldots, X_n = x_n) \qquad (5\text{-}31)$$

This could be used to form an n-dimensional table of probabilities, the joint probability distribution. However, we rarely have use for such a table. The one-dimensional marginal probability function of X_1 is

$$f_1(x_1) = \sum_{x_2} \sum_{x_3} \cdots \sum_{x_n} f(x_1, x_2, \ldots, x_n) \qquad (5\text{-}32)$$

That is, $\Pr(X_1 = x_1, X_2 = x_2, \ldots, X_n = x_n)$ is summed over all values of $x_2, \ldots, x_n$ for which $X_1 = x_1$. The marginal probability functions of $X_2, X_3, \ldots, X_n$, say, $f_2(x_2), f_3(x_3), \ldots, f_n(x_n)$, are similarly defined, the sum in each case being taken over all possible values of the other random variables.

If the random variables are independent, then the joint probability function is the product of the marginal probability functions. That is,

$$f(x_1, x_2, \ldots, x_n) = f_1(x_1)f_2(x_2) \cdots f_n(x_n) \qquad (5\text{-}33)$$

163

Further, if $X_1, X_2, \ldots, X_n$ all have the same marginal probability functions, that is, $f_1(x_1) = g(x_1)$, $f_2(x_2) = g(x_2)$, $\ldots$, $f_n(x_n) = g(x_n)$, then

$$f(x_1, x_2, \ldots, x_n) = g(x_1)g(x_2) \cdots g(x_n) \tag{5-34}$$

When $X_1, X_2, \ldots, X_n$ are independent and all have the same marginal probability function $g(x)$, so that the joint probability function is given by (5-34), the random variables are called a *random sample* from a distribution that has the probability function $g(x)$.

Occasionally we need the marginal distribution of two or more random variables. The marginal probability function of X_1 and X_2, which is also the joint probability function of X_1 and X_2, is

$$f_{12}(x_1, x_2) = \sum_{x_3} \sum_{x_4} \cdots \sum_{x_n} f(x_1, \ldots, x_n) \tag{5-35}$$

That is, $\Pr(X_1 = x_1, X_2 = x_2, \ldots, X_n = x_n)$ is summed over all values of $x_3, \ldots, x_n$ for which $X_1 = x_1$, $X_2 = x_2$. In a similar manner we can write the marginal probability function of any two random variables. The marginal probability function of any three random variables is found by summing over the other $n - 3$ random variables, etc.

The joint conditional probability function of $X_2, \ldots, X_n$ given $X_1 = x_1$ is

$$g(x_2, \ldots, x_n | x_1) = \frac{f(x_1, \ldots, x_n)}{f_1(x_1)} \tag{5-36}$$

Similarly, the joint conditional probability function of the other $n - 1$ random variables given $X_i = x_i$ is obtained by dividing the joint probability function by the marginal probability function $f_i(x_i)$. Also two or more random variables could be given and conditional probability functions are formed in the obvious way. For example, the joint conditional probability function of $X_3, \ldots, X_n$ given $X_1 = x_1$ and $X_2 = x_2$ is

$$h(x_3, \ldots, x_n | x_1, x_2) = \frac{f(x_1, \ldots, x_n)}{f_{12}(x_1, x_2)} \tag{5-37}$$

If

$$Y_1 = u_1(X_1, \ldots, X_n), \ Y_2 = u_2(X_1, \ldots, X_n), \ldots, \\ Y_n = u_n(X_1, \ldots, X_n)$$

and there is a one-to-one correspondence between the n-dimensional random variables $(X_1, \ldots, X_n)$ and $(Y_1, \ldots, Y_n)$, then the joint probability function of $Y_1, Y_2, \ldots, Y_n$ is found by extending the procedure used to get (5-7). We solve the above equations, getting

$$X_1 = w_1(Y_1, \ldots, Y_n), \ X_2 = w_2(Y_1, \ldots, Y_n), \ldots, \\ X_n = w_n(Y_1, \ldots, Y_n)$$

The joint probability function of $Y_1, Y_2, \ldots, Y_n$ is

$$g(y_1, y_2, \ldots, y_n) = f[w_1(y_1, \ldots, y_n), w_2(y_1, \ldots, y_n), \ldots, w_n(y_1, \ldots, y_n)] \quad (5\text{-}38)$$

As in the case of two random variables, the most useful application of derived random variables arises when we seek the probability function of $Y_1 = X_1 + X_2 + \cdots + X_n$. Following the same procedure we used with two random variables, we would let $Y_2 = X_2$, $Y_3 = X_3$, $\ldots$, $Y_n = X_n$ (we need as many Y's as X's). Thus $X_1 = Y_1 - Y_2 - \cdots - Y_n$, $X_2 = Y_2, \ldots, X_n = Y_n$, and (5-38) becomes

$$g(y_1, y_2, \ldots, y_n) = f[y_1 - y_2 - \cdots - y_n, y_2, \ldots, y_n] \quad (5\text{-}39)$$

Then, the marginal probability function of Y_1 is

$$g_1(y_1) = \sum_{y_2} \sum_{y_3} \cdots \sum_{y_n} g(y_1, y_2, \ldots, y_n) \quad (5\text{-}40)$$

Fortunately, this complicated procedure is very seldom needed.

EXAMPLE 5-13

Let $X_1, X_2, \ldots, X_n$ be a random sample from a distribution having the probability function $g(x) = p^x(1 - p)^{1-x}$, $x = 0$, 1. Find the joint probability function of the n random variables and the marginal probability function of $Y_1 = X_1 + X_2 + \cdots + X_n$.

Solution

By (5-34) we have

$$f(x_1, x_2, \ldots, x_n) = [p^{x_1}(1 - p)^{1-x_1}][p^{x_2}(1 - p)^{1-x_2}] \cdots [p^{x_n}(1 - p)^{1-x_n}]$$
$$= p^{(x_1+x_2+\cdots+x_n)}(1 - p)^{n-(x_1+x_2+\cdots+x_n)}$$

where $x_i = 0$, 1, $i = 1, 2, \ldots, n$. If $Y_1 = X_1 + X_2 + \cdots + X_n$, $Y_2 = X_2, \ldots, Y_n = X_n$, the joint probability function of the Y's is

$$h(y_1, y_2, \ldots, y_n) = p^{y_1}(1 - p)^{n-y_1}$$

defined for the 2^n points one gets by converting the points $(x_1, x_2, \ldots, x_n)$ to $(y_1, y_2, \ldots, y_n)$. To get the marginal probability function of Y_1, we would then have to evaluate

$$\sum_{y_2} \sum_{y_3} \cdots \sum_{y_n} h(y_1, y_2, \ldots, y_n)$$

165

Such a complicated procedure is unnecessary if we notice $g(0) = 1 - p$ and $g(1) = p$ and $Y_1 = X_1 + X_2 + \cdots + X_n$ is the sum of 1's (or successes) obtained when n independent experiments are performed, each with probability of success p. Hence, $\Pr(Y_1 = y_1)$ is by (4-2)

$$h_1(y_1) = \binom{n}{y_1} p^{y_1}(1 - p)^{n-y_1} \qquad y_1 = 0, 1, 2, \ldots, n$$

EXAMPLE 5-14

Let $X_1, X_2, \ldots, X_n$ be a random sample from a distribution having the probability function $g(x) = \frac{1}{6}$, $x = 1, 2, 3, 4, 5, 6$. Find the joint probability function of the n random variables.

Solution

By (5-34) we get with no difficulty

$$f(x_1, x_2, \ldots, x_n) = \left(\frac{1}{6}\right)^n \qquad \begin{matrix} x_i = 1, 2, 3, 4, 5, 6 \\ i = 1, 2, \ldots, n \end{matrix}$$

Here we did not ask for the probability function of $Y_1 = X_1 + X_2 + \cdots + X_n$. The derivation of the probability function of Y_1 involves some moderately difficult enumeration problems (see [2], pp. 30–31), a fact that provides some motivation for seeking approximate methods for computing probabilities on Y_1. We shall discuss such an approximation in the next chapter.

In Sec. 5-2 we were able to show that if X_1 and X_2 are independent random variables having Poisson distributions with means μ_1 and μ_2, respectively, then $Y_1 = X_1 + X_2$ has the probability function (5-9). That is, Y_1 has a Poisson distribution with mean $\mu_1 + \mu_2$. This result can be readily extended to n independent random variables $X_1, X_2, \ldots, X_n$ having Poisson distributions with means $\mu_1, \mu_2, \ldots, \mu_n$. The probability function of $Y_1 = X_1 + X_2 + \cdots + X_n$ is

$$g_1(y_1) = \frac{e^{-\mu}\mu^{y_1}}{y_1!} \qquad y_1 = 0, 1, 2, \ldots \qquad (5\text{-}41)$$

where $\mu = \mu_1 + \mu_2 + \cdots + \mu_n$. This result can be established in several ways. First, we can follow the procedure used to get (5-40). To perform the summing required in (5-40) we can apply formula (A2-3) repeatedly instead of just once as we did in deriving (5-9) (see Exercise 5-47). A second method requires that we use an intuitively obvious theorem (which can be proved as an exercise). The theorem is:

If X_1, X_2, X_3 are independent random variables, then so are $X_1 + X_2$ and X_3.

We already know by the derivation of (5-9) that $X_1 + X_2$ has a Poisson distribution with mean $\mu_1 + \mu_2$. By reapplication of the derivation of (5-9) $(X_1 + X_2) + X_3$ has a Poisson distribution with mean $(\mu_1 + \mu_2) + \mu_3$. Next $(X_1 + X_2) + X_3$ is independent of X_4 by the above theorem, and reapplication of the derivation of (5-9) yields the fact that $(X_1 + X_2 + X_3) + X_4$ has a Poisson distribution with mean $(\mu_1 + \mu_2 + \mu_3) + \mu_4$. Continuing in this way yields the desired result.

By using exactly the same argument we used in Sec. 5-2 for two random variables, we can show that if X_1, X_2, . . . , X_k are independent random variables having binomial distributions with the same p but with $n = n_1$, $n = n_2$, . . . , $n = n_k$, respectively, then

$$Y_1 = X_1 + X_2 + \cdot\cdot\cdot + X_k$$

has a binomial distribution with constants $n = n_1 + n_2 + \cdot\cdot\cdot + n_k$ and p. Thus the probability function of Y_1 is

$$g_1(y_1) = \binom{n}{y_1} p^{y_1}(1 - p)^{n-y_1} \qquad y_1 = 0, 1, 2, \ldots, n \qquad (5\text{-}42)$$

The extension of (5-11) is

$$E(Z) = E[w(X_1, X_2, \ldots, X_n)]$$
$$= \sum_{x_n} \cdot\cdot\cdot \sum_{x_1} w(x_1, \ldots, x_n) \Pr(X_1 = x_1, \ldots, X_n = x_n) \qquad (5\text{-}43)$$

The counterparts of (5-12) and its generalization given in Exercise 5-40 are respectively

$$E(X_1 + X_2 + \cdot\cdot\cdot + X_n) = E(X_1) + E(X_2) + \cdot\cdot\cdot + E(X_n) \qquad (5\text{-}44)$$
$$E(a_1X_1 + a_2X_2 + \cdot\cdot\cdot + a_nX_n)$$
$$= a_1E(X_1) + a_2E(X_2) + \cdot\cdot\cdot + a_nE(X_n) \qquad (5\text{-}45)$$

where a_1, a_2, . . . , a_n are constants. Formulas (5-13), (5-14), and (5-15) readily generalize to any number of functions. Following the same scheme used to get (5-20) yields

$$\mathrm{Var}(X_1 + X_2 + \cdot\cdot\cdot + X_n) = \sum_{i=1}^{n} \mathrm{Var}(X_i) + 2 \sum_{\substack{i,j=1 \\ i<j}}^{n} \mathrm{Cov}(X_i, X_j)$$

$$(5\text{-}46)$$

since the square of a multinomial expression is equal to the sum of the squares plus twice all possible cross products. Similarly

$$\text{Var}(a_1X_1 + a_2X_2 + \cdots + a_nX_n) = \sum_{i=1}^{n} a_i^2 \text{Var}(X_i)$$

$$+ 2 \sum_{\substack{i,j=1 \\ i<j}}^{n} a_i a_j \text{Cov}(X_i, X_j) \quad (5\text{-}47)$$

where $a_1, a_2, \ldots, a_n$ are constants. For the special case in which the random variables are independent, (5-46) and (5-47) become

$$\text{Var}(X_1 + X_2 + \cdots + X_n) = \sum_{i=1}^{n} \text{Var}(X_i) \quad (5\text{-}48)$$

and

$$\text{Var}(a_1X_1 + a_2X_2 + \cdots + a_nX_n) = \sum_{i=1}^{n} a_i^2 \text{Var}(X_i) \quad (5\text{-}49)$$

When dealing with a random sample from a distribution whose probability function is $g(x)$, we are sometimes interested in the probability distribution of $Y_1 = X_1 + X_2 + \cdots + X_n$. We can find the mean and variance of Y_1 without its probability distribution. By definition of a random sample, $X_1, X_2, \ldots, X_n$ all have the same probability distribution $g(x)$. Consequently,

$$E(X_1) = E(X_2) = \cdots = E(X_n) = E(X)$$

and

$$\text{Var}(X_1) = \text{Var}(X_2) = \cdots = \text{Var}(X_n) = \text{Var}(X)$$

From (5-44) we see that

$$E(X_1 + X_2 + \cdots + X_n) = nE(X) \quad (5\text{-}50)$$

and from (5-48)

$$\text{Var}(X_1 + X_2 + \cdots + X_n) = n \, \text{Var}(X) \quad (5\text{-}51)$$

If $X_1, X_2, \ldots, X_n$ is a random sample, then

$$\bar{X} = \frac{(X_1 + X_2 + \cdots + X_n)}{n}$$

is called the *sample mean*. Although $Y_1 = X_1 + X_2 + \cdots + X_n$ and $\bar{X}$ differ only in that the latter is the former divided by n, it is sometimes more convenient to use the sample mean. We note that the sample

168

mean can be written

$$\bar{X} = \frac{1}{n} X_1 + \frac{1}{n} X_2 + \cdots + \frac{1}{n} X_n \qquad (5\text{-}52)$$

Using (5-45) and (5-49) with $a_1 = a_2 = \cdots = a_n = \frac{1}{n}$ yields

$$E(\bar{X}) = \frac{1}{n} E(X_1) + \frac{1}{n} E(X_2) + \cdots + \frac{1}{n} E(X_n)$$

$$= \frac{1}{n} [E(X) + E(X) + \cdots + E(X)]$$

$$= \frac{1}{n} [nE(X)]$$

$$= E(X) \qquad (5\text{-}53)$$

$$\operatorname{Var}(\bar{X}) = \frac{1}{n^2} \operatorname{Var}(X_1) + \frac{1}{n^2} \operatorname{Var}(X_2) + \cdots + \frac{1}{n^2} \operatorname{Var}(X_n)$$

$$= \frac{1}{n^2} [\operatorname{Var}(X) + \operatorname{Var}(X) + \cdots + \operatorname{Var}(X)]$$

$$= \frac{1}{n^2} [n \operatorname{Var}(X)]$$

$$= \frac{\operatorname{Var}(X)}{n} \qquad (5\text{-}54)$$

Thus, if $X_1, X_2, \ldots, X_n$ is a random sample from a distribution having probability function $g(x)$, then the derived random variable $\bar{X}$ has the same mean as X and the variance of $\bar{X}$ is only $(1/n)$th as large as the variance of X.

EXAMPLE 5-15

Let $X_1, X_2, \ldots, X_n$ be a random sample from a distribution having the probability function

$$g(x) = p^x(1 - p)^{1-x} \qquad x = 0, 1$$

By using (5-50) and (5-51) find the mean and variance of $Y_1 = X_1 + X_2 + \cdots + X_n$.

Solution

By definition of expected value, we have

$$E(X) = (0)p^0(1 - p)^{1-0} + (1)p^1(1 - p)^{1-1} = p$$
$$E(X^2) = (0^2)p^0(1 - p)^{1-0} + (1^2)p^1(1 - p)^{1-1} = p$$

The variance of X is

$$\operatorname{Var}(X) = p - p^2 = p(1 - p)$$

Hence by (5-50) and (5-51) we have

$$E(Y_1) = np \qquad \text{Var}(Y_1) = np(1 - p)$$

Since we know from Example 5-13 that Y_1 has a binomial distribution, we have verified (4-6) and (4-7), perhaps somewhat more easily than we did in Sec. 4-2.

EXAMPLE 5-16

Let $X_1, X_2, \ldots, X_n$ be a random sample from a distribution having the probability function $g(x) = \frac{1}{6}$, $x = 1, 2, 3, 4, 5, 6$. Find the mean and variance of the derived distribution of $\bar{X}$.

Solution

$$E(X) = 1(\tfrac{1}{6}) + 2(\tfrac{1}{6}) + 3(\tfrac{1}{6}) + 4(\tfrac{1}{6}) + 5(\tfrac{1}{6}) + 6(\tfrac{1}{6})$$
$$= {}^{21}\!\!/_{6} = \tfrac{7}{2} = 3.5$$
$$E(X^2) = 1^2(\tfrac{1}{6}) + 2^2(\tfrac{1}{6}) + 3^2(\tfrac{1}{6}) + 4^2(\tfrac{1}{6}) + 5^2(\tfrac{1}{6}) + 6^2(\tfrac{1}{6}) = {}^{91}\!\!/_{6}$$
$$\text{Var}(X) = {}^{91}\!\!/_{6} - (\tfrac{7}{2})^2 = {}^{35}\!\!/_{12}$$

Then by (5-53) and (5-54) we get

$$E(\bar{X}) = 3.5 \qquad \text{Var}(\bar{X}) = \frac{35}{12n}$$

EXERCISES

5-43   If X_1, X_2, X_3, X_4 is a random sample from a distribution having the probability function

$$g(x) = \frac{x}{10} \qquad x = 1, 2, 3, 4$$

find the joint probability function of the four random variables. Find the mean and variance of $Y_1 = X_1 + X_2 + X_3 + X_4$. Find $E(X_1 X_2 X_3 X_4)$.

5-44   If X_1, X_2, X_3 are independent random variables with probability functions

$$f_1(x_1) = \binom{2}{x_1}\left(\frac{1}{6}\right)^{x_1}\left(\frac{5}{6}\right)^{2-x_1} \qquad x_1 = 0, 1, 2$$

$$f_2(x_2) = \binom{3}{x_2}\left(\frac{1}{6}\right)^{x_2}\left(\frac{5}{6}\right)^{3-x_2} \qquad x_2 = 0, 1, 2, 3$$

$$f_3(x_3) = \binom{4}{x_3}\left(\frac{1}{6}\right)^{x_3}\left(\frac{5}{6}\right)^{4-x_3} \qquad x_3 = 0, 1, 2, 3, 4$$

find the joint probability function of the three random variables. What is the probability function of $Y_1 = X_1 + X_2 + X_3$?

5-45 If $X_1, X_2, \ldots, X_n$ is a random sample from a distribution having probability function

$$g(x) = p(1 - p)^{x-1} \qquad x = 1, 2, \ldots$$

find the joint probability function of the n random variables. Note that $g(x)$ is the negative binomial distribution with $c = 1$, x replacing n. Use the same argument used with the binomial to show that $Y_1 = X_1 + X_2 + \cdots + X_n$ has a negative binomial distribution, with $c = n$.

5-46 One typist makes on the average one error per page. A second typist makes on the average two errors per page. A third typist makes on the average three errors per page. If each of the three types a complete page, what is the probability that the total number of errors on all three pages is more than six? Assume in each case that the number of errors per page follows a Poisson distribution.

5-47 Derive (5-41) for the case $n = 3$ following the first suggested procedure. *Hint:* From (5-39) the joint probability function of Y_1, Y_2, Y_3 is

$$g(y_1, y_2, y_3) = \frac{e^{-(\mu_1+\mu_2+\mu_3)}\mu_1^{y_1-y_2-y_3}\mu_2^{y_2}\mu_3^{y_3}}{(y_1 - y_2 - y_3)!\,y_2!\,y_3!}$$

Since $X_1 = Y_1 - Y_2 - Y_3$ and $X_1 \geq 0$ (since $X_1 = 0, 1, 2, \ldots$), $Y_1 \geq Y_2 + Y_3$ or $Y_3 \leq Y_1 - Y_2$. Hence the joint probability function has nonzero probability at $y_1 = 0, 1, 2, \ldots, y_2 = 0, 1, 2, \ldots, y_3 = 0, 1, 2, \ldots$, but subject to $y_3 \geq y_1 - y_2$. Thus the joint probability function of Y_1 and Y_2 is

$$g_{12}(y_1, y_2) = \frac{e^{-(\mu_1+\mu_2+\mu_3)}\mu_2^{y_2}}{y_2!(y_1 - y_2)!} \sum_{y_3=0}^{y_1-y_2} \frac{(y_1 - y_2)!}{y_3!(y_1 - y_2 - y_3)} \mu_3^{y_3}\mu_1^{y_1-y_2-y_3}$$

$$= \frac{e^{-(\mu_1+\mu_2+\mu_3)}}{y_2!(y_1 - y_2)!} \mu_2^{y_2}(\mu_1 + \mu_3)^{y_1-y_2}$$

where $y_1 = 0, 1, 2, \ldots, y_2 = 0, 1, 2, \ldots$ subject to $y_2 \leq y_1$ (since $0 \leq y_3 \leq y_1 - y_2$). Finally

$$g_1(y_1) = \sum_{y_2=0}^{y_1} g_{12}(y_1, y_2)$$

$$= \frac{e^{-(\mu_1+\mu_2+\mu_3)}}{y_1!} \sum_{y_2=0}^{y_1} \frac{y_1!}{y_2!(y_2 - y_1)!} \mu_2^{y_2}(\mu_1 + \mu_3)^{y_1-y_2}$$

which with the use of formula (A2-3) produces the desired result.

171

5-48 Let X_1, X_2, X_3 be a random sample of size 3 from a distribution whose mean is μ and whose variance is σ^2. We already know that $E(\bar{X}) = \mu$, $\text{Var}(\bar{X}) = \sigma^2/3$, results that were obtained by letting $a_1 = a_2 = \cdots = a_n = 1/n$ in (5-45) and (5-49). Suppose we let $Y_1 = a_1X_1 + a_2X_2 + a_3X_3$ and require that a_1, a_2, a_3 be so chosen as to make $E(Y_1) = \mu$ and $\text{Var}(Y_1)$ a minimum. Show that $a_1 = a_2 = a_3 = \frac{1}{3}$. *Hint:* If $E(Y_1) = \mu$, then $k_1 + k_2 + k_3 = 1$. Also $\text{Var}(Y_1) = (k_1^2 + k_2^2 + k_3^2)\sigma^2$. Eliminate k_3 in $\text{Var}(Y_1)$ by using the first equation. If the square is completed first on k_1, then on k_2, we can write

$$\text{Var}(Y_1) = \left[\frac{(2k_1 + k_2 - 1)^2 + 3(k_2 - \frac{1}{3})^2}{2} + \frac{1}{3} \right]\sigma^2$$

It is obvious that $k_2 = \frac{1}{3}$, $k_1 = \frac{1}{3}$ (and hence $k_3 = \frac{1}{3}$) minimizes $\text{Var}(Y_1)$. The result of the problem generalizes. That is, with a random sample of size n and $Y_1 = a_1X_1 + a_2X_2 + \cdots + a_nX_n$, choose $a_1 = a_2 = \cdots = a_n = 1/n$ to make $E(Y_1) = \mu$ and $\text{Var}(Y_1)$ a minimum.

5-49 Let X_1, X_2, . . . , X_n be a random sample of size n from a distribution whose mean is μ and whose variance is σ^2. From the alternative form of Chebyshev's inequality (3-16) deduce that $\Pr(|\bar{X} - \mu| < a) \geq 1 - \sigma^2/a^2n$. Since $a > 0$ can be chosen arbitrarily small, we conclude that the probability that $\bar{X}$ differs from μ by less than any preassigned number can be made as close to 1 as we like by taking n large enough. Suppose

$$g(x) = \frac{1}{6} \qquad x = 1, 2, 3, 4, 5, 6$$

so that $\mu = 3.5$ and $\sigma^2 = \frac{35}{12}$. How large does n have to be so that $\Pr(|\bar{X} - 3.5| < 1) \geq .9$?

5-5   THE MULTINOMIAL DISTRIBUTION

A fairly obvious generalization of the binomial yields the multinomial distribution. Consider a series of events or experiments which have the following properties:

(a) *The result of each experiment can be classified into one of k categories C_1, C_2, . . . , C_k.*

(b) *The probabilities of falling into these categories are p_1, p_2, . . . , p_k for each experiment.*   (5-55)

(c) *Each experiment is independent of all the others.*

(d)   *The series of experiments is performed a fixed number of times, say, n.*

Thus if $k = 2$, conditions (5-55) are the same as (4-1). We note that $p_1 + p_2 + \cdots + p_k = 1$, since the result of each experiment must fall into one of the k categories. To illustrate a series of events satisfying the above conditions, suppose that a symmetric six-sided die is to be rolled 10 times. Let C_1 be the outcome 1 or 2, C_2 be the outcome 3, 4, or 5, and C_3 be the outcome 6. Thus it is reasonable to take $p_1 = \frac{2}{6}$, $p_2 = \frac{3}{6}$, $p_3 = \frac{1}{6}$. We might like to know the probability that three of the rolls fall in category 1, five in category 2, and two in category 3. One order of outcomes which will yield the desired result is $C_1 C_1 C_1 C_2 C_2 C_2 C_2 C_2 C_3 C_3$. By the multiplication theorem for independent events, the probability of this sequence is

$$(\tfrac{2}{6})(\tfrac{2}{6})(\tfrac{2}{6})(\tfrac{3}{6})(\tfrac{3}{6})(\tfrac{3}{6})(\tfrac{3}{6})(\tfrac{3}{6})(\tfrac{1}{6})(\tfrac{1}{6}) = (\tfrac{2}{6})^3 (\tfrac{3}{6})^5 (\tfrac{1}{6})^2$$

Obviously there are many other orders, each order being an event, which will yield the result under consideration. The probability associated with each sequence is $(\tfrac{2}{6})^3 (\tfrac{3}{6})^5 (\tfrac{1}{6})^2$. The total number of orders that yield the desired result is $10!/3!5!2!$, the number of ways to permute 10 things taken 10 at a time if 3 are alike, 5 are alike, and 2 are alike. Since only one order can occur at a time, the $10!/3!5!2!$ orders represent $10!/3!5!2!$ mutually exclusive events, each with probability of occurrence $(\tfrac{2}{6})^3 (\tfrac{3}{6})^5 (\tfrac{1}{6})^2$. By the addition theorem, the probability that one of these orders occurs is

$$\frac{10!}{3!5!2!} \left(\frac{2}{6}\right)^3 \left(\frac{3}{6}\right)^5 \left(\frac{1}{6}\right)^2 = .081$$

We now generalize the discussion to yield a formula for the probability of obtaining x_1 outcomes in category C_1, x_2 outcomes in category C_2, . . . , x_k outcomes in category C_k, where

$$x_1 + x_2 + \cdots + x_k = n \tag{5-56}$$

One order which produces the result is

$$\underbrace{C_1 \cdots C_1}_{x_1 \text{ times}} \underbrace{C_2 \cdots C_2}_{x_2 \text{ times}} \cdots \underbrace{C_k \cdots C_k}_{x_k \text{ times}}$$

By the multiplication theorem, the probability that this happens is

$$\underbrace{p_1 \cdots p_1}_{x_1 \text{ times}} \underbrace{p_2 \cdots p_2}_{x_2 \text{ times}} \cdots \underbrace{p_k \cdots p_k}_{x_k \text{ times}} = (p_1)^{x_1} (p_2)^{x_2} \cdots (p_k)^{x_k}$$

The total number of orders which produce the result under discussion is

$$\frac{n!}{x_1! x_2! \cdots x_k!}$$

the number of ways to permute n things taken n at a time when x_1 are alike, x_2 are alike, . . . , x_k are alike. Each order corresponds to an event and only one order can occur at a time. Hence the $n!/(x_1! \, x_2! \cdots x_k!)$ orders represent $n!/(x_1! \, x_2! \cdots x_k!)$ mutually exclusive events, each with probability of occurrence $(p_1)^{x_1}(p_2)^{x_2} \cdots (p_k)^{x_k}$. By the addition theorem the probability that one of these orders occurs is

$$f(x_1, x_2, \ldots, x_k) = \frac{n!}{x_1! \cdots x_k!} (p_1)^{x_1} \cdots (p_k)^{x_k} \qquad (5\text{-}57)$$

The x's can take on any of the values $0, 1, 2, \ldots, n$ subject to the restriction (5-56).

Actually, since the sum of the x's is n, there are only $n - 1$ functionally independent random variables, the last being determined when the other $n - 1$ are known. However, $X_1, X_2, \ldots, X_k$ are all perfectly respectable random variables. Thus we could write for (5-57)

$$f(x_1, x_2, \ldots, x_{k-1}) = \frac{n!}{x_1! x_2! \cdots (n - x_1 - \cdots - x_{k-1})!} (p_1)^{x_1}(p_2)^{x_2}$$
$$\cdots (p_{k-1})^{x_{k-1}}(p_k)^{n - x_1 - \cdots - x_{k-1}} \qquad (5\text{-}58)$$

The random variables $X_1, X_2, \ldots, X_{k-1}$ can take any of the values $0, 1, 2, \ldots, n$ subject to the restriction $X_1 + X_2 + \cdots + X_{k-1} \leq n$ [since (5-56) must hold]. The name "multinomial" arises from the fact that the right-hand side of (5-57) is a term in the expansion of $(p_1 + p_2 + \cdots + p_k)^n$.

The computation of (5-57) is a tedious chore when performed by hand calculations. This type of calculation is easily handled by high-speed computing machines. Tabulation would be cumbersome because of the variety of choices possible for the p's, k, and n. In some situations where multinomial probabilities are encountered, satisfactory approximations are available (see [1], chap. 6).

EXAMPLE 5-17

If a die is rolled five times, what is the probability that the results are a 1, a 2, and three other numbers?

Solution

First we check conditions (5-55).

(a) Each throw is a 1, a 2, or not a 1 or 2.
(b) The probabilities $p_1 = \frac{1}{6}$, $p_2 = \frac{1}{6}$, $p_3 = \frac{4}{6}$ remain constant for each throw.
(c) Successive throws are independent.
(d) The die is rolled five times, a fixed number.

Hence the multinomial model seems to be appropriate. Here $x_1 = 1$, $x_2 = 1$, $x_3 = 3$ and the probability of obtaining a 1, a 2, and three other numbers is

$$\frac{5!}{1!1!3!} \left(\frac{1}{6}\right) \left(\frac{1}{6}\right) \left(\frac{4}{6}\right)^3 = \frac{40}{243}$$

Marginal probability functions are obtained by making use of previous arguments. Suppose we want $f_1(x_1)$, the marginal probability function of X_1. We note that X_1 is the number of occurrences of category C_1 for a series of experiments with the following properties:

(a)   The result of each experiment is either C_1 (success) or not C_1 (failure).
(b)   The probability of a success is p_1 for each experiment.
(c)   Each experiment is independent of all the others.
(d)   The series of experiments is performed n times, a fixed number.

Hence X_1 must have a binomial distribution with constants n and p_1 so that

$$f_1(x_1) = \binom{n}{x_1} p_1^{x_1}(1 - p_1)^{n-x_1} \qquad x_1 = 0, 1, 2, \ldots, n \qquad (5\text{-}59)$$

Similarly, the marginal probability function of X_i, $i = 1, 2, \ldots, k$, is

$$f_i(x_i) = \binom{n}{x_i} p_i^{x_i}(1 - p_i)^{n-x_i} \qquad x_i = 0, 1, 2, \ldots, n \qquad (5\text{-}60)$$

The marginal probability function of two random variables can be found by a similar argument. Suppose we want $f_{12}(x_1, x_2)$, the marginal probability function of X_1 and X_2. We now observe that X_1 is the number of occurrences of category C_1, X_2 is the number of occurrences of category C_2 for a series of experiments with the following properties:

(a)   The result of each experiment can be classified into one of three categories C_1, C_2, and not C_1 or C_2.
(b)   The probabilities of falling into these categories are p_1, p_2, and $p_3 = 1 - p_1 - p_2$ for each experiment.
(c)   Each experiment is independent of all the others.
(d)   The series of experiments is performed a fixed number of times n.

Hence X_1 and X_2 have a multinomial distribution with $k = 3$ and

$$f_{12}(x_1, x_2) = \frac{n!}{x_1!x_2!(n - x_1 - x_2)!} (p_1)^{x_1}(p_2)^{x_2}(p_3)^{n-x_1-x_2} \qquad (5\text{-}61)$$

where $x_1 = 0, 1, 2, \ldots, n$, $x_2 = 0, 1, 2, \ldots, n$ subject to $x_1 + x_2 \leq n$. Similarly the marginal probability function of X_i and X_j is

$$f_{ij}(x_i, x_j) = \frac{n!}{x_i! x_j! (n - x_i - x_j)!} (p_i)^{x_i} (p_j)^{x_j} (1 - p_i - p_j)^{n - x_i - x_j} \qquad (5\text{-}62)$$

where $x_i = 0, 1, 2, \ldots, n$, $x_j = 0, 1, 2, \ldots, n$ subject to $x_i + x_j \leq n$.

In Exercise 5-23 we used the derived distribution technique to find the probability function of $Y_1 = X_1 + X_2$, where X_1 and X_2 have a multinomial distribution. This procedure can be used to find the distribution of the sum of any number of X's having a multinomial distribution. It is, however, easier to argue as we have in the preceding two paragraphs. Thus if we want the probability function of $Y_1 = X_1 + X_2$, we can regard old categories C_1 and C_2 as being category C_1' and old categories C_3, C_4, \ldots, C_k as being category C_2', observe that the binomial conditions (4-1) are satisfied, and conclude that Y_1 has a binomial distribution with constants n and $p_1 + p_2$. That is,

$$g(y_1) = \binom{n}{y_1} (p_1 + p_2)^{y_1} (1 - p_1 - p_2)^{n - y_1} \qquad y_1 = 0, 1, 2, \ldots, n$$

$$(5\text{-}63)$$

If $Y_1 = X_1 + X_2 + X_3$, then Y_1 has a binomial distribution with constants n and $p_1 + p_2 + p_3$, etc.

As in the case of marginal distributions, the conditional distributions are multinomial. For example, if

$$f(x_1, x_2, x_3) = \frac{n!}{x_1! x_2! x_3! (n - x_1 - x_2 - x_3)!} (p_1)^{x_1} (p_2)^{x_2} (p_3)^{x_3} (p_4)^{n - x_1 - x_2 - x_3}$$

the marginal probability function of X_1 is given by (5-59). Hence, the conditional probability function of X_2 and X_3 given X_1 is

$$h(x_2, x_3 | x_1) = \frac{f(x_1, x_2, x_3)}{f_1(x_1)}$$

$$= \frac{(n - x_1)!}{x_2! x_3! (n - x_1 - x_2 - x_3)!} \left(\frac{p_2}{1 - p_1} \right)^{x_2} \left(\frac{p_3}{1 - p_1} \right)^{x_3} \left(\frac{p_4}{1 - p_1} \right)^{n - x_1 - x_2 - x_3}$$

$$(5\text{-}64)$$

$$x_2 + x_3 \leq n - x_1$$

since we can write

$$(1 - p_1)^{n - x_1} = (1 - p_1)^{n - x_1 - x_2 - x_3} (1 - p_1)^{x_2} (1 - p_1)^{x_3}$$

We note that (5-64) is the probability function of a multinomial generated by considering $n - x_1$ repetitions of an experiment for which each outcome can be classified into three categories with probabilities $p_2/(1 - p_1)$,

176

$p_3/(1 - p_1)$, and $p_4/(1 - p_1)$. Thus (5-64) is nonzero for $x_2 = 0, 1, 2,$
$\ldots, n - x_1, x_3 = 0, 1, 2, \ldots, n - x_1$, subject to $x_2 + x_3 \leq n - x_1$.

Expected values are easily obtained from previous results. Because of (5-60) each X_i is binomial and we have immediately from (4-6) and (4-7)

$$E(X_i) = np_i \qquad \text{Var}(X_i) = np_i(1 - p_i) \qquad (5\text{-}65)$$

To find $\text{Cov}(X_i, X_j)$ we could use the scheme followed in Exercises 5-38, 5-39, and 5-42; we shall, however, use an alternative method. We know that $Y_1 = X_i + X_j$ has the probability function (5-63) with p_1 and p_2 replaced by p_i and p_j. Hence

$$\text{Var}(Y_1) = \text{Var}(X_i + X_j) = n(p_i + p_j)(1 - p_i - p_j)$$

by (4-7). Also by (5-20)

$$\text{Var}(X_i + X_j) = \text{Var}(X_i) + \text{Var}(X_j) + 2\text{Cov}(X_i, X_j) \qquad (5\text{-}66)$$

The only unknown in (5-66) is $\text{Cov}(X_i, X_j)$. Thus we get

$$\begin{aligned}
\text{Cov}(X_i, X_j) &= \frac{\text{Var}(X_i + X_j) - \text{Var}(X_i) - \text{Var}(X_j)}{2} \\
&= \frac{n(p_i + p_j)(1 - p_i - p_j) - np_i(1 - p_i) - np_j(1 - p_j)}{2} \\
&= -np_ip_j \qquad (5\text{-}67)
\end{aligned}$$

EXAMPLE 5-18

Let X_1 and X_2 have the joint probability function given by (5-61). Find the conditional probability function of X_1 given X_2. Then find the mean and variance of X_1 given X_2.

Solution

The marginal probability function of X_2 is given by (5-60) with $i = 2$. Thus we get

$$h(x_1|x_2) = \frac{f(x_1,x_2)}{f_2(x_2)} = \frac{\dfrac{n!}{x_1!x_2!(n - x_1 - x_2)!}(p_1)^{x_1}(p_2)^{x_2}(p_3)^{n-x_1-x_2}}{\dfrac{n!}{x_2!(n - x_2)!}p_2^{x_2}(1 - p_2)^{n-x_2}}$$

$$= \frac{(n - x_2)!}{x_1!(n - x_2 - x_1)!}\left(\frac{p_1}{1 - p_2}\right)^{x_1}\left(\frac{p_3}{1 - p_2}\right)^{n-x_2-x_1}$$

since $(1 - p_2)^{n-x_2} = (1 - p_2)^{x_1}(1 - p_2)^{n-x_2-x_1}$.

Now

$$\frac{p_3}{1 - p_2} = \frac{1 - p_1 - p_2}{1 - p_2} = 1 - \frac{p_1}{1 - p_2}$$

Hence

$$h(x_1|x_2) = \binom{n - x_2}{x_1} \left(\frac{p_1}{1 - p_2}\right)^{x_1} \left(1 - \frac{p_1}{1 - p_2}\right)^{n - x_2 - x_1}$$

$$x_1 = 0, 1, 2, \ldots, n - x_2$$

so that X_1 given X_2 has a binomial distribution with constants $n - x_2$ and $p_1/(1 - p_2)$. The mean and variance are obtained immediately from (4-6) and (4-7) and we get

$$E(X_1|x_2) = (n - x_2)\left(\frac{p_1}{1 - p_2}\right)$$

$$\text{Var}(X_1|x_2) = (n - x_2)\left(\frac{p_1}{1 - p_2}\right)\left(1 - \frac{p_1}{1 - p_2}\right)$$

EXERCISES

Comment on the reasonableness of the multinomial assumptions when it is appropriate to do so.

5-50　A card is drawn from an ordinary deck of 52 cards. The result is recorded, the card is replaced in the deck, and the deck is shuffled. This is repeated 10 times. Let X_1 be the number of spades drawn and X_2 be the number of hearts drawn. Find $\Pr(X_1 = 2, X_2 = 3)$, the expected number of spades drawn, and $\Pr(X_1 + X_2 \leq 7)$. Suppose we are given the additional information that the 10 draws have produced 2 hearts. Now what is the distribution of the number of spades drawn?

5-51　In a large university it has been determined that 20 percent of the students live in fraternities and sororities, 30 percent live in dormitories, and 50 percent live in private homes. If a committee of five is selected, each person being chosen independently of the others, what is the probability that the committee has one person from a dormitory, one from a private home, and three persons from fraternities and sororities? What is the probability that it contains three or more from fraternities and sororities?

5-52　Suppose that national records reveal that twice as many accidents occur on Saturday and Sunday as on other days of the week. That is, the probability that an accident occurs on Saturday is $\frac{2}{9}$, the probability that an accident occurs on Sunday is $\frac{2}{9}$, and the probability that an accident occurs on each of the other days of the week is $\frac{1}{9}$. From the record file, 50 accident reports are selected independently of one another. The distribution of accidents according to the days of the week is

Sun.	Mon.	Tues.	Wed.	Thurs.	Fri.	Sat.
10	8	2	7	6	3	14

Write down, but do not attempt to evaluate, an expression for computing the probability that the random variables assume this particular set of values.

5-53 In Exercise 5-50 we knew that $p_1 = \frac{1}{4}$, $p_2 = \frac{1}{4}$, $p_3 = \frac{2}{4}$. Suppose we define a new random variable Z that takes on the value 6 if a spade is drawn, 4 if a heart is drawn, and 1 if a diamond or club is drawn. Then

$$E(Z) = 6(\tfrac{1}{4}) + 4(\tfrac{1}{4}) + 1(\tfrac{2}{4}) = 3$$

Construct another random variable

$$Y_1 = 6\left(\frac{X_1}{10}\right) + 4\left(\frac{X_2}{10}\right) + 1\left(\frac{X_3}{10}\right)$$

where X_1, X_2 are defined in Exercise 5-50 and X_3 is the number of diamonds or clubs drawn in ten tries. Show that $E(Y_1) = E(Z)$. Then find $\mathrm{Var}(Y_1)$ by using (5-47).

If p_1, p_2, p_3 were unknown in the card problem, then we could not get a specific number for $E(Z)$ but merely express it in terms of the p's. Since $E(Y_1) = E(Z)$ we could estimate $E(Z)$ by observing $X_1 = x_1$, $X_2 = x_2$, $X_3 = x_3$. Thus if $X_1 = 2$, $X_2 = 3$, $X_3 = 5$,

$$y_1 = 6(\tfrac{2}{10}) + 4(\tfrac{3}{10}) + 1(\tfrac{5}{10}) = 2.9$$

The estimate y_1 has the property that the average value of the random variable Y_1 is $E(Z)$, the quantity which is being estimated. The problem of finding estimates with this property is a typical one in sampling, a topic which will be pursued in Chap. 8.

5-54 Derive the result $\mathrm{Cov}(X_1, X_2) = -np_1p_2$ by using the theorem of Exercise 5-42. *Hint:* $w(X_1, X_2) = (X_1 - np_1)(X_2 - np_2)$ and $\phi(x_2) = E(w|x_2) = (x_2 - np_2)E[(X_1 - np_1)|x_2]$ since by (3-12) the expected value of a constant times a random variable is equal to the constant times the expected value of the random variable. Next use (3-13) to split $E[(X_1 - np_1)|x_2]$ into two expected values, one of which was evaluated in Example 5-18 and the other given by (3-11). Thus, deduce that

$$\phi(X_2) = \frac{-p_1}{1 - p_2}(X_2 - np_2)^2$$

so that $E[\phi(X_2)] = -np_1p_2$.

5-55   The highway patrol stops 100 cars a day selected at random at a roadblock to check the vehicles for unsafe operating conditions. Cars are classified into categories C_1, C_2, C_3, C_4 depending upon the number of violations the officers find. The owners whose cars are classified C_2, C_3, C_4 are fined \$5, \$8, and \$10 respectively while those in C_1 are not fined. Suppose it is known that the percent of automobiles falling into the four categories are respectively .3, .4, .2, .1. If I is the income produced in a day by fines imposed as a result of the roadblock, find $E(I)$ and the standard deviation of I. Use (5-47) to find Var(I).

5-6   THE GENERALIZED HYPERGEOMETRIC DISTRIBUTION

The conditions (4-23) which apply to the hypergeometric probability distribution can be generalized as follows:

(a)   *A drawing is made from N items.*

(b)   *A random sample of size n is selected.*

(c)   *Of the N items k_1 are classified as being in category C_1, k_2*   (5-68)
*are classified as being in category C_2, . . . , k_r are classified
as being in category C_r, where $k_1 + k_2 + \cdots + k_r = N$.*

Thus if $r = 2$, conditions (5-68) are the same as (4-23), those of the hypergeometric probability model. The drawing is done without replacement so that we select a random sample from a finite population (a concept already discussed in Sec. 4-4). Under conditions (5-68) we seek the probability that the sample contains x_1 items in category C_1, x_2 items in category C_2. . . , x_r items in category C_r where

$$x_1 + x_2 + \cdots + x_r = n$$

To illustrate let us again consider the fruit example of Sec. 4-5. Suppose that now the sack contains 6 apples, 4 oranges, and 10 bananas. At random we are going to draw out 6 pieces of fruit and would like to know the probability that we get 1 apple, 2 oranges, and 3 bananas. In terms of the above notation we could let C_1 be apples, C_2 be oranges, and C_3 be bananas. Then we identify $k_1 = 6$, $k_2 = 4$, $k_3 = 10$, $x_1 = 1$, $x_2 = 2$, $x_3 = 3$, $n = 6$, $N = 20$. The desired probability is again computed from our knowledge of combinations. First, the sample can be drawn in $\binom{20}{6}$ ways so that for a random sample we can visualize a sample space with

180

$\binom{20}{6}$ points, each corresponding to one sample, and each being assigned probability $1 \Big/ \binom{20}{6}$. A number of these sample points correspond to the event of interest and our next problem is to determine that number. Now, 1 apple can be selected from 6 in $\binom{6}{1}$ ways, after which 2 oranges can be selected from 4 in $\binom{4}{2}$ ways, after which 3 bananas can be selected from 10 in $\binom{10}{3}$ ways. Hence, by the fundamental counting principle, the number of sample points corresponding to the event of interest is $\binom{6}{1}\binom{4}{2}\binom{10}{3}$. Since all points in the sample space have the same probability $1 \Big/ \binom{20}{6}$, the probability of drawing 1 apple, 2 oranges, and 3 bananas is

$$\frac{\binom{6}{1}\binom{4}{2}\binom{10}{3}}{\binom{20}{6}}$$

The above argument is readily generalized. The total number of ways to draw the random sample is $\binom{N}{n}$ so that the sample space of interest contains $\binom{N}{n}$ points, each assigned probability $1 \Big/ \binom{N}{n}$. Since x_1 items can be selected from k_1 in $\binom{k_1}{x_1}$ ways, after which x_2 items can be selected from k_2 in $\binom{k_2}{x_2}$ ways, . . . , after which x_r items can be selected from k_r in $\binom{k_r}{x_r}$ ways, the probability of drawing x_1 items from category C_1, x_2 items from category C_2, . . . , x_r items from category C_r is

$$f(x_1, x_2, \ldots, x_r) = \frac{\binom{k_1}{x_1}\binom{k_2}{x_2} \cdots \binom{k_r}{x_r}}{\binom{N}{n}} \tag{5-69}$$

The numbers N, n, k_1, k_2, . . . , k_r are constants and x_1, x_2, . . . , x_r are assumed values of random variables, which we shall denote by X_1, X_2, . . . , X_r. We note that $x_i = 0, 1, 2, \ldots, k_i$ for $i = 1, 2, \ldots, r$

subject to the condition $x_1 + x_2 + \cdots + x_r = n$. As in the case of the multinomial, if all but one of the random variables is specified, the last is determined. Consequently we could write the joint probability function as

$$f(x_1, x_2, \ldots, x_{r-1}) = \frac{\binom{k_1}{x_1}\binom{k_2}{x_2} \cdots \binom{k_r}{n - x_1 - x_2 - \cdots - x_{r-1}}}{\binom{N}{n}}$$

(5-70)

but again $X_1, X_2, \ldots, X_r$ are all perfectly respectable random variables.

As in the case of the multinomial, generalized hypergeometric probabilities are tedious to compute but such calculation is easily handled by high-speed computing machines.

EXAMPLE 5-19

From a well-shuffled deck of 52 cards a random sample of 13 cards is drawn. Let X_1 be the number of aces and X_2 the number of kings. Find the joint probability function of X_1 and X_2. Use the result to find the probability that a 13-card hand contains 1 ace and 1 king.

Solution

First we check conditions (5-68).

(a) A drawing is made from $N = 52$ items.
(b) A random sample of size $n = 13$ is selected.
(c) Of the 52 items, $k_1 = 4$ are classified as aces, $k_2 = 4$ are classified as kings, and $k_3 = 44$ are classified otherwise.

Hence the generalized hypergeometric model is appropriate. We get

$$f(x_1, x_2) = \frac{\binom{4}{x_1}\binom{4}{x_2}\binom{44}{13 - x_1 - x_2}}{\binom{52}{13}} \quad \begin{array}{l} x_1 = 0, 1, 2, 3, 4 \\ x_2 = 0, 1, 2, 3, 4 \end{array}$$

The probability that a hand contains 1 ace and 1 king is

$$f(1,1) = \frac{\binom{4}{1}\binom{4}{1}\binom{44}{11}}{\binom{52}{13}} = .195$$

We note that if the drawing is done with replacement, conditions (5-55) and not (5-68) are appropriate. In this situation the probabilities of falling into the various categories remain constant and each experiment (or each draw) is independent of all the others. Hence the multinomial is the correct model to use.

Marginal probability functions are obtained by making use of previous arguments. Suppose we want $f_1(x_1)$, the marginal probability function of X_1. We note that X_1 is the number of items in category C_1 where

(a) The drawing is made of N items.
(b) A random sample of size n is selected.
(c) Of the N items k_1 fall into category C_1, $N - k_1$ fall into a second category C_2.

Hence X_1 must have a hypergeometric distribution with constants N, n, k_1 so that

$$f_1(x_1) = \frac{\binom{k_1}{x_1}\binom{N - k_1}{n - x_1}}{\binom{N}{n}} \tag{5-71}$$

where x_1 can be any nonnegative integer such that the two combination symbols in the numerator make sense (that is, $0 \leq x_1 \leq k_1$, $0 \leq n - x_1 \leq N - k_1$). Similarly, the marginal probability function of X_i, $i = 1, 2, \ldots, k$, is

$$f_i(x_i) = \frac{\binom{k_i}{x_i}\binom{N - k_i}{n - x_i}}{\binom{N}{n}} \tag{5-72}$$

with the corresponding appropriate restriction on x_i.

The marginal probability function of two random variables can be found by a similar argument. Suppose we want $f_{12}(x_1,x_2)$, the marginal probability function of X_1 and X_2. We now observe that X_1 is the number of items in category C_1 and X_2 is the number of items in category C_2 where

(a) The drawing is made of N items.
(b) A random sample of size n is selected.
(c) Of the N items k_1 fall into category C_1, k_2 fall into category C_2, and $N - k_1 - k_2$ fall into a third category C_3.

Hence X_1 and X_2 must have a generalized hypergeometric distribution with constants N, n, k_1, k_2 so that

$$f_{12}(x_1, x_2) = \frac{\binom{k_1}{x_1} \binom{k_2}{x_2} \binom{N - k_1 - k_2}{n - x_1 - x_2}}{\binom{N}{n}} \qquad (5\text{-}73)$$

where x_1, x_2 are any nonnegative integers such that the three combination symbols in the numerator make sense (that is, $0 \leqq x_1 \leqq k_1$, $0 \leqq x_2 \leqq k_2$, $0 \leqq n - x_1 - x_2 \leqq N - k_1 - k_2$). Similarly, the marginal probability function of X_i, X_j is

$$f_{ij}(x_i, x_j) = \frac{\binom{k_i}{x_i} \binom{k_j}{x_j} \binom{N - k_i - k_j}{n - x_i - x_j}}{\binom{N}{n}} \qquad (5\text{-}74)$$

with the corresponding appropriate restrictions on x_i and x_j.

Repeating the same type of argument we can easily get the probability function of $Y = X_1 + X_2$. We can regard old categories C_1 and C_2 as being C_1' and old categories $C_3, \ldots, C_r$ as being category C_2'. The number of items in the two new categories are, respectively, $k_1' = k_1 + k_2$, $k_2' = N - k_1 - k_2$. Since the conditions (4-23) are satisfied, we conclude that Y has a hypergeometric distribution with constants N, n, $k_1 + k_2$. That is,

$$g(y) = \frac{\binom{k_1 + k_2}{y} \binom{N - k_1 - k_2}{n - y}}{\binom{N}{n}} \qquad (5\text{-}75)$$

where y can be any nonnegative integer such that the two combination symbols in the numerator make sense (that is, $0 \leqq y \leqq k_1 + k_2$, $0 \leqq n - y \leqq N - k_1 - k_2$). If $Y = X_1 + X_2 + X_3$, then Y has a hypergeometric distribution with constants N, n, $k_1 + k_2 + k_3$, etc.

As in the case of marginal distributions, the conditional distributions are generalized hypergeometric. For example, if

$$f(x_1, x_2, x_3) = \frac{\binom{k_1}{x_1} \binom{k_2}{x_2} \binom{k_3}{x_3} \binom{N - k_1 - k_2 - k_3}{n - x_1 - x_2 - x_3}}{\binom{N}{n}}$$

the marginal probability function of X_1 is given by (5-71). Hence, the conditional probability function of X_2 and X_3 given X_1 is

$$h(x_2,x_3|x_1) = \frac{f(x_1,x_2,x_3)}{f_1(x_1)} = \frac{\binom{k_2}{x_2}\binom{k_3}{x_3}\binom{N - k_1 - k_2 - k_3}{n - x_1 - x_2 - x_3}}{\binom{N - k_1}{n - x_1}} \qquad (5\text{-}76)$$

where x_1, x_2, x_3 can be any nonnegative integer such that the combination symbols make sense.

Expected values are obtained from the results already available for the hypergeometric distribution. Because of (5-72) each X_i has a hypergeometric distribution and we have immediately from (4-30) and (4-31)

$$E(X_i) = \frac{nk_i}{N} = np_i \qquad (5\text{-}77)$$

$$\text{Var}(X_i) = \frac{nk_i(N - k_i)(N - n)}{N^2(N - 1)} = np_iq_i \frac{N - n}{N - 1} \qquad (5\text{-}78)$$

where $p_i = k_i/N$, $q_i = 1 - p_i$. To find $\text{Cov}(X_i,X_j)$ we could use the theorem of Exercise 5-42; we shall, however, use an alternative method. We know that $Y = X_i + X_j$ has the probability function (5-75) with k_1 and k_2 replaced by k_i and k_j. Hence

$$\text{Var}(Y) = \text{Var}(X_i + X_j) = \frac{n(k_i + k_j)(N - k_i - k_j)(N - n)}{N^2(N - 1)}$$

$$= n(p_i + p_j)(1 - p_i - p_j) \frac{N - n}{N - 1}$$

by (4-31). Also, from the derivation of (5-67) we had

$$\text{Cov}(X_i,X_j) = \frac{\text{Var}(X_i + X_j) - \text{Var}(X_i) - \text{Var}(X_j)}{2}$$

which now becomes

$$\text{Cov}(X_i,X_j) = \frac{n}{2}\left(\frac{N - n}{N - 1}\right)[(p_i + p_j)(1 - p_i - p_j) - p_i(1 - p_i)$$
$$- p_j(1 - p_j)]$$
$$= -n\frac{N - n}{N - 1}p_ip_j \qquad (5\text{-}79)$$

EXAMPLE 5-20

Let X_1 and X_2 have the joint probability function given by (5-73). Find the conditional probability function of X_1 given X_2. Then find the mean and variance of X_1 given X_2.

Solution

The marginal probability function of X_2 is given by (5-72) with $i = 2$. Thus we get

$$h(x_1|x_2) = \frac{f(x_1,x_2)}{f_2(x_2)} = \frac{\binom{k_1}{x_1}\binom{N - k_1 - k_2}{n - x_1 - x_2}}{\binom{N - k_2}{n - x_2}}$$

so that X_1 given X_2 has a hypergeometric distribution with constants $N - k_2$, $n - x_2$, k_1. The mean and variance are obtained from (4-30) and (4-31). We get

$$E(X_1|x_2) = \frac{(n - x_2)k_1}{N - k_2} = (n - x_2)\frac{p_1}{1 - p_2}$$

$$\begin{aligned}
\mathrm{Var}(X_1|x_2) &= \frac{(n - x_2)k_1(N - k_1 - k_2)(N - k_2 - n + x_2)}{(N - k_2)^2(N - k_2 - 1)} \\
&= \frac{(n - x_2)p_1(1 - p_1 - p_2)}{(1 - p_2)^2}\left(\frac{N - k_2 - n + x_2}{N - k_2 - 1}\right) \\
&= (n - x_2)\left(\frac{p_1}{1 - p_2}\right)\left(1 - \frac{p_1}{1 - p_2}\right)\left(\frac{N - k_2 - n + x_2}{N - k_2 - 1}\right)
\end{aligned}$$

EXERCISES

5-56   From a well-shuffled deck of 52 cards a random sample of 10 cards is drawn. Let X_1 be the number of spades drawn and X_2 be the number of hearts drawn. Use the joint probability function of X_1 and X_2 to evaluate $\Pr(X_1 = 2, X_2 = 3)$. Find the expected number of spades drawn. Write down a single sum that, if evaluated, would give $\Pr(X_1 + X_2 \leq 7)$. Suppose we are given the additional information that the 10 draws have produced 2 hearts. Now what is the distribution of the number of spades drawn?

5-57   Ten cans of fruit and vegetables, all the same size, have lost their labels. It is known, however, that three contain tomatoes, two contain corn, and five contain pears. Six cans are selected at random. Let X_1 be the number of cans of tomatoes and X_2 be the number of cans of corn in the sample. Write down the joint probability function of X_1 and X_2 and use it to evaluate $\Pr(X_1 = 2, X_2 = 1)$. Let $Y = $ the number of cans of vegetables in the sample. Find the probability function of Y and $\Pr(Y \leq 3)$.

5-58   A university committee is to be selected from 15 faculty members, 6 from arts and sciences, 4 from education, 2 from engineering, and 3 from the business school. The committee is to contain 7 members, selected at random from the 15. Let X_1, X_2, X_3 represent the number in the sample from arts and sciences, education, and engineering, respectively. Find the joint probability function of X_1, X_2, X_3. Find the mean and variance of X_1.

5-59 If each $k_i = 1$, that is, every category contains only one item, show that every X_i has the same marginal distribution

$$f_i(x_i) = \frac{N-n}{N} \qquad x_i = 0$$

$$= \frac{n}{N} \qquad x_i = 1$$

Find the mean and variance of X_i. If each $k_i = 1$ show that the joint probability distribution of X_i and X_j is given by

x_j \\ x_i	0	1	Row totals
0	$\dfrac{(N-n)(N-n-1)}{N(N-1)}$	$\dfrac{n(N-n)}{N(N-1)}$	$\dfrac{N-n}{N}$
1	$\dfrac{n(N-n)}{N(N-1)}$	$\dfrac{n(n-1)}{N(N-1)}$	$\dfrac{n}{N}$
Column totals	$\dfrac{N-n}{N}$	$\dfrac{n}{N}$	1

Hence every pair of random variables has this same probability distribution. Find $\text{Cov}(X_i, X_j)$.

5-60 In Exercise 5-56 we knew that $k_1 = 13$, $k_2 = 13$, $k_3 = 26$. Suppose we define a new random variable Z that takes on the value 6 if a spade is drawn, 4 if a heart is drawn, and 1 if a diamond or club is drawn. Then for a single draw

$$E(Z) = 6(^{13}\!/_{52}) + 4(^{13}\!/_{52}) + 1(^{26}\!/_{52}) = 3$$

Construct another random variable

$$Y_2 = 6\left(\frac{X_1}{10}\right) + 4\left(\frac{X_2}{10}\right) + 1\left(\frac{X_3}{10}\right)$$

where X_1 and X_2 are defined in Exercise 5-56 and X_3 is the number of diamonds or clubs drawn in ten tries. Show $E(Y_2) = E(Z)$. Then find $\text{Var}(Y_2)$ by using (5-47). Note that $\text{Var}(Y_2) < \text{Var}(Y_1)$ where Y_1 is defined in Exercise 5-53.

If k_1, k_2, k_3 were unknown, then we could not get a specific number for $E(Z)$ but merely express it in terms of the k's. Since $E(Y_2) = E(Z)$ we could estimate $E(Z)$ by observing $X_1 = x_1$, $X_2 = x_2$, $X_3 = x_3$. Thus if $X_1 = 2$, $X_2 = 3$, $X_3 = 5$

$$y_2 = 6(\tfrac{2}{10}) + 4(\tfrac{3}{10}) + 1(\tfrac{5}{10}) = 2.9$$

The estimate y_2 has the property that the average value of the random variable Y_2 is $E(Z)$; also, Y_2 has a smaller variance than the random variable Y_1. If several random variables can be used to estimate the same quantity, intuitively it seems desirable to use the one with the smaller variance.

5-61 Derive the result $\mathrm{Cov}(X_1, X_2) = -np_1p_2(N - n)/(N - 1)$ by using the theorm of Exercise 5-42. Observe the hint of Exercise 5-54.

5-62 A stock of canned goods contains 100 cans, all the same size, which have lost their labels. It is known that 30 of the cans contain tomatoes, 20 contain corn, and 50 contain pears. A buyer wants to purchase 10 cans selected at random from the stock. The price is 30 cents for tomatoes, 20 cents for corn, and 40 cents for pears. The buyer could open all 10 cans to determine the contents and pay the exact value. This seems undesirable since he is not ready to use the contents of all 10 cans. Consequently, he agrees to pay the expected cost of 10 cans. If C represents that cost, find $E(C)$ and the standard deviation of C.

5-7 THE DISTRIBUTION OF THE MINIMUM AND MAXIMUM OF A RANDOM SAMPLE

Let $X_1, X_2, \ldots, X_n$ be a random sample from a distribution that has probability function $f(x)$. That is, $X_1, X_2, \ldots, X_n$ are independent random variables all having the same marginal probability function $f(x)$. We seek the probability functions of $Y_1 = \min X_i$ and $Y_n = \max X_i$ where $\min X_i$ and $\max X_i$ are respectively the smallest and largest X in the sample of size n. We have already considered problems of this type for the case $n = 2$ in Examples 5-4 and 5-5 (in the former X_1, X_2 did not constitute a random sample).

First we shall consider a specific example. Let X have the uniform distribution

$$f(x) = \frac{1}{6} \qquad x = 1, 2, 3, 4, 5, 6$$

a random variable that characterizes the behavior of a symmetric six-sided die. Suppose that the die is rolled $n = 3$ times and we want

$\Pr(Y_1 = 1)$. Obviously $\Pr(Y_1 = 1) + \Pr(Y_1 > 1) = 1$ since Y_1 can take on the values 1, 2, 3, 4, 5 6. Thus, if it is profitable to do so, we could calculate $\Pr(Y_1 > 1)$ and obtain $\Pr(Y_1 = 1)$ by subtraction. It is easy to see that $Y_1 > 1$ requires that X_1, X_2, X_3 all be greater than 1. Hence

$$\Pr(Y_1 > 1) = \Pr(X_1 > 1)\Pr(X_2 > 1)\Pr(X_3 > 1)$$

since the three events $X_1 > 1$, $X_2 > 1$, $X_3 > 1$ are independent. But $\Pr(X > 1) = \frac{5}{6}$ so that $\Pr(Y_1 > 1) = (\frac{5}{6})^3$. We get

$$\Pr(Y_1 \leq 1) = \Pr(Y_1 = 1) = 1 - \Pr(Y_1 > 1) = 1 - \left(\frac{5}{6}\right)^3$$

Next let us find $\Pr(Y_1 = 2) = \Pr(Y_1 \leq 2) - \Pr(Y_1 \leq 1)$. The same argument yields

$$\Pr(Y_1 \leq 2) = 1 - \Pr(Y_1 > 2)$$

$$= 1 - \Pr(X_1 > 2)\Pr(X_2 > 2)\Pr(X_3 > 2) = 1 - \left(\frac{4}{6}\right)^3$$

so that

$$\Pr(Y_1 = 2) = \left[1 - \left(\frac{4}{6}\right)^3\right] - \left[1 - \left(\frac{5}{6}\right)^3\right] = \left(\frac{5}{6}\right)^3 - \left(\frac{4}{6}\right)^3$$

Continuing in a similar manner we find that the probability distribution of Y_1 is

$$
\begin{aligned}
g_1(y_1) &= 1 - \left(\frac{5}{6}\right)^3 & y_1 &= 1 \\
&= \left(\frac{5}{6}\right)^3 - \left(\frac{4}{6}\right)^3 & y_1 &= 2 \\
&= \left(\frac{4}{6}\right)^3 - \left(\frac{3}{6}\right)^3 & y_1 &= 3 \\
&= \left(\frac{3}{6}\right)^3 - \left(\frac{2}{6}\right)^3 & y_1 &= 4 \\
&= \left(\frac{2}{6}\right)^3 - \left(\frac{1}{6}\right)^3 & y_1 &= 5 \\
&= \left(\frac{1}{6}\right)^3 & y_1 &= 6
\end{aligned}
$$

which can be written as

$$g_1(y_1) = \left(\frac{7 - y_1}{6}\right)^3 - \left(\frac{6 - y_1}{6}\right)^3 \qquad y_1 = 1, 2, 3, 4, 5, 6$$

If the die is rolled n times, replace the exponent 3 by n.

In generalizing the above discussion we shall use the notation

$$G_1(y_1) = \Pr(Y_1 \leq y_1) \qquad F(x) = \Pr(X \leq x)$$

Then

$$G_1(y_1) = \Pr(Y_1 \leq y_1) = 1 - \Pr(Y_1 > y_1)$$
$$= 1 - \Pr(X_1 > y_1)\Pr(X_2 > y_1) \cdots \Pr(X_n > y_1)$$
$$= 1 - [\Pr(X > y_1)]^n$$
$$= 1 - [1 - \Pr(X \leq y_1)]^n$$

Thus we have

$$G_1(y_1) = 1 - [1 - F(y_1)]^n \tag{5-80}$$

Next, if Y_1 assumes consecutive integer values, then the probability function of Y_1 is

$$g_1(y_1) = \Pr(Y_1 = y_1) = \Pr(Y_1 \leq y_1) - \Pr(Y_1 \leq y_1 - 1)$$
$$= G_1(y_1) - G_1(y_1 - 1)$$
$$= \{1 - [1 - F(y_1)]^n\} - \{1 - [1 - F(y_1 - 1)]^n\}$$
$$= [1 - F(y_1 - 1)]^n - [1 - F(y_1)]^n \tag{5-81}$$

Similarly, we can get corresponding results for Y_n. We have

$$G_n(y_n) = \Pr(Y_n \leq y_n) = \Pr(X_1 \leq y_n)\Pr(X_2 \leq y_n) \cdots \Pr(X_n \leq y_n)$$
$$= [\Pr(X \leq y_n)]^n$$

since Y_n will be less than or equal to y_n if and only if each X_i is less than or equal to y_n. Consequently,

$$G_n(y_n) = [F(y_n)]^n \tag{5-82}$$

and if Y_n assumes consecutive integer values the probability function of Y_n is

$$g_n(y_n) = [F(y_n)]^n - [F(y_n - 1)]^n \tag{5-83}$$

EXAMPLE 5-21

Suppose that five days a week two people toss a fair coin to determine who will buy coffee. Find the probability that for a given individual the maximum number of losses during a week will be four or more if a three-week period is considered.

Solution

Let X be the number of times the given individual loses in a week. Then X has a binomial distribution with $n = 5$, $p = \frac{1}{2}$, and

$$f(x) = \binom{5}{x}(.5)^x(.5)^{5-x} \qquad x = 0, 1, 2, 3, 4, 5$$

We have a random sample X_1, X_2, X_3. Let $Y_3 = \max X_i$. We seek $\Pr(Y_3 \geq 4)$. Now

$$\Pr(Y_3 \geq 4) = 1 - \Pr(Y_3 \leq 3) = 1 - G_3(3) = 1 - [F(3)]^3$$

190

by using (5-82).   But

$$F(3) = \sum_{x=0}^{3} b(x;5,.5) = .81250 \qquad \text{by Appendix B1}$$

Thus

$$\Pr(Y_3 \geq 4) = 1 - (.81250)^3 = 1 - .536 = .464$$

EXAMPLE 5-22

A secretary averages one error per typed page.   If five pages are selected at random from her work, find the probability that the maximum number of errors found on any one page exceeds three.   Assume that the errors per page follow a Poisson distribution.

Solution

If X is the number of errors on a page, then

$$f(x) = \frac{e^{-1}(1)^x}{x!} \qquad x = 0, 1, 2, \ldots$$

We have a random sample X_1, X_2, X_3, X_4, X_5.   If we let $Y_5 = \max X_i$, we want $\Pr(Y_5 \geq 4)$.   But

$$\Pr(Y_5 \geq 4) = 1 - \Pr(Y_5 \leq 3) = 1 - G_5(3) = 1 - [F(3)]^5$$

by using (5-82).   Also

$$F(3) = \sum_{x=0}^{3} \frac{e^{-1}(1)^x}{x!} = .98101 \qquad \text{by Appendix B4}$$

Thus

$$\Pr(Y_5 \geq 4) = 1 - (.98101)^5 = 1 - .909 = .091$$

EXERCISES

5-63   A symmetric six-sided die is rolled three times.   Find the distribution function and the probability function of Y_3, the maximum number showing in three throws.   Then find $E(Y_3)$.

5-64   Five numbers are selected at random (by using a table of random numbers) from the distribution whose probability function is $f(x) = \frac{1}{10}$, $x = 1, 2, 3, 4, 5, 6, 7, 8, 9, 10$.   Find the probability distribution of the minimum of the five numbers. Use the result to find $\Pr(Y_1 \leqq 3)$.

5-65 A student takes four multiple-choice examinations during the semester. Each examination has 25 questions and each question has 5 possible answers. If the student only guesses on each question, what is the probability that his minimum score is 5 or less correct out of 25?

5-66 To determine who pays for coffee, three people each toss a coin and the odd man pays. If all coins show heads or all show tails, they are tossed again. Let N be the number of tosses by each individual required to reach a decision. If the three people repeat this procedure five times during the week, what is the probability that the maximum number of tosses required on any one day during the week exceeds three? (See Example 4-8 for the distribution of N.)

5-67 In Exercise 5-66 find the probability that the minimum number of tosses occurring during the five-day period is one.

5-68 In Example 5-22 find the probability that the minimum number of errors found on one of the five pages exceeds one.

5-69 Let Y_r be the rth smallest of the random variables $X_1, X_2, \ldots, X_n$, which constitute a random sample from the distribution whose probability function is $f(x)$. Hence, Y_1 is the smallest and Y_n is the largest of the n random variables. Show that

$$G(y_r) = \Pr(Y_r \leqq y_r) = \sum_{k=r}^{n} \binom{n}{k} [F(y_r)]^k [1 - F(y_r)]^{n-k}$$

$$= \sum_{k=r}^{n} b(k;n,p)$$

where $b(k;n,p)$ is defined by (4-2) and $p = F(y_r)$. Verify that $r = 1$ and $r = n$ give (5-80) and (5-82), respectively. Suppose

$$f(x) = \frac{1}{10} \qquad x = 1, 2, \ldots, 10$$

and $n = 20$. Verify that $\Pr(Y_{10} \leqq 5) = .58810$. Make a table giving the probability distributions of Y_{10}.

REFERENCES

1 GUENTHER, WILLIAM C.: "Concepts of Statistical Inference," McGraw-Hill Book Company, New York, 1965.

2 MOOD, ALEXANDER M., and FRANKLIN A. GRAYBILL: "Introduction to the Theory of Statistics," 2d-ed., McGraw-Hill Book Company, New York, 1963.

5-1 If X_1 and X_2 are random variables with joint probability function $f(x_1,x_2)$, then the marginal probability functions of X_1 and X_2 are

$$f_1(x_1) = \sum_{x_2} f(x_1,x_2) \tag{5-1}$$

$$f_2(x_2) = \sum_{x_1} f(x_1,x_2) \tag{5-2}$$

and the conditional probability functions of X_1 given X_2 and of X_2 given X_1 are

$$g(x_1|x_2) = \frac{f(x_1,x_2)}{f_2(x_2)} \qquad [f_2(x_2) > 0] \tag{5-3}$$

$$h(x_2|x_1) = \frac{f(x_1,x_2)}{f_1(x_1)} \qquad [f_1(x_1) > 0] \tag{5-4}$$

If

$$f(x_1,x_2) = f_1(x_1)f_2(x_2) \tag{5-5}$$

then X_1 and X_2 are said to be independent random variables.

5-2 Let X_1 and X_2 be random variables with joint probability function $f(x_1,x_2)$, and let $Y_1 = u_1(X_1,X_2)$ and $Y_2 = u_2(X_1,X_2)$ be solvable for X_1 and X_2, giving one set of values $X_1 = w_1(Y_1,Y_2)$, $X_2 = w_2(Y_1,Y_2)$ for each Y_1, Y_2. Then

$$g(y_1,y_2) = \Pr(Y_1 = y_1,\ Y_2 = y_2) = f[w_1(y_1,y_2),\ w_2(y_1,y_2)] \tag{5-7}$$

If X_1 and X_2 are independent random variables both having a Poisson distribution with means μ_1 and μ_2, then $Y_1 = X_1 + X_2$ has Poisson distribution with mean $\mu_1 + \mu_2$ [see (5-9)]. If X_1 and X_2 are independent random variables both having binomial distributions with probability of success on a single trial p, and $n = n_1$, $n = n_2$, respectively, then $Y = X_1 + X_2$ has a binomial distribution with probability of success p, and $n = n_1 + n_2$ [see (5-10)].

5-3 Let $Z = w(X_1,X_2)$. Then

$$E(Z) = \sum_{x_2} \sum_{x_1} w(x_1,x_2)f(x_1,x_2) \tag{5-11}$$

where $f(x_1,x_2)$ is the joint probability function of X_1 and X_2. For any two random variables X_1 and X_2,

$$E(X_1 + X_2) = E(X_1) + E(X_2) \tag{5-12}$$

If X_1 and X_2 are independent random variables, then

$$E[u_1(X_1)u_2(X_2)] = E[u_1(X_1)]E[u_2(X_2)] \tag{5-15}$$

193

If X_1 and X_2 are random variables with means μ_1 and μ_2, then

$$\text{Cov}(X_1, X_2) = E[(X_1 - \mu_1)(X_2 - \mu_2)] \tag{5-19}$$
$$= E(X_1, X_2) - \mu_1\mu_2 \tag{5-22}$$

is the covariance of X_1 and X_2. If X_1 and X_2 are independent, then

$$\text{Cov}(X_1, X_2) = 0$$

For any two random variables X_1 and X_2

$$\text{Var}(X_1 + X_2) = \text{Var}(X_1) + \text{Var}(X_2) + 2\text{Cov}(X_1, X_2) \tag{5-20}$$

5-4 If $X_1, X_2, \ldots, X_n$ are random variables with joint probability function $f(x_1, x_2, \ldots, x_n)$, then the one-dimensional marginal probability function of X_1 is

$$f_1(x_1) = \sum_{x_2} \sum_{x_3} \cdots \sum_{x_n} f(x_1, x_2, \ldots, x_n) \tag{5-32}$$

with a similar expression holding for other marginals. If
$X_1, X_2, \ldots, X_n$
are independent, then

$$f(x_1, x_2, \ldots, x_n) = f_1(x_1)f_2(x_2) \cdots f_n(x_n) \tag{5-33}$$

If in addition each marginal probability function is the same, then $X_1, X_2, \ldots, X_n$ is called a random sample, and

$$f(x_1, x_2, \ldots, x_n) = g(x_1)g(x_2) \cdots g(x_n) \tag{5-34}$$

For any random variables $X_1, X_2, \ldots, X_n$

$$E(X_1 + X_2 + \cdots + X_n) = E(X_1) + E(X_2) + \cdots + E(X_n) \tag{5-44}$$

$$\text{Var}(X_1 + X_2 + \cdots + X_n) = \sum_{i=1}^{n} \text{Var}(X_i) + 2 \sum_{\substack{i,j=1 \\ i<j}}^{n} \text{Cov}(X_i, X_j) \tag{5-46}$$

If $X_1, X_2, \ldots, X_n$ is a random sample, the sample mean is

$$\bar{X} = \frac{X_1 + X_2 + \cdots + X_n}{n}$$

and

$$E(\bar{X}) = E(X) \tag{5-53}$$

$$\text{Var}(\bar{X}) = \frac{\text{Var}(X)}{n} \tag{5-54}$$

5-5 If $X_1, X_2, \ldots, X_k$ are random variables associated with the number of outcomes in categories $C_1, C_2, \ldots, C_k$ for a series of experiments satisfying (5-55), then their joint probability function is

$$f(x_1, x_2, \ldots, x_k) = \frac{n!}{x_1! x_2! \cdots x_k!} p_1^{x_1} \cdots p_k^{x_k} \qquad (5\text{-}57)$$

and the random variables are said to have a multinomial distribution. One-dimensional marginals are binomial distributions (5-60). Two-dimensional marginals are multinomial distributions (5-62).

5-6 If $X_1, X_2, \ldots, X_r$ are random variables associated with the number of outcomes in categories $C_1, C_2, \ldots, C_r$ for a series of experiments satisfying (5-68), then their joint probability function is

$$f(x_1, x_2, \ldots, x_r) = \frac{\binom{k_1}{x_1} \binom{k_2}{x_2} \cdots \binom{k_r}{x_r}}{\binom{N}{n}} \qquad (5\text{-}69)$$

and the random variables are said to have a generalized hypergeometric distribution. One-dimensional marginals are hypergeometric distributions (5-72). Two-dimensional marginals are generalized hypergeometric distributions (5-74). Conditional distributions are generalized hypergeometric distributions (5-76). If X_i and X_j are any two of the random variables $X_1, X_2, \ldots, X_r$ that have a generalized hypergeometric distribution, then

$$\mathrm{Cov}(X_i, X_j) = -n \frac{N-n}{N-1} p_i p_j \qquad (5\text{-}79)$$

where $p_i = k_i / N$.

5-7 Let $X_1, X_2, \ldots, X_n$ be a random sample from a distribution that has probability function $f(x)$. If $Y_1 = \min X_i$, $Y_n = \max X_i$, then

$$G_1(y_1) = 1 - [1 - F(y_1)]^n \qquad (5\text{-}80)$$

$$G_n(y_n) = [F(y_n)]^n \qquad (5\text{-}82)$$

where $G_1(y_1)$, $G_n(y_n)$, and $F(x)$ are the distribution functions of Y_1, Y_n, and X, respectively. If Y_r is the rth smallest of the X_i, then

$$G_r(y_r) = \sum_{k=r}^{n} b(k; n, p)$$

where $p = F(y_r)$.

CHAPTER SIX

Continuous Random Variables and Large-sample Approximations

6-1   CONTINUOUS RANDOM VARIABLES

Our discussion in preceding chapters has been concerned entirely with discrete sample spaces.   The sample space which we have encountered for every situation has consisted of either a finite number of points or a countably infinite number of points.   A random variable associated with a discrete sample space is called a *discrete random variable*.   Thus every random variable which we have considered has been a discrete random variable.

In many interesting applications of probability the random variable under consideration can take on any value within an interval (or perhaps several intervals).   Let us turn to a simple example.   Consider the spinner pictured in Fig. 6-1 with the outside scale labeled from 0 to 1.

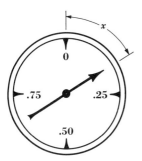

Figure 6-1 A balanced spinner.

Assume that the spinner is perfectly balanced, that is, as likely to stop at one place as another if spun. Let X denote the distance on the circular scale from 0 to the end of the spinner. When the spinner stops after being spun we observe $X = x$. Suppose that A_1 is the event that X is between .50 and .75. Our intuition tells us that .25 would be a good choice for $\Pr(A_1)$, the probability that A_1 happens. We would like to construct a method of assigning weights (or probabilities) to an event A that is harmonious with our notions of probability, particularly definition 1-5. In this example it is easy. We define

$\Pr(A) = \Pr(X$ belongs to an interval or several intervals comprising $A)$
$\quad\quad\;\; =$ length of the interval or several intervals

We note that $\Pr(A)$ satisfies the conditions

(a) $\Pr(A) \geqq 0$ for any A

(b) $\Pr(S) = 1$ $\hspace{5cm}$ (6-1)

where S is the entire sample space, that is, the interval 0 to 1 which has length 1. Conditions (6-1) roughly correspond to (1-4). If the interval corresponding to the event A has no length, then $\Pr(A) = 0$. Hence, the probability that X takes on any specific value, say, $X = .50$, is 0, that is, $\Pr(X = .50) = 0$, a fact that is true for any continuous random variable.

We can regard $\Pr(A)$ of the spinner example as the area above the horizontal axis and under the curve

$f(x) = 1$ $0 \leqq x < 1$
$\quad\;\; = 0$ elsewhere

between the end points of the interval that corresponds to event A. Thus if A corresponds to the interval $(.50 < X < .75)$, the area under the curve between .50 and .75 is .25 (see Fig. 6-2).

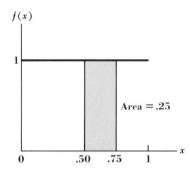

Area = .25

Figure 6-2 Probability interpreted
as area under a curve.

The function $f(x)$ is called a density function and has the properties:

(a) *It is nonnegative.*
(b) *The total area under the curve is 1.* (6-2)

Any function $f(x)$ satisfying the properties (6-2) can be used as a density function and areas under $f(x)$ above the horizontal axis and within intervals can be interpreted as probabilities. It is easy to construct density functions. Any nonnegative $f(x)$ can be used if the area under the curve is finite. If the area is C instead of 1, construct a new $f(x)$ by dividing the old one by C. Constructing density functions that associate realistic probabilities with events arising from continuous random variables is, in general, not so easy as it was for the spinner problem, where the choice was rather obvious. Even if we are satisfied that a particular $f(x)$ is appropriate, the evaluation of areas under the curve may be a difficult numerical problem. Fortunately, tables that make the evaluation easy are available for most important density functions.

The preceding discussion suggests that we make the following definition of probability in the continuous case:

Let $f(x)$ be a density function with properties (6-2), and let A be (6-3)
an event corresponding to an interval or several intervals in the
sample space S. Then $Pr(A)$ is the area under the curve $f(x)$ in
the interval (or intervals) corresponding to A.

In comparing the continuous situation with the discrete we have observed that (1) intervals are used instead of sample points, (2) weights (or probabilities) are associated with intervals instead of points, (3) probabilities are evaluated by finding areas under curves rather than by summing weights.

In this book we are not interested in random variables that have a continuous distribution. Our sole purpose in discussing the continuous case is to demonstrate an approximation that may be used to evaluate certain probabilities associated with discrete random variables.

198

The random variable Z whose density function is

$$\phi(z) = \frac{1}{\sqrt{2\pi}}\, e^{-z^2/2} \qquad -\infty < z < \infty \tag{6-4}$$

is said to have a *standard normal distribution*. Here $e = 2.71828 \cdots$ and $\pi = 3.14159 \cdots$. It is instructive to note some of the geometric properties of the density function (6-4). The graph of $\phi(z)$ appears in Fig. 6-3. The curve approaches, but never touches, the horizontal axis. The random variable Z can take on any value between minus infinity and plus infinity. However, the tables we shall discuss in the next paragraph reveal that practically all the area under the curve is between -3 and $+3$ so that the probability that Z is less than -3 or greater than $+3$ is very small. The curve is symmetrical about the vertical axis through $z = 0$. In other words, the curve to the right of the vertical axis is exactly the same as the curve to the left of it. Since the total area under the curve is 1, the area on each side of the vertical axis is $\frac{1}{2}$. Because of symmetry it is necessary to tabulate areas on only one side of the vertical axis. However, to make the use of Appendix B5 a little easier, entries are given for both positive and negative values of z_p.

Appendix B5 contains values of p such that

$$\Pr(Z < z_p) = p \tag{6-5}$$

This means that the probability that the random variable Z is smaller than some specified z_p is equal to p. The geometrical relationship between z_p and p is illustrated in Fig. 6-3. As an example, suppose we want to know the probability that Z is less than -1.12, that is, $\Pr(Z < -1.12)$. Proceed down the left-hand column of the first page of Appendix B5 to $z_p = -1.10$, then across the row to the column under .02. We find $p = .1314$, so that $\Pr(Z < -1.12) = .1314$.

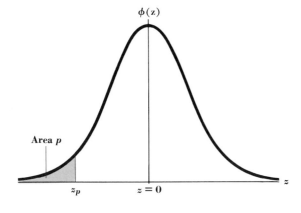

Figure 6-3   Standard normal curve showing tabled entry p. The total area under the curve is 1.

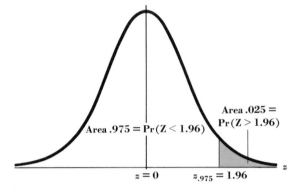

Area .025 = Pr(Z > 1.96)

Area .975 = Pr(Z < 1.96)

$z = 0$ $z_{.975} = 1.96$

Figure 6-4   Standard normal curve showing Pr(Z < 1.96) and Pr(Z > 1.96) as areas (Example 6-1).

EXAMPLE 6-1

Find the probability that Z is less than 1.96.

Solution

We look on the second page of Appendix B5 opposite the row 1.9 and under column .06. The table yields $\Pr(Z < 1.96) = .97500$. We note that $\Pr(Z < 1.96) + \Pr(Z > 1.96) = 1$, since $Z > 1.96$ and $Z < 1.96$ are mutually exclusive events, one of which must happen $[\Pr(Z = 1.96) = 0]$. Consequently, $\Pr(Z > 1.96) = .02500$ by subtraction. These area relationships are indicated in Fig. 6-4.

EXAMPLE 6-2

Find the probability that Z is between -1.12 and 1.96.

Solution

We seek the shaded area pictured in Fig. 6-5. Since the events $Z < -1.12$ and $-1.12 < Z < 1.96$ are mutually exclusive,

$$\Pr(Z < -1.12) + \Pr(-1.12 < Z < 1.96) = \Pr(Z < 1.96)$$

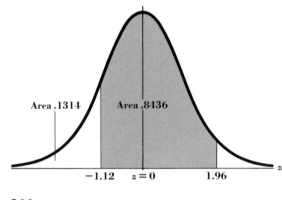

Area .1314   Area .8436

-1.12 $z = 0$ 1.96

Figure 6-5   Standard normal curve showing Pr(-1.12 < Z < 1.96) as an area (Example 6-2).

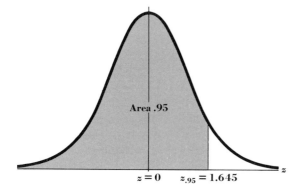

Area .95

$z = 0$ $z_{.95} = 1.645$ z

Figure 6-6   Standard
normal curve showing
$z_{.95}$ (Example 6-3).

In other words

$$Pr(-1.12 < Z < 1.96) = Pr(Z < 1.96) - Pr(Z < -1.12)$$
$$= .9750 - .1314$$
$$= .8436$$

It is geometrically obvious that the shaded area is obtained by subtraction.

EXAMPLE 6-3

Find $z_{.95}$.

Solution

We want to find the value of z below which lies .95 of the area under the curve as indicated in Fig. 6-6.   From Appendix B5 we find

$$Pr(Z < 1.64) = .9495   Pr(Z < 1.65) = .9505$$

Since .95 is halfway between .9405 and .9505, we use $z_{.95} = 1.645$.

EXAMPLE 6-4

Find a value of Z, say, z, such that $Pr(-z < Z < z) = .95$.

Solution

We seek the shaded area in Fig. 6-7.   Since the curve is symmetrical, the area is .025 under each tail.   Hence the area below z is .975 and $z = z_{.975} = 1.96$ from Example 6-1.

The density function (6-4) can be generalized to

$$f(x) = \frac{1}{\sigma \sqrt{2\pi}} e^{-(x-\mu)^2/2\sigma^2}   -\infty < x < \infty \tag{6-6}$$

201

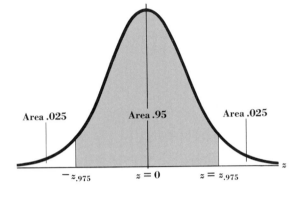

Area .025   Area .95   Area .025

$-z_{.975}$ $z = 0$ $z = z_{.975}$

Figure 6-7 Standard normal curve showing z such that $Pr(-z < Z < z) = .95$ (Example 6-4).

A random variable X having the density function (6-6) is said to have a *normal distribution* with mean μ and variance σ^2. If the definitions of $E(X) = \mu$ and $\sigma^2 = E[(X - \mu)^2]$ are extended so as to apply to continuous random variables, then it is possible to show that an X characterized by the density $f(x)$ does have mean μ and variance σ^2 so that the meaning of these symbols is consistent with previous usage. It is immediately apparent that if we let $\mu = 0$, $\sigma = 1$, and replace x by z, then the normal density function (6-6) reduces to the standard normal density (6-4).

The graph of the density function (6-6) is shown in Fig. 6-8. It is very similar to the one for the standard normal. The curve is symmetric about the vertical axis through $x = \mu$. Practically all the area is between $x = \mu - 3\sigma$ and $x = \mu + 3\sigma$; thus σ controls the shape of the curve. If σ is small, the curve has most of its area near μ. The larger σ, the more the area is spread out (see Fig. 6-9).

Fortunately no new tables are needed to evaluate areas (probabilities) under the curve. It is an elementary calculus problem to show that

$$Pr(x_1 < X < x_2) = Pr\left(\frac{x_1 - \mu}{\sigma} < Z < \frac{x_2 - \mu}{\sigma}\right) \tag{6-7}$$

$f(x)$

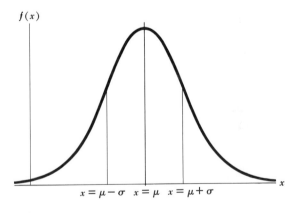

$x = \mu - \sigma$ $x = \mu$ $x = \mu + \sigma$

x

Figure 6-8 Normal distribution with mean μ, standard deviation σ.

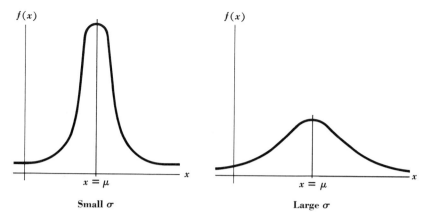

Small σ | Large σ

Figure 6-9 Effect of σ on the shape of the normal curve.

where Z has the standard normal distribution. Similarly, it can be shown

$$\Pr(X < x) = \Pr\left(Z < \frac{x - \mu}{\sigma}\right) \qquad (6\text{-}8)$$

EXAMPLE 6-5

Suppose that a random variable X has a normal distribution with $\mu = 100$, $\sigma = 10$. Find the probability that X is between 90 and 120.

Solution

According to (6-7) we have

$$\Pr(90 < X < 120) = \Pr\left(\frac{90 - 100}{10} < Z < \frac{120 - 100}{10}\right)$$
$$= \Pr(-1 < Z < 2)$$
$$= \Pr(Z < 2) - \Pr(Z < -1)$$
$$= .9773 - .1587 \qquad \textit{from Appendix B5}$$
$$= .8186$$

EXERCISES

6-1 Use Appendix B5 to find (a) $\Pr(Z < 1.37)$, (b) $\Pr(-.67 < Z < 1.37)$, (c) $\Pr(Z > 1.00)$, and (d) the probability that Z differs from 0 by at least 1.

6-2 Use Appendix B5 to find (a) $z_{.99}$, (b) $z_{.90}$, (c) $z_{.025}$, and (d) $z_{.05}$.

6-3 Use Appendix B5 to find z such that $\Pr(-z < Z < z) = $ (a) .90, (b) .98, (c) .99.

6-4 Find (a) $\Pr(.32 < Z < 1.65)$, (b) $\Pr(-1.57 < Z < .68)$, and (c) $\Pr(-2.32 < Z < -1.64)$.

6-5 With the information given in Example 6-5, find (a) $\Pr(X < 90)$, (b) $\Pr(X > 120)$.

6-6 If a random variable X has a normal distribution with mean $\mu = \frac{1}{2}$ and $\sigma = .01$, find the probability that X is between .48 and .52. Find the probability that X is either less than .47 or greater than .53.

6-7 If a random variable X has a normal distribution with mean $\mu = 35$ and standard deviation $\sigma = 8$, find the probability that X will exceed 50.

6-3 THE DISTRIBUTION OF SAMPLE MEANS AND SAMPLE SUMS

If $X_1, X_2, \ldots, X_n$ is a random sample from a distribution having probability function $g(x)$, then we call

$$Y = X_1 + X_2 + \cdots + X_n \tag{6-9}$$

the sample sum and

$$\bar{X} = \frac{Y}{n} = \frac{X_1 + X_2 + \cdots + X_n}{n} \tag{6-10}$$

the sample mean. From (5-50) and (5-51) we already know that

$$E(Y) = nE(X) \tag{6-11}$$

$$\mathrm{Var}(Y) = n\,\mathrm{Var}(X) \tag{6-12}$$

and from (5-53) and (5-54) we have

$$E(\bar{X}) = E(X) \tag{6-13}$$

$$\mathrm{Var}(\bar{X}) = \frac{\mathrm{Var}(X)}{n} \tag{6-14}$$

These facts were derived without finding the probability distribution of either Y or $\bar{X}$.

Frequently we are interested in making a probability statement about Y or $\bar{X}$. For example, we may want to know $\Pr(Y \leq y)$ or $\Pr(\bar{X} \leq \bar{x})$. The standard procedure for evaluating a probability of this type consists of two steps:

1 Find the derived distribution of Y or $\bar{X}$.
2 Using the derived distribution, add together the terms whose sum yields the desired probability.

204

Either or both of the above steps may lead to difficulties. The derived distribution may be difficult to obtain if n is very large, requiring either tedious enumeration or clever use of counting devices (as mentioned in Example 5-14). Even when it is easy to write down the derived distribution, the calculation of probabilities may be a considerable chore. For example, in Example 5-13 it was found that the sample sum had a binomial distribution. As long as n is in the range of existing tables of the binomial the calculation of probabilities is a trivial exercise. This will probably be the case in the majority of applications. However, if $n = 2,000$ considerable calculation may be required to provide the desired answer. Of course, high-speed computers can be used to surmount both types of difficulties. Our objective here is to discuss a procedure that provides a reasonably accurate approximation when the above difficulties are encountered.

For illustrative purposes let X be a random variable with probability function

$$g(x) = \tfrac{1}{6} \qquad x = 1, 2, 3, 4, 5. 6$$

Hence X could be the number of spots that show when a six-sided symmetric die is rolled. We have seen in Example 5-14 that if $X_1, X_2, \ldots, X_n$ is a random sample from this distribution, then the joint probability function is

$$f(x_1, x_2, \ldots, x_n) = (\tfrac{1}{6})^n \qquad \begin{matrix} x_i = 1, 2, 3, 4, 5, 6 \\ i = 1, 2, \ldots, n \end{matrix}$$

Now let us construct the probability distribution of $\bar{X}$ for $n = 2$. The 36 possible outcomes for (X_1, X_2) appear in Table 6-1. Each has probability of occurrence $(\tfrac{1}{6})^2 = \tfrac{1}{36}$. From each pair we can compute a value of $\bar{X}$, obtaining the 36 values found in Table 6-2. Since each sample had probability of occurrence $\tfrac{1}{36}$, each sample mean appearing in Table 6-2

Table 6-1 **Enumeration of all possible samples of size 2 obtained from a die random variable**

1,1	2,1	3,1	4,1	5,1	6,1
1,2	2,2	3,2	4,2	5,2	6,2
1,3	2,3	3,3	4,3	5,3	6,3
1,4	2,4	3,4	4,4	5,4	6,4
1,5	2,5	3,5	4,5	5,5	6,5
1,6	2,6	3,6	4,6	5,6	6,6

Table 6-2 Sample means computed from
           the 36 samples of Table 6-1

1.0	1.5	2.0	2.5	3.0	3.5
1.5	2.0	2.5	3.0	3.5	4.0
2.0	2.5	3.0	3.5	4.0	4.5
2.5	3.0	3.5	4.0	4.5	5.0
3.0	3.5	4.0	4.5	5.0	5.5
3.5	4.0	4.5	5.0	5.5	6.0

also has probability of occurrence $\frac{1}{36}$. From the latter table we can construct the probability distribution of $\bar{X}$ which is presented in Table 6-3.

It is easy to verify by direct calculation that $E(\bar{X}) = E(X)$ and $\text{Var}(\bar{X}) = \text{Var}(X)/2$. We have

$$E(X) = 1(\tfrac{1}{6}) + 2(\tfrac{1}{6}) + 3(\tfrac{1}{6}) + 4(\tfrac{1}{6}) + 5(\tfrac{1}{6}) + 6(\tfrac{1}{6}) = 3.5$$

$$E(\bar{X}) = 1.0(\tfrac{1}{36}) + 1.5(\tfrac{2}{36}) + 2.0(\tfrac{3}{36}) + 2.5(\tfrac{4}{36}) + 3.0(\tfrac{5}{36}) + 3.5(\tfrac{6}{36})$$
$$\qquad + 4.0(\tfrac{5}{36}) + 4.5(\tfrac{4}{36}) + 5.0(\tfrac{3}{36}) + 5.5(\tfrac{2}{36}) + 6.0(\tfrac{1}{36})$$

$$= \frac{126}{36} = 3.5$$

$$\text{Var}(X) = (1 - 3.5)^2(\tfrac{1}{6}) + (2 - 3.5)^2(\tfrac{1}{6}) + (3 - 3.5)^2(\tfrac{1}{6})$$
$$\qquad + (4 - 3.5)^2(\tfrac{1}{6}) + (5 - 3.5)^2(\tfrac{1}{6}) + (6 - 3.5)^2(\tfrac{1}{6}) = \frac{35}{12}$$

$$\text{Var}(\bar{X}) = (1.0 - 3.5)^2(\tfrac{1}{36}) + (1.5 - 3.5)^2(\tfrac{2}{36}) + (2.0 - 3.5)^2(\tfrac{3}{36})$$
$$\qquad + (2.5 - 3.5)^2(\tfrac{4}{36}) + (3.0 - 3.5)^2(\tfrac{5}{36}) + (3.5 - 3.5)^2(\tfrac{6}{36})$$
$$\qquad + (4.0 - 3.5)^2(\tfrac{5}{36}) + (4.5 - 3.5)^2(\tfrac{4}{36}) + (5.0 - 3.5)^2(\tfrac{3}{36})$$
$$\qquad + (5.5 - 3.5)^2(\tfrac{2}{36}) + (6.0 - 3.5)^2(\tfrac{1}{36})$$

$$= \frac{35}{24} = \frac{35}{12}\left(\frac{1}{2}\right)$$

For the case $n = 3$ the enumeration is a little more complicated. The counterparts of Tables 6-1 and 6-2 contain 216 entries instead of 36.

Table 6-3 Probability distribution of $\bar{X}$ for die random
           variable with $n = 2$

$\bar{x}$	1.0	1.5	2.0	2.5	3.0	3.5	4.0	4.5	5.0	5.5	6.0
$h_2(\bar{x})$	$\tfrac{1}{36}$	$\tfrac{2}{36}$	$\tfrac{3}{36}$	$\tfrac{4}{36}$	$\tfrac{5}{36}$	$\tfrac{6}{36}$	$\tfrac{5}{36}$	$\tfrac{4}{36}$	$\tfrac{3}{36}$	$\tfrac{2}{36}$	$\tfrac{1}{36}$

Table 6-4 Probability distribution of $\bar{X}$ for die random variable with $n = 3$

$\bar{x}$	1	$\frac{4}{3}$	$\frac{5}{3}$	2	$\frac{7}{3}$	$\frac{8}{3}$	3	$\frac{10}{3}$
$h_3(\bar{x})$	$\frac{1}{216}$	$\frac{3}{216}$	$\frac{6}{216}$	$\frac{10}{216}$	$\frac{15}{216}$	$\frac{21}{216}$	$\frac{25}{216}$	$\frac{27}{216}$
$\bar{x}$	$\frac{11}{3}$	4	$\frac{13}{3}$	$\frac{14}{3}$	5	$\frac{16}{3}$	$\frac{17}{3}$	6
$h_3(\bar{x})$	$\frac{27}{216}$	$\frac{25}{216}$	$\frac{21}{216}$	$\frac{15}{216}$	$\frac{10}{216}$	$\frac{6}{216}$	$\frac{3}{216}$	$\frac{1}{216}$

However, it is not too difficult to verify that the probability distribution of $\bar{X}$ is the one given in Table 6-4. It can be verified that $E(\bar{X}) = 3.5$ and $\text{Var}(\bar{X}) = (\frac{35}{12})(\frac{1}{3})$.

It is interesting to examine the graphs of $g(x)$, $h_2(\bar{x})$, and $h_3(\bar{x})$. These appear in Figs. 6-10, 6-11, and 6-12. With a little effort we could draw the corresponding graphs for $n = 4$, $n = 5$, etc. We would find that as n increases, (1) the number of points on the graph increases, (2) the graph looks more and more like the points that lie on a normal curve. We might suspect that probabilities for $\bar{X}$ could be approximated by finding an area under a normal curve if n is not too small. This is indeed the case even though $\bar{X}$ is a discrete random variable which has nonzero

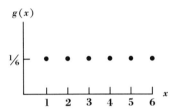

Figure 6-10 Graphical representation of $g(x)$.

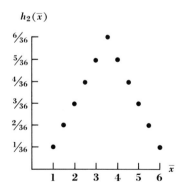

Figure 6-11 Graphical representation of distribution of $\bar{X}$ for samples of size 2.

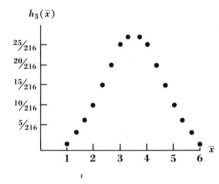

$h_3(\bar{x})$

$^{25}\!/_{216}$
$^{20}\!/_{216}$
$^{15}\!/_{216}$
$^{10}\!/_{216}$
$^{5}\!/_{216}$

1 2 3 4 5 6 $\bar{x}$

Figure 6-12   Graphical representation of distribution of $\bar{X}$ for samples of size 3.

probability only at isolated points while the normal random variable associates probabilities with intervals.   Our suspicions are confirmed by one of the most remarkable theorems in mathematical literature called the *central limit theorem.*   The theorem is as follows:

Let X_1, X_2, . . . , X_n be a random sample from a distribution with mean μ and variance σ^2.   Then the distribution of

$$Z' = \frac{\bar{X} - \mu}{\sigma/\sqrt{n}} \tag{6-15}$$

$$= \frac{Y - n\mu}{\sigma\sqrt{n}} \tag{6-16}$$

approaches the standard normal distribution as n increases.

We already knew from the discussion at the beginning of the section that $E(\bar{X}) = \mu$, $\text{Var}(\bar{X}) = \sigma^2/n$.   These facts together with the theorem imply that the distribution of $\bar{X}$ is approximately normal with mean μ and variance σ^2/n for moderate or large n.   Equivalently, Y is approximately normal with mean $n\mu$ and variance $n\sigma^2$.

Let us now consider a couple of examples that illustrate the usefulness of the theorem.

EXAMPLE 6-6

A six-sided symmetric die is thrown 105 times.   Approximate the probability that $\bar{X}$ is between 3 and 4 inclusive.

Solution

We have already found $E(X) = 3.5$, $\text{Var}(X) = {}^{35}\!/_{12}$.   Hence $\text{Var}(\bar{X}) = \text{Var}(X)/n = ({}^{35}\!/_{12})({}^{1}\!/_{105}) = \frac{1}{36}$.   The theorem implies that $\bar{X}$ is

208

approximately normal with mean 3.5 and variance $\frac{1}{36}$. Thus

$$\Pr(3 \leq \bar{X} \leq 4) = \Pr\left(\frac{3 - 3.5}{\frac{1}{6}} \leq \frac{\bar{X} - 3.5}{\frac{1}{6}} \leq \frac{4 - 3.5}{\frac{1}{6}}\right)$$

$$= \Pr(-3 \leq Z' \leq 3)$$
$$\cong \Pr(-3 < Z < 3)$$
$$= \Pr(Z < 3) - \Pr(Z < -3)$$
$$= .998650 - .001350 \quad \text{from Appendix B5}$$
$$= .997300$$

We note that $\Pr(3 < \bar{X} < 4) = \Pr(315 < Y < 420)$.

EXAMPLE 6-7

A six-sided symmetric die is thrown 1,000 times. What is the probability that 170 or more 6's occur?

Solution

Let X be a random variable taking on the value 0 if a 6 does not occur and 1 if a 6 occurs on a single throw. Then the probability function of X is

$$f(x) = \frac{5}{6} \quad x = 0$$
$$= \frac{1}{6} \quad x = 1$$

and

$$E(X) = 0(\tfrac{5}{6}) + 1(\tfrac{1}{6}) = \tfrac{1}{6}$$

$$\text{Var}(X) = \left(0 - \frac{1}{6}\right)^2 \frac{5}{6} + \left(1 - \frac{1}{6}\right)^2 \frac{1}{6} = \frac{5}{36}$$

If we let $Y = X_1 + X_2 + \cdots + X_{1,000}$ we are seeking $\Pr(170 \leq Y \leq 1,000)$. The central limit theorem implies that Y is approximately normal with mean $1,000(\tfrac{1}{6})$ and variance $1,000(\tfrac{5}{36})$. Hence

$$\Pr(170 \leq Y \leq 1,000) = \Pr\left[\frac{170 - 1,000(\tfrac{1}{6})}{\sqrt{1,000(\tfrac{5}{36})}} \leq \frac{Y - 1,000(\tfrac{1}{6})}{\sqrt{1,000(\tfrac{5}{36})}} \right.$$
$$\left. \leq \frac{1,000 - 1,000(\tfrac{1}{6})}{\sqrt{1,000(\tfrac{5}{36})}}\right]$$

$$= \Pr(.283 \leq Z' \leq 70.7)$$
$$\cong \Pr(.283 < Z < 70.7)$$
$$= \Pr(Z < 70.7) - \Pr(Z < .283)$$
$$= 1 - \Pr(Z < .283)$$

From Appendix B5 we find

$$\Pr(Z < .28) = .6103 \qquad \Pr(Z < .29) = .6141$$

Thus, to two decimal places $\Pr(Z < .283) = .61$ and

$$\Pr(170 \leqq Y \leqq 1,000) = .39$$

We have already discussed the adequacy of the binomial model in Example 4-1 for throwing a die a fixed number of times. In addition we found in Example 5-13 that Y has a binomial distribution. We identify $n = 1000$, $p = \frac{1}{6}$ and the exact probability is

$$\Pr(170 \leqq Y \leqq 1,000) = \sum_{y=170}^{1,000} b(y;1,000,\tfrac{1}{6})$$

$$= .40158$$

by using the Harvard University table mentioned in Chap. 4. Here we can compare the approximation with the exact probability.

The die random variable of Example 6-6 has a uniform distribution and its probability function is a special case of

$$f(x;k) = \frac{1}{k} \qquad x = 1, 2, \ldots, k \qquad (6\text{-}17)$$

with $k = 6$. In Exercise 3-45 we found that if X has probability function (6-17), then $E(X) = (k + 1)/2$ and $\mathrm{Var}(X) = (k^2 - 1)/12$. Hence, if $X_1, X_2, \ldots, X_n$ is a random sample from a distribution having probability function (6-17), the distribution of $\bar{X}$ has mean $E(\bar{X}) = (k + 1)/2$ and variance $\mathrm{Var}(\bar{X}) = (k^2 - 1)/12n$. Thus

$$\Pr(\bar{x}_1 \leqq \bar{X} \leqq \bar{x}_2) = \Pr\left[\frac{\bar{x}_1 - (k+1)/2}{\sqrt{(k^2-1)/12n}} \leqq \frac{\bar{X} - \mu}{\sigma/\sqrt{n}} \leqq \frac{\bar{x}_2 - (k+1)/2}{\sqrt{(k^2-1)/12n}} \right]$$

$$= \Pr\left[\frac{\bar{x}_1 - (k+1)/2}{\sqrt{(k^2-1)/12n}} \leqq Z' \leqq \frac{\bar{x}_2 - (k+1)/2}{\sqrt{(k^2-1)/12n}} \right]$$

and by the central limit theorem

$$\Pr(\bar{x}_1 \leqq \bar{X} \leqq \bar{x}_2) \cong \Pr\left[\frac{\bar{x}_1 - (k+1)/2}{\sqrt{(k^2-1)/12n}} \leqq Z \leqq \frac{\bar{x}_2 - (k+1)/2}{\sqrt{(k^2-1)/12n}} \right]$$

$$(6\text{-}18)$$

Formula (6-18) is a generalization of the approximation used in Example 6-6.

It is also easy to generalize the approximation used in Example 6-7. Let $X_1, X_2, \ldots, X_n$ be a random sample from a distribution having probability function $f(x) = p^x(1 - p)^{1-x}$, $x = 0, 1, 0 < p < 1$. Then $E(X) = p$, $\text{Var}(X) = p(1 - p)$, $E(Y) = np$, $\text{Var}(Y) = np(1 - p)$, and

$$\Pr(y_1 \leq Y \leq y_2) = \Pr\left[\frac{y_1 - np}{\sqrt{np(1 - p)}} \leq \frac{Y - \mu}{\sigma\sqrt{n}} \leq \frac{y_2 - np}{\sqrt{np(1 - p)}}\right]$$

$$= \Pr\left[\frac{y_1 - np}{\sqrt{np(1 - p)}} \leq Z' \leq \frac{y_2 - np}{\sqrt{np(1 - p)}}\right]$$

and by the central limit theorem

$$\Pr(y_1 \leq Y \leq y_2) \cong \Pr\left[\frac{y_1 - np}{\sqrt{np(1 - p)}} \leq Z \leq \frac{y_2 - np}{\sqrt{np(1 - p)}}\right] \qquad (6\text{-}19)$$

Since we know from Example 5-13 that Y has a binomial distribution with constants n and p, (6-19) can also be regarded as an approximation for binomial probabilities.

From (6-19) we can get an approximation for a sum of negative binomial probabilities. Recall from formula (4-10) that if N has a negative binomial distribution, then

$$\sum_{n=c}^{r} b^*(n;c,p) = \sum_{x=c}^{r} b(x;r,p)$$

Hence

$$\Pr(c \leq N \leq r) = \Pr(c \leq X \leq r)$$

$$\cong \Pr\left[\frac{c - rp}{\sqrt{rp(1 - p)}} \leq Z \leq \frac{r - rp}{\sqrt{rp(1 - p)}}\right] \qquad (6\text{-}20)$$

which follows immediately from (6-19) with $y_1 = c$, $y_2 = r$, $n = r$.

It is also worthwhile to write down the approximation for the Poisson distribution. If $X_1, X_2, \ldots, X_n$ is a random sample from a distribution having probability function

$$f(x) = \frac{e^{-\mu}\mu^x}{x!} \qquad x = 0, 1, 2, \ldots \qquad \mu > 0$$

then from (4-36) and (4-37) we have $E(X) = \text{Var}(X) = \mu$ so that $E(Y) = n\mu$, $\text{Var}(Y) = n\mu$. Thus we can write

$$\Pr(y_1 \leq Y \leq y_2) = \Pr\left(\frac{y_1 - n\mu}{\sqrt{n\mu}} \leq \frac{Y - n\mu}{\sqrt{n\mu}} \leq \frac{y_2 - n\mu}{\sqrt{n\mu}}\right)$$

$$= \Pr\left(\frac{y_1 - n\mu}{\sqrt{n\mu}} \leq Z' \leq \frac{y_2 - n\mu}{\sqrt{n\mu}}\right)$$

and by the central limit theorem

$$\Pr(y_1 \leqq Y \leqq y_2) \cong \Pr\left(\frac{y_1 - n\mu}{\sqrt{n\mu}} \leqq Z \leqq \frac{y_2 - n\mu}{\sqrt{n\mu}}\right) \tag{6-21}$$

Since we know from (5-41) that Y has a Poisson distribution with constant $n\mu$, (6-21) can also be regarded as approximation for Poisson probabilities.

EXAMPLE 6-8

If the random variable X has a Poisson distribution with $\mu = 100$, approximate the probability that $X \leqq 95$.

Solution

Let $Y = X$, $n = 1$ and use (6-21). We get

$$\Pr(0 \leqq Y \leqq 95) \cong \Pr\left(\frac{0 - 100}{\sqrt{100}} \leqq Z \leqq \frac{95 - 100}{\sqrt{100}}\right)$$
$$= \Pr(-10 \leqq Z \leqq -.5)$$
$$= \Pr(Z \leqq -.5) - \Pr(Z \leqq -10)$$
$$= .3085 - 0$$
$$= .3085$$

Since the Molina tables described in Chap. 4 contain entries for $\mu = 100$, we find using that table

$$\Pr(0 \leqq X \leqq 95) = .331192$$

Hence, for this probability, the approximation is off by a little over .02. We note that we could look upon the random variable X as having a Poisson distribution with $\mu = 1$ and let $Y = X_1 + X_2 + \cdots + X_{100}$, the sum of 100 independent random variables. Or, just as well, Y could be the sum of 1,000 independent Poisson random variables with $\mu = .1$. The probability is the same as the one already obtained. With the Poisson it is the size of $n\mu$, not n, that determines the accuracy of the approximation.

It has been found by actual experience that the normal approximation can be improved by replacing y_1 by $y_1 - .5$ and y_2 by $y_2 + .5$. This is sometimes referred to as the "continuity correction" and applies equally as well to the uniform, binomial, Poisson, or any other discrete distribution that has nonzero probabilities for a consecutive set of integers. In terms of limits on $\bar{X}$, say, $\bar{x}_1$ and $\bar{x}_2$, this amounts to replacing $\bar{x}_1$ and $\bar{x}_2$ by $\bar{x}_1 - 1/2n$ and $\bar{x}_2 + 1/2n$, respectively.

EXAMPLE 6-9 ━━━

Recompute the approximations used in Examples 6-6, 6-7, and 6-8 using the continuity correction.

Solution

In Example 6-6 we replace $\Pr(3 \leq \bar{X} \leq 4)$ by $\Pr[3 - (\frac{1}{2}10) \leq \bar{X} \leq 4 + (\frac{1}{2}10)]$ and proceed as before, subtracting 3.5 and dividing by $\frac{1}{6}$. We get

$$\Pr(3 \leq \bar{X} \leq 4) \cong \Pr(-3.03 < Z < 3.03) = .998$$

which differs by very little from the previous answer.

In Example 6-7 we write

$$\Pr(170 \leq Y \leq 1{,}000) = \Pr(169.5 \leq Y \leq 1{,}000.5)$$

Then, as before, subtract $1{,}000(\frac{1}{6})$ from the two limits and divide the results by $1{,}000(\frac{5}{36})$. We get

$$\Pr(170 \leq Y \leq 1{,}000) \cong \Pr(.240 < Z < 70.72) = 1 - .595 = .405$$

which is slightly closer to the exact value.

In Example 6-8 we write

$$\Pr(0 \leq Y \leq 95) = \Pr(-.5 \leq Y \leq 95.5)$$

$$= \Pr\left(\frac{-.5 - 100}{\sqrt{100}} \leq Z' \leq \frac{95.5 - 100}{\sqrt{100}}\right)$$

$$= \Pr(-10.05 \leq Z' \leq -.45)$$

$$\cong \Pr(-10.05 < Z < -.45)$$

$$= .3264$$

which is somewhat closer to the exact probability, agreeing to two decimal places when rounded off to the nearest hundredth.

━━━

We would, of course, not use the normal approximation in cases where the exact probability can be read directly from tables. This means that we are unlikely to use the approximation on the binomial unless $n > 1{,}000$ (the upper limit of the Harvard table) and on the Poisson unless $n\mu > 100$ (the upper limit on the Molina table). For the binomial case, if $p < .05$ or $p > .95$ the Poisson approximation described in Sec. 4-6, and illustrated in Example 4-18, is apt to give better results.

When the normal approximation is used we might just as well use the continuity correction since the arithmetic is about the same as when it is not used. The change produced in the probability is greatest when either y_1 or y_2 is near $E(Y)$ since then the limit on Z is near 0, the region where normal probabilities increase the greatest.

213

EXERCISES

6-8   The random variable X has a binomial distribution with $n = 100$, $p = .10$. Approximate $\Pr(X \leq 12)$ both with and without the continuity correction and compare with the exact probability given in Appendix B1.

6-9   The random variable X has a binomial distribution with $n = 100$, $p = .50$. Approximate $\Pr(X \leq 53)$ both with and without the continuity correction and compare with the exact probability given in Appendix B1.

6-10   Let $X_1, X_2, \ldots , X_n$ be a random sample from a distribution having probability function $f(x) = p^x(1 - p)^{1-x}, x = 0, 1, 0 < p < 1$, and $Y = X_1 + X_2 + \cdots + X_n$. If $n = 2,500, p = .10$, approximate $\Pr(Y \geq 277)$ both with and without the continuity correction.

6-11   The random variable X has a Poisson distribution with $\mu = 2$. Let $X_1, X_2, \ldots , X_n$ be a random sample with $n = 200$. If $Y = X_1 + X_2 + \cdots + X_n$, approximate $\Pr(Y \leq 364)$ both with and without the continuity correction.

6-12   From a table of random numbers, 33 numbers are selected from those between 1 and 10 inclusive. Hence, the probability function of X, the number appearing on any given selection, is (6-17) with $k = 10$. Approximate the probability that $\bar{X}$ is greater than or equal to 6.

6-13   A cup contains 15 six-sided symmetric dice. The cup is turned over so that the dice fall onto a table. Use the normal approximation with the continuity correction to approximate $\Pr(40 \leq Y \leq 60)$ where Y is the sum of the spots on the 15 dice.

6-14   Consider again Example 4-7. Suppose that 900 good potatoes are needed. Using the continuity correction, approximate the probability that more than 1,000 potatoes need be cut open.

6-1 A continuous random variable can assume any value within an interval or several intervals.

Associated with a continuous random variable X is a density function $f(x)$ where

(a) $f(x)$ is nonnegative.

(b) The total area under the graph of $f(x)$ and above the x axis is 1.

Areas under density functions are interpreted as probabilities. The choice of the correct density function to yield probabilities for a random variable X is not always easily made and is influenced by experience.

6-2 One of the most important density functions is the standard normal. Areas under the curve (probabilities) are well tabulated.

6-3 Let $X_1, X_2, \ldots, X_n$ be a random sample from a distribution with mean μ and variance σ^2. Then the distribution of

$$Z' = \frac{\bar{X} - \mu}{\sigma/\sqrt{n}} = \frac{Y - n\mu}{\sigma\sqrt{n}} \qquad \text{(6-15), (6-16)}$$

approaches the standard normal as n increases. If X has probability function

$$f(x) = p^x(1 - p)^{1-x} \qquad x = 0, 1$$
$$0 < p < 1$$

then

$$\Pr(y_1 \leqq Y \leqq y_2) \cong \Pr\left[\frac{y_1 - np}{\sqrt{np(1 - p)}} \leqq Z \leqq \frac{y_2 - np}{\sqrt{np(1 - p)}}\right] \qquad \text{(6-19)}$$

where Z has a standard normal distribution. If N has a negative binomial distribution with constants c and p, then

$$\Pr(c \leqq N \leqq r) \cong \Pr\left[\frac{c - rp}{\sqrt{rp(1 - p)}} \leqq Z \leqq \frac{r - rp}{\sqrt{rp(1 - p)}}\right] \qquad \text{(6-20)}$$

If X has a Poisson distribution with mean μ, then

$$\Pr(y_1 \leqq Y \leqq y_2) \cong \Pr\left(\frac{y_1 - n\mu}{\sqrt{n\mu}} \leqq Z \leqq \frac{y_2 - n\mu}{\sqrt{n\mu}}\right) \qquad \text{(6-21)}$$

215

CHAPTER SEVEN

Statistical Applications of Probability: Estimation and Testing Hypotheses

7-1 INTRODUCTION

In Chap. 4 we considered some important probability distributions. These distributions, which were used to compute probabilities for one-dimensional random variables, include the binomial, negative binomial, uniform, hypergeometric, and Poisson. Before making any computations, we not only attempted to justify the use of the model but also had to identify certain constants. These constants are called *parameters*. For any given problem they are fixed numbers, but they may take on different values when another problem is considered. The parameters encountered in Chap. 4 were n and p with the binomial; c and p with the negative binomial; k with the uniform; N, n, and k with the hypergeometric; μ with the Poisson. In Chap. 5 parameters associated with the multinomial were n, p_1, p_2, . . . , p_k and with the generalized hypergeometric were N, n, k_1, k_2, . . . , k_r. We shall define a *parameter* as a constant entering into a probability function.

For the probability problems of Chaps. 4 and 5 all parameters were known. In many situations that arise this is not the case. Let us reconsider Example 4-5 concerning potatoes. We were given that $p = .90$ and $n = 25$. Knowing that 90 percent of the potatoes were good, we were able to write down the probability function of the random variable X, the number of good potatoes in a random sample of 25. That probability function is

$$f(x) = b(x;25,.90) = \binom{25}{x} (.90)^x (.10)^{25-x} \qquad x = 0, 1, 2, . . . , 25$$

In a more practical situation we would not know p and it is quite likely that we would want to make some inference about it.

Suppose that a potato wholesaler claims that his product is 90 percent good. We might believe him, but it is possible that we would require some evidence to support his claim before we accepted it as the truth. In order to make a decision, suppose that we count the number of good potatoes in a random sample of size 25 and accept his statement unless the number of good potatoes found in the sample is too small. Which results are "too small" could be determined on a probability basis computed under the assumptions that the binomial model is correct (which we are willing to believe) and $p = .90$ (about which we are undecided). This procedure is called *testing a hypothesis* and will be discussed later in the chapter. As we shall use the terms, a *hypothesis* is an assumption concerning the parameters of a probability function and a *test* is a rule indicating when to reject the hypothesis. In the potato example the hypothesis is $p = .90$. The test could be: Reject $p = .90$ if 19 or fewer good potatoes are observed in the sample.

In the potato example we may not be interested in testing a hypothesis about the proportion of good potatoes. We are stuck with the crop no matter what p happens to be. We may, however, want to estimate p from a random sample to make some sort of price adjustment. Two kinds of estimates are in common usage. The first, called a point estimate, yields one value—say, $\hat{p} = .81$—which is our best guess for the proportion of good potatoes. The second, called an interval estimate, specifies a range such as .76 to .85, which we think includes or captures the true but unknown p. Both methods of estimation will be considered.

The potato example not only motivates a definition for statistics but illustrates the main difference between probability and statistics. As we shall use the term, *statistics* is a field of endeavor concerned with drawing inferences about unknown parameters of a probability distribution by observing one or more random variables having the given probability distribution whose parameters are unknown. Probability theory enables us to derive distributions for random variables. Statistics, on the other hand, uses the information furnished by an observed value of a random variable to draw inferences about unknown parameters of a probability distribution.

The random variable that is observed in a statistical problem is usually called a *statistic*. Two random variables, or statistics, that arise frequently when dealing with discrete probability distributions are the sample mean $\bar{X}$ and the sample sum $Y = \sum_{i=1}^{n} X_i = n\bar{X}$. A statistic is a random variable which can be computed from a random sample. Thus as soon as a sample is observed, a statistic assumes a specific known value. Other possible statistics include Y_1, the smallest observation in a random sample, Y_n, the largest observation in a random sample, or $R = Y_n - Y_1$, the range of the observations of a random sample.

The probability distribution of a statistic is sometimes called a *sampling distribution*. The binomial can be regarded as the sampling distribution of the number of successes in n trials when the probability of a success remains constant and trials are independent. The negative binomial can be regarded as the sampling distribution of the number of trials required to achieve c successes when the probability of a success is constant from trial to trial and trials are independent. The hypergeometric can be considered as the sampling distribution of the number of good items in a sample of size n drawn without replacement from a population of N items containing k good ones. The Poisson can be regarded as the sampling distribution of the number of times an event happens in a given time interval (area or volume of space) given that the average number of times the event happens is μ and assuming that the Poisson is a reasonable model.

7-2 POINT ESTIMATION

Suppose that a random sample of 25 potatoes yields 20 good ones. A reasonable and obvious estimate to use for p, the fraction of good potatoes, is $^{20}/_{25} = .8$. In general, to estimate p for a series of experiments satisfying the binomial conditions, we would perform n experiments, count the number of successes x, and use the fraction x/n as an estimate of p. That is, not knowing the real value of p it seems natural to estimate p by using the observed fraction of success in a sample. Since x is an observed value of a random value X, x/n is an observed value of the random variable X/n. Further, since X has a binomial distribution we can study the behavior of X/n, observing any good or bad properties the estimate may have.

In order to distinguish between the random variable and its observed value, we shall refer to the former as an *estimator* and to the latter as an *estimate*. Thus in the previous paragraph X/n is an estimator, x/n is an estimate.

It is easy to write down the probability function of $T = X/n$. We know that the probability function of X is

$$f(x) = \binom{n}{x} p^x (1 - p)^{n-x} \qquad x = 0, 1, 2, \ldots, n \tag{7-1}$$

Since $X = nT$, it follows immediately that

$$g(t) = \binom{n}{nt} p^{nt} (1 - p)^{n-nt} \qquad t = 0, \frac{1}{n}, \frac{2}{n}, \ldots, \frac{n-1}{n}, 1 \tag{7-2}$$

There is, however, no useful purpose served in having (7-2). All questions of interest can be answered by using (7-1), a tabulated distribution whose mean and variance we have already derived. Since

$$\Pr(T = t) = \Pr\left(\frac{X}{n} = \frac{x}{n}\right) = \Pr(X = x)$$

any probability concerning T can be obtained from Appendix B1. Since $E(X) = np$, we have

$$E\left(\frac{X}{n}\right) = \frac{1}{n} E(X) = \frac{np}{n} = p \tag{7-3}$$

That is, the average value of the random variable T is p, the quantity we are estimating. Intuitively, this seems like a very desirable property for an estimator to have. A random variable is said to be an *unbiased estimator* of a parameter if the average or expected value of the random variable is equal to the parameter.

In addition to being unbiased, a good estimate should tend to be close to the parameter that is being estimated. That is, the probability that the estimator is near the parameter should be large. The measure of closeness most commonly used is the variance. In general, we usually prefer to use unbiased estimators with variance as small as possible. These two concepts form the basis of the definition we shall use for a "best" estimator.

Now let us formalize the previous discussion. Suppose we know that a random variable X has a given probability function depending upon an unknown parameter θ. Thus, we may know that X has a binomial distribution but do not know the correct value of p. To indicate this dependence upon θ we shall use the notation $f(x;\theta)$ instead of $f(x)$ to denote the probability function of X. Let $X_1, X_2, \ldots, X_n$ denote a random sample from the distribution having probability function $f(x;\theta)$ and let T be a statistic that is a function of $X_1, X_2, \ldots, X_n$ (i.e., we might have $T = X_1 + X_2 + \cdots + X_n$). If $E(T) = \theta$, then T is an *unbiased estimator* of θ. If, in addition to being unbiased, $\text{Var}(T) = E[(T - \theta)^2]$ is less than or equal to the variance of any other unbiased estimator, then T is called a "best" estimator of θ. The quotation marks, which will be dropped henceforth, are used to indicate the arbitrariness of our definition. Others may prefer to define best in a different way. We could, for example, say that an estimator is best if it has a smaller variance than any other statistic used as an estimator for the unknown parameter, thus not requiring the estimator to be unbiased.

For all estimators that we shall study, we shall be able to compute the mean and variance. However, showing that an unbiased estimator has a variance that is no larger than the variance of any other unbiased estimator of the same parameter requires sophisticated mathematics. Consequently, we shall merely state that an estimator is best and leave the proof for more advanced courses.

Now let us return to the binomial situation. Let X have probability function

$$f(x;p) = p^x(1 - p)^{1-x} \qquad x = 0, 1 \tag{7-4}$$
$$0 < p < 1$$

We already know that $Y = X_1 + X_2 + \cdots + X_n$ has a binomial distribution with parameters n and p (Example 5-13) so that $E(Y) = np$. Hence it seems logical to use

$$T = \bar{X} = \frac{X_1 + X_2 + \cdots + X_n}{n} = \frac{Y}{n} \tag{7-5}$$

as an estimator of p. Since $E(\bar{X}) = E(X) = p$, $\bar{X}$ is an unbiased estimator. Its variance, which is

$$\text{Var}(\bar{X}) = \frac{\text{Var}(X)}{n} = \frac{p(1 - p)}{n} \tag{7-6}$$

is the smallest possible for unbiased estimators of p. Consequently, $\bar{X}$ (which is nothing more than the fraction of successes in a sample of size n) is the best estimator of p.

Next, suppose that X has probability function

$$f(x;p) = p(1 - p)^{x-1} \qquad x = 1, 2, 3, \ldots \tag{7-7}$$
$$0 < p < 1$$

Thus X has a negative binomial distribution with parameters $c = 1$ and p. From Exercise 5-45 we know that $Y = X_1 + X_2 + \cdots + X_n$ has a negative binomial distribution with parameters $c = n$ and p. From (4-12) we know that $E(Y) = n/p$. This might lead one to suspect that n/Y is an unbiased estimator of p, a conjecture that is false. The correct statistic to use is

$$T = \frac{n - 1}{Y - 1} \tag{7-8}$$

To show that T is unbiased we evaluate

$$E(T) = \sum_{y=n}^{\infty} \left(\frac{n-1}{y-1}\right)\left(\frac{y-1}{n-1}\right) p^n (1-p)^{y-n}$$

$$= \sum_{y=n}^{\infty} \left(\frac{y-2}{n-2}\right) p^n (1-p)^{y-n}$$

Now let $y - 1 = u$. We get

$$E(T) = p \sum_{u=n-1}^{\infty} \left(\frac{u-1}{n-1-1}\right) p^{n-1}(1-p)^{u-(n-1)} = p(1) = p$$

The sum on u is 1 since this represents the sum of all the probabilities for a negative binomial distribution with parameters $c = n - 1$ and p. It can be shown that the variance of T is as small as possible for unbiased estimators of p. Thus T is the best estimator of p if X has the negative binomial distribution given by (7-7).

If X has a Poisson distribution with probability function

$$f(x;\mu) = \frac{e^{-\mu}\mu^x}{x!} \qquad x = 0, 1, 2, \ldots \tag{7-9}$$

we know from (5-41) that $Y = X_1 + X_2 + \cdots + X_n$ has a Poisson distribution with parameter $n\mu$ so that $E(Y) = n\mu$. Hence it seems logical to use

$$T = \bar{X} = \frac{X_1 + X_2 + \cdots + X_n}{n} = \frac{Y}{n} \tag{7-10}$$

as an estimator of μ. Since $E(\bar{X}) = E(X) = \mu$, $\bar{X}$ is unbiased with

$$\mathrm{Var}(\bar{X}) = \frac{\mathrm{Var}(X)}{n} = \frac{\mu}{n} \tag{7-11}$$

It can be shown that no other unbiased estimator of μ has smaller variance. Hence $\bar{X}$ is the best estimator of μ.

EXAMPLE 7-1

A coin is tossed 100 times and 44 heads are recorded. Find the best estimate p, the probability of getting a head on a single throw. Find the probability that the best estimator will differ from p by no more than .08 if $p = .40$, if $p = .50$, if $p = .60$.

Solution

The result on each throw is a binomial random variable X with probability function (7-4). Hence the best estimate of p is $\bar{x} = y/n$ where $y = 44$, $n = 100$, so that $\bar{x} = .44$. The probability we seek is

$$\Pr(p - .08 \leq \frac{Y}{100} \leq p + .08) = \Pr(100p - 8 \leq Y \leq 100p + 8)$$

where Y has a binomial distribution with parameters $n = 100$ and p. If $p = .40, .50, .60$ we get, respectively,

$$\Pr(32 \leq Y \leq 48) = \sum_{y=32}^{48} b(y;100,.40) = .95770 - .03985 = .91785$$

$$\Pr(42 \leq Y \leq 58) = \sum_{y=42}^{58} b(y;100,.50) = .95569 - .04431 = .91138$$

$$\Pr(52 \leq Y \leq 68) = \sum_{y=52}^{68} b(y;100,.60) = \sum_{y=32}^{48} b(x;100,.40) = .91785$$

EXAMPLE 7-2

Find the best estimate of p, the probability that a missile is successful, using the information of Exercise 4-13. If the actual value of p is .30, find in terms of unevaluated binomial probabilities

$$\Pr(.1 \leq T \leq .5)$$

using $n = 4$.

Solution

The random variable X, the number of trials to get one success, has probability function (7-7). Thus the best estimator of p is given by (7-8) with $n = 4$, $y = 11$, and $t = (4 - 1)/(11 - 1) = .3$. Next, we want to evaluate

$$\Pr\left(.1 \leq \frac{3}{Y-1} \leq .5\right)$$

222

computed with $p = .3$ and $n = 4$. The inequality $.1 \leq 3/(Y-1)$ is equivalent to $Y \leq 31$ and $3/(Y-1) \leq .5$ reduces to $7 \leq Y$. Hence

$$\Pr\left(.1 \leq \frac{3}{Y-1} \leq .5\right) = \Pr(7 \leq Y \leq 31)$$

$$= \sum_{y=7}^{31} b^*(y;4,.3)$$

$$= \sum_{y=4}^{31} b^*(y;4,.3) - \sum_{y=4}^{6} b^*(y;4,.3)$$

$$= \sum_{x=4}^{31} b(x;31,.3) - \sum_{x=4}^{6} b(x;6,.3)$$

These sums cannot be evaluated from our binomial table, but the Harvard University table [5] yields $.99284 - .07047 = .92237$.

EXAMPLE 7-3

A random sample of 10 typed pages is selected from the work turned out by a secretary. A total of 6 errors is found on the 10 pages. Find the best estimate of the average number of errors that the secretary makes per page. Find the probability that a best estimator differs from true value of the parameter by no more than .3 if the true value is .5. If the true value is .4. If the true value is .3.

Solution

The random variable X, the number of errors per typed page, can be assumed to have a Poisson distribution (recall Example 4-17). Hence the best estimate is $\bar{x}$. We have $y = 6$, $n = 10$, $\bar{x} = .6$. The probability we seek is

$$\Pr\left(\mu - .3 \leq \frac{Y}{10} \leq \mu + .3\right) = \Pr(10\mu - 3 \leq Y \leq 10\mu + 3)$$

where Y has a Poisson distribution with parameter 10μ. If $\mu = .5$

$$\Pr(2 \leq Y \leq 8) = \sum_{y=0}^{8} f(y;5) - \sum_{y=0}^{1} f(y;5) = .93191 - .04043 = .89148$$

If $\mu = .4$

$$\Pr(1 \leq Y \leq 7) = \sum_{y=0}^{7} f(y;4) - \sum_{y=0}^{0} f(y;4) = .94887 - .01832 = .93755$$

223

If $\mu = .3$

$$\Pr(0 \le Y \le 6) = \sum_{y=0}^{6} f(y;.3) = .96649$$

all sums being evaluated from Appendix B4.

In Examples 7-1, 7-2, 7-3, we computed the probability that Y was in an interval whose end points were integers. To evaluate something like $\Pr(3.2 \le Y \le 7.8)$, this is the same as $\Pr(4 \le Y \le 7)$ when Y has non-zero probability only at nonnegative integer values.

EXERCISES

7-1 A four-sided unsymmetric die is rolled 100 times to estimate the probability p that the side labeled 1 falls on the bottom. If a 1 is counted 28 times, what is the best estimate of the unknown probability p? Find the probability that the best estimator will differ from p by no more than .10 if $p = .20$. If $p = .25$. If $p = .30$.

7-2 A production line process turns out bolts, a fraction of which require rework. To estimate this fraction bolts are selected at random until 10 requiring rework are obtained. If the 10th defective bolt is found on the 73rd selection, what is the best estimate of p? If the real value of $p = .10$, find the probability that the estimator T differs from p by no more than .05, that is, $\Pr(.05 \le T \le .15)$. Express the result as unevaluated binomial sums.

7-3 A manufacturer inspects a random sample of 20 refrigerators and finds a total of 6 defects. Assume that X, the number of defects found in a refrigerator, has a Poisson distribution. Find the best estimate of μ. If the true value of μ is .50, find $\Pr(.30 \le \bar{X} \le .70)$.

7-4 Suppose that the random variable X has the probability function

$$f(x;k) = \frac{1}{k} \qquad x = 1, 2, \ldots, k$$

On the basis of a random sample $X_1, X_2, \ldots, X_n$ we would like to estimate k. Let $Y_1 = \min X_i$, $Y_n = \max X_i$ as in Sec. 5-7. Use (5-81) and (5-82) to show that $T = Y_1 + Y_n - 1$ is an unbiased estimator of k.

7-5 Suppose that the random variable X has a hypergeometric distribution with probability function given by (4-24). From (4-30) we have $E(X) = nk/N$, from which it follows that $E(X/n) = k/N$ and $E(NX/n) = k$. Hence, if a random sample of size n is drawn without replacement from N items, k of which have some characteristic, then X/n, the fraction of the sample having the characteristic, is an

224

unbiased estimator of k/N, the fraction of all N items having the characteristic. Observe that one does not need to know N to use this estimator. Further NX/n is an unbiased estimator of k. (a) A box contains 1,000 camera flash bulbs. A random sample of 25 is selected and tried and yields 2 defectives. Find unbiased estimates of the fraction of defectives in the box and the number of defectives in the box. Express as a hypergeometric probability (but do not evaluate) the probability that the estimator X/n differs by no more than .01 from k/N if $k/N = .05$. (b) A random sample (drawn without replacement) of 200 voters in a community yields 110 Democrats. Find an unbiased estimate of the proportion of Democrats in the community.

7-6   Suppose $(X_1, X_2, \ldots, X_k)$ is a multinomial random variable having probability function (5-57). Find unbiased estimates of $p_1, p_2, \ldots, p_k$. [*Hint:* Consider the marginal distribution of X_i whose probability function is given by (5-60).] Consider again Exercise 5-52 and assume that $p_1, p_2, \ldots, p_7$ are unknown. Give unbiased estimates of the p's based upon the given sample of size 50.

7-3   INTERVAL ESTIMATION

In Example 7-3 in which we considered a Poisson random variable with unknown mean μ we found that

$$\Pr(\mu - .3 \leq T \leq \mu + .3) = .93755 \qquad (7\text{-}12)$$

if $\mu = .4$. Consider the two inequalities in (7-12), that is, $\mu - .3 \leq T$ and $T \leq \mu + .3$. Since the same number can be added to or subtracted from each side of an inequality without disturbing the sense of the inequality, the two inequalities are equivalent to $\mu \leq T + .3$ and $T - .3 \leq \mu$, respectively. Thus, if $\mu - .3 \leq T \leq \mu + .3$, then

$$T - .3 \leq \mu \leq T + .3$$

and vice versa. This fact enables us to rewrite (7-12) as

$$\Pr(T - .3 \leq \mu \leq T + .3) = .93755 \qquad (7\text{-}13)$$

The probability statement (7-13) means that the probability is .93755 that the random interval $(T - .3, T + .3)$ contains (or captures) μ if $\mu = .4$. Once we observe $T = t = .6$, the observed value from Example 7-3, we can calculate a specific interval $(.6 - .3, .6 + .3)$ or $(.3, .9)$. The

specific interval is called a *confidence interval* for μ, and the probability
.93755 is called the confidence coefficient.

If we knew that $\mu = .4$, then from a practical point of view it makes no
sense to obtain an interval that we hope captures μ. However, in applied
problems the parameter is usually unknown and an investigator may wish
to be able to compute two numbers $u_1(t)$, $u_2(t)$ such that the random
interval $[u_1(T), u_2(T)]$ contains the true value of the parameter with
fairly high probability. The statistician usually begins by selecting
arbitrarily the confidence coefficient that we shall designate by $1 - \alpha$.
Then he tries to find $u_1(T)$, $u_2(T)$ such that

$$\Pr[u_1(T) < \mu < u_2(T)] = 1 - \alpha \qquad (7\text{-}14)$$

That is, $u_1(T)$ and $u_2(T)$ are so chosen that the probability is $1 - \alpha$
that they contain the true value of the parameter. Regardless of how
$u_1(T)$ and $u_2(T)$ are determined, $[u_1(t), u_2(t)]$ is a confidence interval for μ
with confidence coefficient $1 - \alpha$. (Actually, when T is a discrete random variable, the probability (7-14) is made $\geq 1 - \alpha$ unless an auxiliary
random variable is also used as described under 2 below. The addition
of the second random variable allows equality to be achieved.)

The discussion of the procedure for finding $[u_1(t), u_2(t)]$ will be left for
more advanced courses. Usually a fairly involved numerical problem is
encountered requiring the use of high-speed computing machines. With
the Poisson parameter in mind, we shall now describe briefly the three
types of confidence intervals that one may wish to use.

1 The end points of the intervals can be chosen so that the length of the
interval $u_2(t) - u_1(t)$ is a minimum for all values of t. The interval
$[u_1(t), u_2(t)]$ is then called the *shortest* confidence interval with con-

Table 7-1 **Short confidence intervals for
the Poisson parameter μ,
confidence coefficient .95**

Observed value of Y, y	$nu_1(y)$	$nu_2(y)$
5	1.970	11.177
6	2.613	12.817
7	3.285	13.765
8	3.285	14.921
9	4.460	16.768
10	5.323	17.633

fidence coefficient $1 - \alpha$. It would be desirable to have a table from which one could read the end points of the shortest interval for values of t. Apparently, no such table exists. However, Crow and Gardner [3] have prepared a table that yields short intervals for Poisson μ (shorter, at least, than intervals described below under 3). A sample from these results appears in Table 7-1. To enter the table one needs y (and not $t = \bar{x}$). The complete table contains entries for $y = 0(1)300$ and confidence coefficients .80, .90, .95, .99, .999.

2   It seems desirable to have

$$\Pr[u_1(T) < \mu' < u_2(T)] < 1 - \alpha \qquad (7\text{-}15)$$

if $\mu' \neq \mu$. In other words it is reasonable to require that the probability of capturing the wrong values of μ be less than the probability of capturing the correct value of μ. An interval $[u_1(t), u_2(t)]$ so chosen is called an unbiased confidence interval. If from all unbiased confidence intervals we obtain the shortest one, then the resulting interval $[u_1(t), u_2(t)]$ is called the *shortest unbiased* confidence interval with confidence coefficient $1 - \alpha$. (The shortest interval and the short interval described under 1 are not unbiased.)

Some tables of shortest unbiased intervals for Poisson μ have been published by Blyth and Hutchinson [2]. To enter their tables one needs $y + x$, where X is the random variable associated with the balanced spinner (Fig. 6-1). Their table gives $nu_1(y + x)$, $nu_2(y + x)$ to two and three significant figures for confidence coefficients .95 and .99 and $y = 0(1)250$. A sample from these results appears in Table 7-2.

Table 7-2   *Shortest unbiased confidence intervals for the Poisson parameter μ, confidence coefficient .95*

Observed value of $Y + X$, $y + x$	$nu_1(y + x)$	$nu_2(y + x)$
5.7	1.7	11.2
5.8	1.8	11.3
5.9	1.9	11.4
6.0	2.0	11.5
6.1	2.0	11.7
6.2	2.1	11.8

Table 7-3 *Equal-tail confidence in-*
tervals for the Poisson
parameter μ, confidence
coefficient .95

Observed value of Y, y	$nu_1(y)$	$nu_2(y)$
5	1.6	11.7
6	2.2	13.1
7	2.8	14.4
8	3.5	15.8
9	4.1	17.1
10	4.8	18.4

3 The most frequently used confidence intervals are determined so that

$$\Pr[u_1(T) < \mu] = \Pr[u_2(T) > \mu] = \frac{\alpha}{2}$$

The resulting interval $[u_1(t), u_2(t)]$ is called an *equal-tail* confidence interval with coefficient $1 - \alpha$. The main virtues of equal-tail intervals are (a) they can sometimes be calculated quickly from other tables which are readily available and (b) the intervals described under *1* and *2* differ very little from the equal-tail interval when the sample size is moderately large.

Table 7-3 yields some equal-tail intervals. Equal-tail intervals for the Poisson parameter are easily computed from chi-square tables (see Guenther [4], p. 172).

EXAMPLE 7-4

Using the observed results in Example 7-3, find the short, the shortest unbiased, and the equal-tail confidence intervals for μ with confidence coefficient .95. Assume that the balanced spinner produces an observed result approximately equal to .1.

Solution

We had $y = 6$, $n = 10$. From Table 7-1, $10u_1(y) = 2.613$, $10u_2(y) = 12.817$, short interval is $(.26, 1.28)$. With $y + x = 6 + .1 = 6.1$, Table 7-2 yields $10u_1(y + x) = 2.0$, $10u_2(y + x) = 11.7$ so that the shortest unbiased interval is $(.20, 1.17)$. From Table 7-3, $10u_1(y) = 2.2$, $10u_2(y) = 13.1$, equal-tail interval is $(.22, 1.31)$.

Several comments concerning confidence intervals seem to be appropriate. First, since there may be considerable variation in point estimates, it may be more satisfying from a practical point of view to have an interval for the parameter with which we associate a high degree of confidence. Second, having obtained a confidence interval such as $(.26, 1.28)$, the short interval of Example 7-4, we may be inclined to write

$$\Pr(.26 < \mu < 1.28) = .95$$

This is, of course, not correct since μ, being a constant, such as $.55$, is either in the interval or it is not. Consequently

$$\Pr(.26 < \mu < 1.28) = 1 \qquad \text{if } \mu \text{ is in } (.26, 1.28)$$
$$= 0 \qquad \text{if } \mu \text{ is not in } (.26, 1.28)$$

The interval $(.26, 1.28)$ is one of many such intervals that might be obtained. If thousands of such intervals were calculated in the same way, each based on a different random sample of the same size, then in the long run we would expect that about 95 out of every 100 intervals would include μ. The procedure is somewhat analogous to throwing intervals at a point target in such a way that 95 percent of the time the interval will cover the point.

A number of independent tabulations have been made for equal-tail confidence intervals for the binomial parameter p. Tables give $u_1(y)$,

Table 7-4 *Equal-tail confidence intervals for the binomial parameter p, confidence coefficient .95, n = 10*

Observed value of Y, y	$u_1(y)$	$u_2(y)$
0	.0000	.4113
1	.0025	.4450
2	.0252	.5561
3	.0667	.6525
4	.1216	.7376
5	.1871	.8129
6	.2624	.8784
7	.3475	.9333
8	.4439	.9748
9	.5550	.9975
10	.5887	1.0000

$u_2(y)$ such that $\Pr[u_1(Y) < p] = \Pr[p < u_2(Y)] = \alpha/2$. A sample table is given in Table 7-4. A bibliography of some existing equal-tail tables is given by Owen ([8], p. 273).

EXAMPLE 7-5

In the potato example mentioned in Sec. 7-1 suppose that a random sample of 10 potatoes yields 8 good ones. Find an equal-tail confidence interval for p, the fraction of good potatoes in the crop, with confidence coefficient .95.

Solution

We have $y = 8$, $n = 10$. From Table 7-4 we find $u_1(y) = .4439$, $u_2(y) = .9748$. Hence, the desired interval is $(.4439, .9748)$. The interval is very wide, a situation we should expect with such a small n. To narrow the interval we can either (1) decrease the confidence coefficient or (2) increase n. For example, with $n = 10$, confidence coefficient .50, $y = 8$ we find from Owen's table [8] that the interval is $(.6446, .9036)$; with $n = 100$, confidence coefficient .95, $y = 80$ the interval is shortened to approximately $(.72, .88)$. (The result is from Guenther [4], p. 330.)

Some tables of shortest unbiased interval for binomial p have been prepared by Blyth and Hutchinson [1]. To enter their tables one needs $y + x$, where X is the random variable associated with the balanced spinner (Fig. 6-1). Their table gives $u_1(y + x)$, $u_2(y + x)$ to two decimal places for confidence coefficients .95, .99 and for $n = 2(1)24(2)50$.

EXERCISES

7-7 Using the information of Exercise 7-3, find (a) a short, (b) a shortest unbiased, (c) an equal-tail confidence interval for μ with confidence coefficient .95. Assume that the balanced spinner produces an observed result approximately equal to .2.

7-8 An individual keeps a record of the number of telephone calls he receives per day. From his records he picks a random sample of five days during which the following number of calls were received: 1, 2, 1, 0, 1. Assuming that X, the number of calls received in a day, has a Poisson distribution with mean μ, find (a) a short, (b) a shortest unbiased, and (c) an equal-tail confidence interval for μ with confidence coefficient .95. Assume that the balanced spinner produces an observed result approximately equal to .8.

7-9 A random sample of 10 radish seeds of a certain brand are planted and 9 germinate. Find an equal-tail confidence interval for p, the fraction of seeds of this brand that germinate.

230

7-10　A four-sided unsymmetric die is rolled 10 times and the side labeled 1 falls face down 2 times. Find a confidence interval with coefficient .95 for p, the probability of rolling a 1.

7-11　The graphs reproduced by Guenther ([4], pp. 329–331) enable one to read an equal-tail confidence interval, sufficiently accurate for practical purposes, up to $n = 1,000$. For n's larger than this (and much smaller too) an equal-tail confidence interval with confidence coefficient approximately $1 - \alpha$ is given by

$$\left(\bar{x} - z_{1-\alpha/2} \sqrt{\frac{\bar{x}(1 - \bar{x})}{n}}, \; \bar{x} + z_{1-\alpha/2} \sqrt{\frac{\bar{x}(1 - \bar{x})}{n}} \right)$$

where $z_{1-\alpha/2}$ is defined by Eq. (6-5). We commented in Example 7-5 that for $n = 100$, $y = 80$ graphs gave the equal-tail confidence interval with coefficient .95 as (.72,.88). Find the interval given by the approximation. Observe that since $1 - \alpha = .95$, $\alpha = .05$, $\alpha/2 = .025$.

7-12　A machine makes bolts, and a fraction p of them require rework. A random sample of 2,500 bolts from those produced by the machine includes 277 bolts needing rework. If the random variable X has probability function (7-4), where $X = 0$ if no rework is needed, $X = 1$ if rework is required, use the result given in Exercise 7-11 to find an equal-tail confidence interval for p with confidence coefficient approximately equal to .95.

7-13　We previously commented that equal-tail confidence intervals for Poisson μ can be computed from chi-square tables. This is possible even with very large values of y. For large values of y an equal-tail confidence interval with confidence coefficient approximately equal to $1 - \alpha$ is given by

$$\left(\bar{x} - z_{1-\alpha/2} \sqrt{\frac{\bar{x}}{n}}, \; \bar{x} + z_{1-\alpha/2} \sqrt{\frac{\bar{x}}{n}} \right)$$

Suppose that the individual in Exercise 7-8 uses a random sample of 200 days and records a total of 324 telephone calls. Find the interval given by the approximation using a confidence coefficient of .95.

7-14　A random sample of 225 typed pages is selected from the work turned out by a secretary. A total of 256 errors is found on the 225 pages. Using the result given in Exercise 7-13, find an equal-tail confidence interval with confidence coefficient approximately .95 for the average number of errors per page.

Let us again return to the potato problem discussed in Example 4-5 and Sec. 7-1. We have already decided that the probability function of X, the number of good potatoes in a random sample of size n, is

$$f(x;p) = \binom{n}{x} p^x (1-p)^{n-x} \qquad x = 0, 1, 2, \ldots, n$$

We are undecided about p but wish to investigate the claim that $p = .90$. Thus $p = .90$ is the hypothesis. As we shall use the term, a *hypothesis* is a statement about a parameter (or parameters) of a probability function. The other parameter n has yet to be chosen. Suppose we decide to look at a random sample of $n = 25$ potatoes and base our decision about p on x. Now if the hypothesis is true, the probability function is completely determined, being

$$f(x;.90) = \binom{25}{x} (.90)^x (.10)^{25-x} \qquad x = 0, 1, 2, \ldots, 25$$

The next issue to be decided is the selection of the values of x that cast doubt upon the hypothesis. Hardly anyone would believe $p = .90$ if 0, 1, or 2 good potatoes are in the sample. Not many people would be willing to believe the hypothesis if $x = 10$. Nearly everyone would feel that $p = .90$ is not unreasonable if $x = 23$. We must make a statistical judgment as to what we are going to regard as unreasonable results. To be specific, suppose that we decide that we shall reject $p = .90$ if X takes on a value of 19 or less. In other words our test is: Reject $p = .90$ if $x \leq 19$. Let us examine some of the consequences of our test rule.

First, if $p = .90$

$$\Pr(X \leq 19) = \sum_{x=0}^{19} b(x;25,.90) = \sum_{x=6}^{25} b(x;25,.10) = 1 - \sum_{x=0}^{5} b(x;25,.10)$$

$$= 1 - .96660 \qquad \textit{from Appendix B1}$$
$$= .03340$$

Thus, the probability of rejecting the hypothesis when it is true is .03340. In other words, about 3 out of every 100 samples would lead to an incorrect decision. A hypothesis would be rejected when it is true. This kind of mistake is known as a *Type I error*. The probability of committing a Type I error is called the *level of significance* and is denoted by the Greek letter α. Here $\alpha = .03340$. The set of outcomes for the experiment that lead to rejection of the hypothesis ($x = 0, 1, 2, \ldots , 19$) is called the *critical region*.

Next suppose that the true value of $p = .80$, not $.90$. Then

$$\Pr(X \leq 19) = \sum_{x=0}^{19} b(x;25,.80) = \sum_{x=6}^{25} b(x;25,.20) = 1 - \sum_{x=0}^{5} b(x;25,.20)$$

$$= 1 - .61669$$
$$= .38331$$

Of course, since the hypothesis is now false, we would like to reject with a very high probability. We would much prefer to have $\Pr(X \leq 19)$ near 1 instead of equal to $.38331$. The probability that we accept the hypothesis $p = .90$ using the test rule is $.61669$. By accepting a false hypothesis, we commit another kind of mistake called a *Type II error*. The probability of making a Type II error is designated by β (here $\beta = .61669$ if $p = .80$).

The various possibilities regarding the hypothesis and the decision are summarized in Table 7-5.

If the hypothesis $p = .90$ is rejected by obtaining an x less than or equal to 19, then we conclude $p < .90$. The statement $p < .90$ is referred to as the *alternative hypothesis*. We shall designate the hypothesis by H_0 and the alternative hypothesis by H_1. Thus, in our example we choose between H_0: $p = .90$ and H_1: $p < .90$, the choice being made on the basis of the observed value of a binomial random variable. In all hypothesis-testing problems that we shall consider, a random variable is observed in order to make a choice between a hypothesis H_0 and an alternative hypothesis H_1, both of which are suggested by the particular problem at hand. As we shall see, it is important to formulate the alternative hypothesis at the time the hypothesis is selected and before the statistic used to reach a decision is observed.

Another important concept is *power*. The power of the test is defined to be the probability of rejecting the hypothesis. In our example the power is $.03340$ if $p = .90$ and $.38331$ if $p = .80$. We can compute

Table 7-5 Hypothesis versus decision

	Decision	
	Accept	Reject
Hypothesis true	Correct decision	Type I error
Hypothesis false	Type II error	Correct decision

$\Pr(X \leq 19)$ for any p given in the binomial table. Appendix B1 yields

$$
\begin{aligned}
\text{Power} &= B(19;25,.75) = 1 - B(5;25,.25) = 1 - .37828 = .62172 \\
&\hspace{7.5cm} p = .75 \\
&= B(19;25,.70) = 1 - B(5;25,.30) = 1 - .19349 = .80651 \\
&\hspace{7.5cm} p = .70 \\
&= B(19;25,.60) = 1 - B(5;25,.40) = 1 - .02936 = .97064 \\
&\hspace{7.5cm} p = .60 \\
&= B(19;25,.50) = .99796 \qquad p = .50
\end{aligned}
$$

Of course, power could be obtained for many more values of p by consulting more extensive tables. Plotting power against p yields the graph of Fig. 7-1. An ideal power curve would be of height 1 for all values of the parameter specified by H_1 and of height 0 for values specified by H_0. That is, if the hypothesis is true we would always like to accept, and if the hypothesis is false we would always like to reject. Unfortunately, it is usually impossible to achieve this perfect state of affairs. Much of the literature of mathematical statistics is concerned with finding tests for a given H_0 with the largest power when H_1 is true.

In constructing a test for the potato example it appears that we arbitrarily selected the critical region to be $x = 0, 1, 2, \ldots, 19$. Usually the statistician selects α instead and finds the corresponding critical region. Common choices for α are .05 and .01, although consideration of the consequences of a Type I error may lead to some other choice. Suppose we had selected $\alpha = .05$ for the potato problem. The binomial table gives (with $p = .90$) $\Pr(X \leq 19) = .03340$ and $\Pr(X \leq 20) = .09799$, showing that this was an impractical choice for α. If we want α to be

*Figure 7-1 Power curve for test of H_0: $p = .90$
against H_1: $p < .90$ with $n = 25$.*

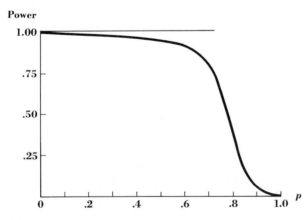

234

.05, or something close to it, we can either

1 reject when $x = 0, 1, 2, \ldots, 19$ with $\alpha = .03340$
2 reject when $x = 0, 1, 2, \ldots, 20$ with $\alpha = .09799$
3 reject when $x = 0, 1, 2, \ldots, 19$ and reject with probability $(.05 - .03340)/(.09799 - .03340) = .26$ when $x = 20$, in which case $\alpha = .05$. To make a decision when $x = 20$ we could use a balanced spinner or a table of random numbers.

In practice one usually follows procedure 1. That is, if we aim at a significance level α_0 (here $\alpha_0 = .05$) we select a sensible critical region such that $\alpha \leqq \alpha_0$ but with α as close to α_0 as possible.

It should be pointed out that an observation of the test statistic that does not fall in the critical region does not prove that the hypothesis is true. For example $x = 23$ may be a reasonable value of X for $p = .85$, $p = .87$, or for many values of p other than $p = .90$. Consequently, the rejection of a hypothesis is a more satisfying statistical decision than acceptance.

EXERCISES

7-15 Suppose we had selected as a critical region $x = 0, 1, 2, \ldots, 18$ for the potato example. Verify that α has been reduced to .00948. With the new critical region, what is the power of the test if $p = .80, .75, .70, .60, .50$? What would you say is the effect on β when α is decreased?

7-16 Suppose that we had selected a sample size of $n = 100$ for the potato example. Verify that the critical region $x = 0, 1, 2, \ldots, 84$ has associated with it an $\alpha = .03989$. If $p = .90$ is rejected when $x = 0, 1, 2, \ldots, 84$, find the power of the test when $p = .80, .75, .70, .60, .50$. Compare with the values obtained when $n = 25$. What does increasing n appear to do to the power curve?

7-5   A TWO-SIDED ALTERNATIVE

In the potato example only small values of the statistic led to rejection of the hypothesis. The alternative to the hypothesis, H_1: $p < .90$, is called a one-sided alternative. A number of problems require two-sided alternatives with both small and large values of the statistic leading to rejection. Let us consider an example.

According to a genetic theory, 25 percent of a species have a certain characteristic. If p is the fraction of the species that has the characteristic, theory suggests H_0: $p = .25$. If the theory is not true, then we will believe either $p < .25$ or $p > .25$. The alternative can be written more compactly as H_1: $p \neq .25$ and is said to be two-sided. In the potato problem we did not have to worry about the case $p > .90$ since our main concern was directed toward guarding against poor quality, and if $p > .90$, so much the better. With the genetic problem, sample results indicating either $p < .25$ or $p > .25$ will contradict the theory $p = .25$.

Suppose that, in order to test the theory, a random sample of 100 seeds is observed and 36 have the characteristic. We can justify the use of the binomial model by the same argument we used in the potato example (Example 4-5). Hence, if the theory is true, X, the number of seeds in a random sample of 100, has the probability function

$$f(x) = \binom{100}{x} (.25)^x (.75)^{100-x} \qquad x = 0, 1, 2, \ldots, 100$$

According to the theory $E(X) = np = 100(.25) = 25$. If the observed value of X, in this case $x = 36$, is sufficiently far away from 25, we shall not believe the theory.

In order to make a decision, one way to proceed would be to compute the probability that X deviates from 25 by as much as or more than the observed deviation $36 - 25 = 11$. Thus, we would find

$$\Pr(X \leq 14) + \Pr(X \geq 36)$$

assuming X has a binomial distribution with $n = 100$, $p = .25$. From Appendix B1 the sum of these probabilities is $.00542 + .00041 = .00583$. The most obvious conclusion is that (1) $p = .25$ and an unusually large deviation has occurred or (2) $p \neq .25$ but is larger than .25. In view of the small probability $(.00583)$, (2) is the more reasonable.

Regarding the example as a hypothesis-testing problem with H_0: $p = .25$ and H_1: $p \neq .25$, we would arbitrarily select an aimed-at level of significance, say, $\alpha_0 = .05$. Then we would attempt to find a sensible critical region with actual level of significance close to .05. Intuitively, both large and small values of x should be included in the region. Since a large x contradicts the theory equally as well as a small one, it seems reasonable to associate approximately half of the level of significance with small x's and half with large x's. A practical way to proceed is to find x_1 and x_2 so that $\Pr(X \leq x_1)$ and $\Pr(X \geq x_2)$ are each as close to

$$\alpha_0/2 = .025$$

as possible and use $x \leq x_1$ and $x \geq x_2$ as the critical region. To be consistent we shall always take x_1 to be the maximum x such that

$$\Pr(X \leq x) \leq \alpha_0/2$$

and x_2 to be the minimum x such that $\Pr(X \geq x) \leq \alpha_0/2$. With $n = 100$, $p = .25$ it is easy to verify that

$$\Pr(X \leq 16) \leq .02111 \qquad \Pr(X \leq 17) = .03763$$
$$\Pr(X \geq 35) = .01643 \qquad \Pr(X \geq 34) = .02759$$

Hence, with $\alpha_0 = .05$, we reject H_0: $p = .25$ if $x \leq 16$ or $x \geq 35$ and the significance level is $.02111 + .01643 = .03754$. Since the actual observed x was 36, H_0 is rejected.

The calculation of power for the genetic example is easy with binomial tables. We need

$$\text{Power} = \Pr(X \leq 16) + \Pr(X \geq 35) \tag{7-16}$$

evaluated for various values of p. We already know that if $p = .25$, Power $= .03754$. From Appendix B1 we get with

$p = .10$ Power $= .97940 + .00000 = .97940$
$p = .20$ Power $= .19234 + .00034 = .19268$
$p = .30$ Power $= .00097 + .16286 = .16383$
$p = .40$ Power $= .00000 + .86966 = .86966$
$p = .50$ Power $= .00000 + .99911 = .99911$

Of course, power could be obtained for many more values of p by consulting more extensive tables. Plotting the points on a graph yields Fig. 7-2.

Let us suppose that we observe a random sample of 100 seeds before we formulate our alternative and count 34 seeds having the characteristic. With $\alpha_0 = .05$, we would accept H_0: $p = .25$ when tested against H_1: $p \neq .25$ since $x = 34$ does not fall in the critical region previously found. However, suppose we test H_0: $p = .25$ against the alternative suggested by the sample, that is, H_1: $p > .25$, with the same α_0. Now,

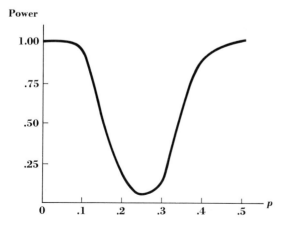

Figure 7-2 Power curve for test of H_0: $p = .25$ against H_1: $p \neq .25$ with $n = 100$.

237

only large values of x support H_1 and lead to rejection. The binomial table yields (with $p = .25$)

$$\Pr(X \geq 33) = .04460 \qquad \Pr(X \geq 32) = .06935$$

If we follow the convention of making α as close to α_0 as possible, yet less than α_0, then the critical region is $x \geq 33$ with $\alpha = .04460$. Now x falls in the critical region and the sample "proves" that $p > .25$. Thus if an alternative suggested by the sample is chosen, an unwarranted conclusion is apt to be drawn. In general, it is not an acceptable procedure to use the sample results to formulate a hypothesis (or an alternative) and then use the same sample for the test.

EXERCISES

7-17 Suppose we had chosen $\alpha_0 = .01$ in the genetic example. Verify that the critical region is $x \leq 13$ and $x \geq 38$ and $\alpha = .00521$. With this critical region find the power of the test if $p = .10, .20, .30, .40$. What does decreasing α do to the power?

7-18 Suppose we had selected a sample of size $n = 50$ seeds for the genetic example and had used $\alpha_0 = .05$. Verify that the critical region is $x \leq 6$ and $x \geq 20$ and $\alpha = .03331$. Find the power of this test for $p = .10, .20, .30, .40$. What does decreasing n appear to do to the power?

7-6 TESTING HYPOTHESES ABOUT THE BINOMIAL PARAMETER p

In the previous two sections we have used specific examples to introduce the ideas and language of hypothesis testing. Now we shall make a more systematic study of a few of the standard problems. Although the choice of statistics and critical regions will be based on intuitive appeal, they are consistent with results obtained by using sound mathematical principles (and discussed in more advanced courses). In particular, most of the tests have the property that their power curves are in general better than power curves associated with other tests of the same hypothesis.

First we shall consider a random variable X that has probability function

$$f(x;p) = p^x(1 - p)^{1-x} \qquad x = 0, 1 \tag{7-17}$$
$$0 < p < 1$$

where p is unknown. A specific problem may suggest that p takes on a particular value, say, p_0. We then hypothesize that $p = p_0$ and, on the basis of a random sample from (7-17), try to decide whether the hypothesis is reasonable or unreasonable. We have already encountered the standard hypothesis-testing situations. With the potato example we illustrated testing

$$H_0: \quad p = p_0 \qquad \text{against} \qquad H_1: \quad p < p_0 \qquad\qquad (7\text{-}18)$$

with $p_0 = .90$. The incorrect analysis in the genetics problem led to testing

$$H_0: \quad p = p_0 \qquad \text{against} \qquad H_1: \quad p > p_0 \qquad\qquad (7\text{-}19)$$

with $p_0 = .25$. Finally, the correct analysis with the genetics problem resulted in testing

$$H_0: \quad p = p_0 \qquad \text{against} \qquad H_1: \quad p \neq p_0 \qquad\qquad (7\text{-}20)$$

again with $p_0 = .25$. Although other hypotheses and alternatives can be formulated, these are the ones that arise most frequently.

The statistic used to test the hypothesis about p is

$$Y = X_1 + X_2 + \cdots + X_n$$

where n is the sample size (which is chosen arbitrarily or to meet some power requirement). We already know from Example 5-13 that Y, the number of successes in n trials, has a binomial distribution with parameters n and p. Further, if the H_0 of (7-18), (7-19), or (7-20) is true, then Y is binomial with parameters n and p_0.

When (7-18) is tested, small values of y support H_1 and lead to rejection of H_0. As we have previously observed, a desired significance level α_0 arbitrarily chosen in advance is unlikely to be achieved. Consequently the critical region usually chosen for (7-18) is $y \leq y_1$ where y_1 is the largest y such that $\Pr(Y \leq y) \leq \alpha_0$. The actual significance level is $\Pr(Y \leq y_1) = \alpha$. When we attempted to achieve $\alpha_0 = .05$ with $n = 25$, the critical region for testing $H_0: \ p = .90$ against $H_1: \ p < .90$ turned out to be $y \leq 19$ with $\alpha = .03340$. Similarly, when we test (7-19) only large values of y support H_1. Consequently, H_0 is rejected when $y \geq y_1$ where y_1 is the smallest y such that $\Pr(Y \geq y) \leq \alpha_0$ and the significance level is $\Pr(Y \geq y_1) = \alpha$. For the two-sided situation of (7-20) both small and large y contradict H_0 and lead to rejection. If we aim at a significance level α_0, the critical region usually selected is $y \leq y_1$ and $y \geq y_2$, where y_1 is the largest value of y such that $\Pr(Y \leq y) \leq \alpha_0/2$ and y_2 is the smallest y such that $\Pr(Y \geq y) \leq \alpha_0/2$. The true significance level is

$$\alpha = \Pr(Y \leq y_1) + \Pr(Y \geq y_2)$$

Of course, all calculations for determining the critical regions and the significance level are performed with $p = p_0$.

EXAMPLE 7-6

A coin is tossed 100 times and 43 heads are recorded. Does this result indicate that the coin is biased? Use $\alpha_0 = .05$.

Solution

The hypothesis and alternative that the problem suggests before any results are observed is H_0: $p = \frac{1}{2}$ against H_1: $p \neq \frac{1}{2}$. It is reasonable to assume that the observations are a random sample from a distribution with probability function (7-17) where $X = 0$ if a tail is observed and $X = 1$ if a head is observed. We would expect p to remain the same and the independence condition to be satisfied if the condition of this coin does not change and each toss is performed in the same way. Thus Y, the number of heads obtained in 100 tosses, has a binomial distribution with $n = 100$ and $p = \frac{1}{2}$ if H_0 is true. From Appendix B1 we find

$$\Pr(Y \leq 39) = .01760 \qquad \Pr(Y \leq 40) = .02844$$
$$\Pr(Y \geq 61) = .01760 \qquad \Pr(Y \geq 60) = .02844$$

Thus the critical region is $y \leq 39$, $y \geq 61$ and

$$\alpha = .01760 + .01760 = .03520$$

Since the sample produced $y = 43$, H_0 is not rejected.

EXAMPLE 7-7

Suppose we are considering for purchase a new machine that makes bolts. We shall buy the machine if the fraction of bolts requiring rework is .10 (or less). A random sample of 25 bolts produced by the machine is examined, and 4 require rework. With $\alpha_0 = .05$ what decision should we make?

Solution

The hypothesis and alternative the problem suggests is H_0: $p = .10$ against H_1: $p > .10$. We shall buy the machine unless we conclude that H_1 is true. Let $X = 0$ if a bolt does not require rework, $X = 1$ if it does. Unless the machine shows some sign of wear or a setting changes, it is not unreasonable to assume that for each bolt X has the distribution with probability function (7-17), that is, p remains constant. To make the independence part of the random sample conditions seem reasonable, perhaps one should avoid including in the sample bolts produced in succession. If the machine is inclined to produce several poor bolts in a row, then the independence condition is more realistic if every fifth (or perhaps every tenth) bolt is taken for the sample. Thus, it is not unreasonable to assume that Y, the number of bolts requiring rework, has a binomial distribution with $n = 25$ and, if H_0 is true, $p = .10$. From Appendix B1 we find

$$\Pr(Y \geq 6) = \sum_{y=6}^{25} b(y;25,.10) = .03340$$

$$\Pr(Y \geq 5) = \sum_{y=5}^{25} b(y;25,.10) = .09796$$

Hence the critical region is $y \geq 6$ and $\alpha = .03340$. Since the sample contained $y = 4$ bolts requiring rework, H_0 is accepted and the machine will be purchased.

As we have previously demonstrated, power calculations are routine with binomial tables.

EXAMPLE 7-8

Find the power of test used in Example 7-7 if $p = .20$. If $p = .30$.

Solution

We need $\Pr(Y \geq 6)$ if $p = .20$.

$$\Pr(Y \geq 6) = \sum_{y=6}^{25} b(y;25,.20)$$

$$= 1 - B(5;25,.20)$$
$$= 1 - .61669 = .38331$$

Similarly, if $p = .30$

$$\Pr(Y \geq 6) = 1 - B(5;25,.30)$$
$$= 1 - .19349 = .80651$$

EXAMPLE 7-9

Suppose that $n = 100$ bolts is used in Example 7-7. Find the power when $p = .20$. When $p = .30$.

Solution

If $p = .10$, $\Pr(Y \geq 16) = .03989$, $\Pr(Y \geq 15) = .07257$. Hence the critical region is $y \geq 16$ with $\alpha = .03989$. If $p = .20$,

$$\Pr(Y \geq 16) = 1 - B(15;100,.20)$$
$$= 1 - .12851 = .87149$$

If $p = .30$,

$$\Pr(Y \geq 16) = 1 - B(15;100,.30)$$
$$= 1 - .00040 = .99960$$

EXAMPLE 7-10

How large does n have to be in Example 7-7 to raise the power to .95 or greater if $p = .20$?

Solution

Since $n = 100$ does not achieve the desired power, our tables in Appendix B1 are inadequate. Even with more extensive tables some trial and error is involved. For $n = 130$, $p = .10$, the Harvard University table [5] gives $\Pr(Y \geq 19) = .05957$, $\Pr(Y \geq 20) = .03443$. Hence the critical region is $y \geq 20$. If $n = 130$, $p = .20$, $\Pr(Y \geq 20) = .92676$.

Next try $n = 140$. With $p = .10$ we find $\Pr(Y \geq 20) = .06590$, $\Pr(Y \geq 21) = .03919$. Thus the critical region is $y \geq 21$. If $p = .20$, $\Pr(Y \geq 21) = .94766$.

With $n = 150$, $p = .10$ we get $\Pr(Y \geq 21) = .07209$, $\Pr(Y \geq 22) = .04396$ so that the critical region is $y \geq 22$. If $p = .20$, $\Pr(Y \geq 22) = .96278$.

Thus it appears that n is cornered between 140 and 150. If the Ordnance Corps table [10] were used, the exact minimum n could be found. However, for practical purposes, any n between 140 and 150 could be used. Certainly $n = 150$ is on the safe side. (Using the procedure illustrated following Example 7-11, it can be shown that the exact minimum n is 135 and the critical region is $y \geq 20$.)

EXAMPLE 7-11

Find the power of test used in Example 7-6 if $p = .40$.

Solution

We need $\Pr(Y \leq 39) + \Pr(Y \geq 61)$ computed with $p = .40$, $n = 100$, and we get

$$\Pr(Y \leq 39) = \sum_{y=0}^{39} b(y;100,.40) = .46208$$

$$\Pr(Y \geq 61) = \sum_{y=61}^{100} b(y;100,.40) = 1 - .99998 = .00002$$

Thus, the required power is $.46208 + .00002 = .46210$.

Example 7-10 illustrates a standard type of statistical problem. We wish to test H_0: $p = p_0$ against H_1: $p > p_0$ and choose n so that the power curve satisfies two requirements. First, if H_0 is true we do not want to reject more than a fraction α_0 of the time. Hence, one equation which y_1 and n must satisfy is

$$\sum_{y=y_1}^{n} b(y;n,p_0) \leqq \alpha_0 \tag{7-21}$$

On the other hand, if p is actually $p = p_1 > p_0$, we would like to reject H_0 at least a fraction $1 - \beta$ of the time. Thus y_1 and n must also satisfy

$$\sum_{y=y_1}^{n} b(y;n,p_1) \geqq 1 - \beta \tag{7-22}$$

In other words, if $p = p_0$ we desire that the power will be no greater than α_0 but if $p = p_1$ we insist that the power will be no less than $1 - \beta$. In Example 7-10 $p_0 = .10$, $p_1 = .20$, $\alpha_0 = .05$, $1 - \beta = .95$. Example 7-10 also demonstrates that (7-21) and (7-22) will be satisfied once n exceeds some minimum value. In applications, it would be good procedure to find the exact minimum n (if available tables make this possible) and use that sample size. Our sample table of the binomial is not well adapted to finding the minimum n but either of the tables mentioned in Example 7-10 works quite well for this problem.

The solution of (7-21) and (7-22) for y_1 and minimum n is often required in statistical quality control. Let us illustrate by an example. Suppose that a production process (such as a machine producing bolts) mass produces a certain item. Usually, not every item will satisfy prescribed specifications, and those which fail to do so are referred to as defectives. For a number of such processes it is not unreasonable to assume that as long as the process is operating satisfactorily (i.e., a machine does not wear out), a fraction p of the items will be defective and the number of defectives in a sample of size n has a binomial distribution with parameters n and p. A potential customer for the product may be well satisfied and will buy a very large order if $p = p_0$ (say, .01). On the other hand, if $p = p_1$ (say, .12) the number of defectives may be so large as to seriously interfere with profit-making potential. In order to make a decision as to whether or not he should sign a contract with the producer of the item, the customer looks at a random sample of size n. If $p = p_0$, he would like to be reasonably sure that he does not reject the contract and decides that the probability of rejection should be less than or equal to α_0 (say, .05). On the other hand, if $p = p_1$, he would like to reject with fairly high probability, at least $1 - \beta$ (say, .90). Hence y_1 and n satisfying (7-21) and (7-22) are required. To obtain the minimum n we need more extensive binomial tables and a standard procedure would be helpful.

Suppose we start with $y_1 = 1$, the smallest possible value of y_1. The Harvard table [5] yields

$$\sum_{y=1}^{5} b(y;5,.01) = .04901 \qquad \sum_{y=1}^{6} b(y;6,.01) = .05852$$

so that $n = 5$ is the largest n which permits (7-21) to be satisfied. However, since

$$\sum_{y=1}^{5} b(y;5,.12) = .47227$$

Eq. (7-22) cannot be satisfied with $y_1 = 1$. Next try $y_1 = 2$. We find

$$\sum_{y=2}^{35} b(y;35,.01) = .04786 \qquad \sum_{y=2}^{36} b(y;36,.01) = .05035$$

so that $n = 35$ is the largest n that permits (7-21) to be satisfied. Now

$$\sum_{y=2}^{35} b(y;35,.12) = .93419$$

so that (7-22) is also satisfied. In fact, some smaller n's will also satisfy both inequalities. Backing down the table with $p = .12$ we find

$$\sum_{y=2}^{30} b(y;30,.12) = .89003 \qquad \sum_{y=2}^{31} b(y;31,.12) = .90063$$

Hence the minimum n to satisfy both inequalities is $n = 31$ and the accompanying $y_1 = 2$. Thus, a random sample of size 31 would be observed and the contract rejected if $y = 2$ or more defectives are found. Had a solution for (7-21) and (7-22) been impossible with $y_1 = 2$, we would have tried $y_1 = 3$, then $y_1 = 4$, etc. This procedure of increasing y_1 until a solution is found prevents bypassing the minimum n as was done in Example 7-10.

EXAMPLE 7-12

In the quality-control example just discussed, suppose $p_0 = .01$, $p_1 = .10$, $\alpha_0 = .05$, $\beta = .20$. Use Table 7-6 to find y_1 and the minimum n required to meet conditions (7-21) and (7-22).

Solution

The inequalities cannot be satisfied with $y_1 = 1$. With $y_1 = 2$, $n = 29$ is the minimum n to satisfy both inequalities.

Table 7-6   Abbreviated binomial table for Example 7-12

| | $\Pr(Y \geq 1)$ | | | $\Pr(Y \geq 2)$ | |
	$p = .01$	$p = .10$		$p = .01$	$p = .10$
$n = 3$	.02970	.27100	$n = 28$	.03182	.78485
4	.03940	.34390	29	.03396	.80113
5	.04901	.40951	30	.03615	.81630
6	.05852	.46856	35	.04786	.87762
7	.06793	.52170	36	.05035	.88736

We have just seen how (7-21) and (7-22) can be used to obtain the critical region and the minimum sample size for testing H_0: $p = p_0$ against H_1: $p > p_0$. The power of the test is such that

Power $\leq \alpha_0$ $p = p_0$

Power $\geq 1 - \beta$ $p = p_1 > p_0$

If observations are expensive we can alter the procedure to cut down on the sample size needed to meet the given power requirements. Suppose that $n = 30$, $y_1 = 20$, so that we reject if 20 or more successes are obtained from a sample of 30. If the items are observed one at a time, it is obvious that we could quit sampling and reject H_0 if the 20th success occurs before the 30th observation. Likewise, once we have observed 11 failures we could quit and accept H_0 since it is then impossible to get more than 19 successes, a result that does not fall in the critical region. Hence, it is possible that a decision will be reached with less than the 30 observations which the test seems to require. In general, if the above procedure is adopted, we:

(a) Determine n and y_1 such that (7-21) and (7-22) are satis-
     fied and n is a minimum.
(b) Stop sampling and   (7-23)
       (i) reject H_0 as soon as y_1 successes are observed
       (ii) accept H_0 as soon as $n - y_1 + 1$ failures are
       observed (whichever occurs first).

The number of observations required to reach a decision, say, X, is now a random variable, and it can be shown that

$$E(X) = \frac{y_1}{p} [1 - B(y_1; n + 1, p)] + \frac{n - y_1 + 1}{1 - p} [B(y_1 - 1; n + 1, p)]$$
$$(7-24)$$

where $B(r;n,p)$ is the binomial sum (4-3). Procedure (7-23) can be used to test H_0: $p = p_0$ against H_1: $p < p_0$. Observe that the latter hypothesis and alternative can be written as H_0: $q = q_0$ against H_1: $q > q_0$, where $q = 1 - p$. Then proceed as before using q in place of p.

EXAMPLE 7-13

Describe procedure (7-23) for Example 7-12. Find the expected number of observations required to reach a decision if H_0 is true. If H_1 is true.

Solution

We found $n = 29$, $y_1 = 2$ so that $n - y_1 + 1 = 28$. Hence we would:

(i) stop sampling and reject as soon as 2 defectives are obtained
(ii) stop sampling and accept as soon as 28 nondefectives are obtained.

According to (7-24) the expected number of observations required to reach a decision if H_0 is true is

$$E(X) = \frac{2}{.01}[1 - B(2;30,.01)] + \frac{28}{.99}[B(1;30,.01)]$$

With the aid of the Harvard University table [5] we find that

$$1 - B(2;30,.01) = .00332 \qquad B(1;30,.01) = .96385$$

Thus, if H_0 is true,

$$E(X) = \frac{2}{.01}(.00332) + \frac{28}{.99}(.96385)$$

$$= .664 + 27.260 = 27.924$$

If H_1 is true,

$$E(X) = \frac{2}{.10}[1 - B(2;30,.10)] + \frac{28}{.90}[B(1;30,.10)]$$

The Harvard table yields

$$1 - B(2;30,.10) = .58865 \qquad B(1;30,.10) = .18370$$

and

$$E(X) = \frac{2}{.10} (.58865) + \frac{28}{.90} (.18370)$$

$$= 11.773 + 5.715 = 17.448$$

Patil [9] has observed that a good procedure for testing H_0: $p = p_0$ against H_1: $p > p_0$ based upon the negative binomial distribution is identical with (7-23). Hence, we do not have to consider the negative binomial in hopes of deriving a better test than we already have.

EXERCISES

7-19 A seed company claims that 90 percent of its radish seeds germinate. Fifty seeds are planted and eight fail to germinate. Test a hypothesis formulated from the company's claim. Use $\alpha_0 = .05$.

7-20 Find the power of the test used in Exercise 7-19 if $p = .75$. If $p = .60$.

7-21 According to a genetic theory, 25 percent of a species have a certain characteristic. A random sample of 20 of the species contains 9 having the characteristic. If $\alpha_0 = .05$, is the theory contradicted?

7-22 Find the power of test used in Exercise 7-21 if 10 percent of the species have the characteristic. If 40 percent have the characteristic.

7-23 Suppose we wish to test H_0: $p = .20$ against H_1: $p > .20$ for a binomial random variable using $\alpha_0 = .05$. If, further, we wish the power to be at least .90 if $p = .50$, what is the smallest n of those listed in Appendix B1 that meets the requirement? What is the critical region?

7-24 Suppose we wish to test H_0: $p = .50$ against H_1: $p < .50$ for a binomial random variable using $\alpha_0 = .01$. If, further, we wish the power to be at least .95 if $p = .10$, what is the smallest n of those listed in Appendix B1 that meets the requirement? What is the critical region?

7-25 The standard cure for tuberculosis is successful 30 percent of the time. A new cure is tried on a group of 50 patients and is successful in 29 cases. The new cure is more expensive and we do not want to use it unless we can prove at the .01 level of significance that it is better. Recommend the appropriate action.

7-26 A fruit wholesaler inspects a random sample of one hundred grapefruit from each carload lot he received. He would like to reject the carload if 10 percent are bad but he is afraid to reject wrongly more than 1 percent of the time. On the other hand, he cannot afford to accept lots with 25 percent (or more) bad more than 10 percent of the time. Can he meet both requirements with the sample he has?

7-27 An entrance examination to a state university is considered satisfactory if 10 percent fail. A new examination is to be tried on a random sample of 50 students. If the failure rate is .10, we would like to have the probability of rejecting the examination be .05 or less. In addition, if the failure rate is .30 we would like to reject the examination with probability at least .95. Is a sample of size 50 sufficiently large? If 9 out of 50 fail the examination, what decision would be made?

7-28 We wish to test H_0: $p = .10$ against H_1: $p > .10$ with $\alpha_0 = .05$ and a power of at least .90 if $p = .30$. Use the Harvard table (or some other more extensive binomial table) to find y_1 and the minimum n so that (7-21) and (7-22) are satisfied. Describe procedure (7-23) for this problem and find the expected number of observations when $p = .10$. When $p = .20$. When $p = .30$.

7-29 Let X be the number of observations required to reach a decision using procedure (7-23). Assume $y_1 < n - y_1 + 1$, or $y_1 < (n + 1)/2$ (as in Example 7-13 and Exercise 7-28). Then, if $y_1 \leq x \leq n - y_1$ the probability that a decision is reached on the xth observation is the probability that y_1 successes are obtained in x observations, that is, the negative binomial probability $b^*(x;y_1,p)$. If $n - y_1 + 1 \leq x \leq n$, then again a decision is reached on the xth observation if the y_1th success is obtained on the xth observation. However, for x in the latter range, a decision is also reached if the $(n - y_1 + 1)$st failure is observed on the xth observation, the probability of so doing being $b^*(x; n - y_1 + 1, 1 - p)$. Hence, the probability function of X is

$$
\begin{aligned}
f(x) &= b^*(x;y_1,p) & y_1 \leq x \leq n - y_1 \\
&= b^*(x;y_1,p) + b^*(x;n - y_1 + 1, 1 - p) & n - y_1 + 1 \leq x \leq n
\end{aligned}
$$

(To make it easier to follow the above derivation, it may help to rewrite the discussion using the specific y_1 and n found in Exercise 7-28.)

(a) Show by making use of the relationship between negative binomial and binomial sums that the sum of the probabilities for all values of X is 1.

(b) Express the probability of rejecting H_0: $p = p_0$ in terms of a sum on X and show that this is equal to the power of the test.

(c) Show that $E(X)$ is as given by formula (7-24). Hint: Follow the steps used to obtain the mean of a negative binomial random variable. This yields negative binomial sums that can be converted to binomial sums. [If $n - y_1 + 1 \leq y_1$, or $y_1 \geq (n + 1)/2$, only minor modifications are needed to derive the probability function of x.]

7-7 TESTING HYPOTHESES ABOUT THE POISSON PARAMETER μ

The hypothesis-testing situations we shall discuss are similar to those encountered with the binomial parameter p. If an individual decides that he will have his telephone taken out if his average number of calls is less than two per day, he will be interested in testing

$$H_0: \quad \mu = \mu_0 \qquad \text{against} \qquad H_1: \quad \mu < \mu_0 \tag{7-25}$$

(with $\mu_0 = 2$) and he will remove the telephone if H_0 is rejected. If an executive is willing to hire a secretary unless the secretary averages more than one error per typed page, the executive will probably test

$$H_0: \quad \mu = \mu_0 \qquad \text{against} \qquad H_1: \quad \mu > \mu_0 \tag{7-26}$$

(with $\mu_0 = 1$) and hire the secretary unless H_0 is rejected. If a given small volume of blood contains on the average five red cells in a healthy individual, a laboratory technician may wish to test

$$H_0: \quad \mu = \mu_0 \qquad \text{against} \qquad H_1: \quad \mu \neq \mu_0 \tag{7-27}$$

(with $\mu_0 = 5$) if either too many or too few red cells indicate trouble.
Suppose that a random variable X has the probability function

$$f(x;\mu) = \frac{e^{-\mu}\mu^x}{x!} \qquad x = 0, 1, 2, \ldots \tag{7-28}$$

where μ is unknown. Then, if $X_1, X_2, \ldots, X_n$ is a random sample of size n, we know that $Y = X_1 + X_2 + \cdots + X_n$ has a Poisson distribution with parameter $n\mu$ (see the derivation associated with formula (5-41)). That is,

$$g(y) = \frac{e^{-n\mu}(n\mu)^y}{y!} \qquad y = 0, 1, 2, \ldots \tag{7-29}$$

Further, if $\mu = \mu_0$, then Y has the probability function given by (7-29) with parameter $n\mu_0$.
When (7-25) is tested, small values of y support H_1 and lead to rejection of H_0. The critical region usually selected is $y \leq y_1$ where y_1 is the

largest y such that $\Pr(Y \leq y) \leq \alpha_0$ and α_0 is the prechosen aimed-at significance level. The actual significance level is $\Pr(Y \leq y_1) = \alpha$. Similarly, when we test (7-26) only large values of y support H_1. Consequently, H_1 is rejected when $y \geq y_1$ where y_1 is the smallest y such that $\Pr(Y \geq y) = \alpha \leq \alpha_0$. For the two-sided situation of (7-27) both small and large y contradict H_0 and lead to rejection. If we aim at a significance level α_0, the critical region usually selected is $y \leq y_1$ and $y \geq y_2$ where y_1 is the largest value of y such that $\Pr(Y \leq y) \leq \alpha_0/2$ and y_2 is the smallest y such that $\Pr(Y \geq y) \leq \alpha_0/2$. The true significance level is

$$\alpha = \Pr(Y \leq y_1) + \Pr(Y \geq y_2)$$

Of course all calculations for determining the critical region are performed with the Poisson parameter $n\mu_0$.

EXAMPLE 7-14

Suppose that an individual wants to have his telephone disconnected if his average number of calls per day is less than two. He selects five days at random and records the following number of calls on these days: 0, 2, 1, 1, 1. Using $\alpha_0 = .05$, should he have his telephone removed?

Solution

We have previously indicated that this problem suggests we test H_0: $\mu = 2$ against H_1: $\mu < 2$. In Chap. 4 we observed that the Poisson is a reasonable model to use for calculating probabilities about X, the number of calls per day. Hence, if H_0 is true, Y has a Poisson distribution with parameter $n\mu_0 = 5(2) = 10$. From Appendix B4 we find

$$\Pr(Y \leq 4) = .02925 \qquad \Pr(Y \leq 5) = .06709$$

so that the critical region is $y \leq 4$. Since the random sample produced $y = 5$, H_0 is not rejected. The average number of calls per day may be less than two but this sample has not supported the alternative.

EXAMPLE 7-15

During a 12-month period, the number of twin births per month recorded in a hospital are 2, 0, 1, 1, 0, 0, 3, 2, 1, 1, 0, 1. Do these results contradict the hypothesis that the average number of twin births is .5 per month? Use $\alpha_0 = .05$.

Solution

The statement of the problem suggests that we test H_0: $\mu = .5$ against H_1: $\mu \neq .5$. We should attempt to convince ourselves that X, the number of twin births in a month, has a Poisson distribution. Let us consider conditions (4-32). Births that occur in one time interval are in no way related to births that occur in another nonoverlapping time interval. It is not unreasonable to assume that the probability that a twin birth occurs in a small time interval is proportional to the length of the interval. The probability that two or more twin births occur in a small time interval is obviously very small. This, of course, does not prove that the Poisson is the correct model, but it does make it seem reasonable. Because of the independence of births, the randomness condition of sampling is satisfied. Thus, Y has a Poisson distribution with parameter $n\mu_0 = 12(.5) = 6$. From Appendix B4 we find

$$\Pr(Y \le 1) = .01735 \qquad \Pr(Y \le 2) = .06197$$
$$\Pr(Y \ge 12) = .02009 \qquad \Pr(Y \ge 11) = .04262$$

so that the critical region is $y \le 1$ and $y \ge 12$, and $\alpha = .01735 + .02009 = .03744$. Since the sample produced $y = 12$, H_0 is rejected. We would undoubtedly conclude that the average number of twin births per month at the hospital is greater than .5.

As in the binomial case, power calculations are fairly routine if the necessary tables are available.

EXAMPLE 7-16

Find the power of the test used in Example 7-14 if $\mu = 1$.

Solution

Now Y has a Poisson distribution with parameter $n\mu = 5(1) = 5$. We get $\Pr(Y \le 4) = .44049$ from Appendix B4.

EXAMPLE 7-17

Find the power of the test used in Example 7-15 if $\mu = .75$.

Solution

Now Y has a Poisson distribution with parameter $n\mu = 12(.75) = 9$. Appendix B5 yields

$$\Pr(Y \le 1) + \Pr(Y \ge 12) = .00123 + .19699 = .19822$$

With more extensive tables of the Poisson (i.e., the Molina tables [6]) we could find by trial and error the smallest sample size n that makes the power at least $1 - \beta$ at some specified alternative value of μ.

EXAMPLE 7-18

Suppose that in Example 7-14 we wish to take a sample large enough so that when $\mu = 1$ the power is at least .90. How large must n be?

Solution

The minimum n is quickly found with the Molina table. We find: If $n = 13$, $n\mu_0 = 26$, $\Pr(Y \leq 17) = .04111$, $\Pr(Y \leq 18) = .06463$, the critical region is $y \leq 17$. If $\mu = 1$, $n\mu = 13$, $\Pr(Y \leq 17) = .89046$. If $n = 14$, $n\mu_0 = 28$, $\Pr(Y \leq 19) = .04781$, $\Pr(Y \leq 20) = .07274$, the critical region is $y \leq 19$. If $\mu = 1$, $n\mu = 14$, $\Pr(Y \leq 19) = .92349$. Hence, $n = 14$ is the minimum n that meets the requirement.

EXERCISES

7-30 An executive is willing to hire a secretary unless she averages more than one error per typed page. A random sample of five pages is selected from some prepared by the secretary. The errors per page are 3, 3, 4, 1, 2. Using $\alpha_0 = .05$, what decision should be made?

7-31 Find the power of the test used in Exercise 7-30 if $\mu = 1.6$. If $\mu = 2$.

7-32 A bakery would like a certain kind of cookie that they make to contain two raisins on the average. A random sample of four cookies yields a total of five raisins. Test the appropriate hypothesis using $\alpha_0 = .05$.

7-33 What is the power of the test used in Exercise 7-32 if the average number of raisins is 2.5? If the average is 1?

7-34 Suppose we test H_0: $\mu = .01$ against H_1: $\mu > .01$ for a Poisson random variable using $\alpha_0 = .05$. If we wish the power to be at least .90 if $\mu = .10$, is $n = 60$ large enough to meet the requirement?

7-35 Suppose we test H_0: $\mu = .5$ against H_1: $\mu < .5$ for a Poisson random variable using $\alpha_0 = .05$. If we wish the power to be at least .90 when $\mu = .1$, is $n = 20$ large enough to meet the requirement?

7-36 The number of defects per yard of cloth is known to have a Poisson distribution. A wholesaler has received a large order of cloth from a manufacturer. To determine whether or not he should accept the shipment, the wholesaler inspects 100 square yards, selected at random a yard at a time. If the average number of defects per yard is .02, then he does not want to reject such shipments more than one time in a hundred. On the other hand, if the average number of defects per yard is .10, he cannot afford to accept such shipments more than 15 percent of the time. Can this be accomplished with the sample size that is used? What decision should be made if the sample yields a total of five defects?

When $n > 1,000$ so that existing binomial tables are of no assistance, then instead of Y we can use the statistic

$$Z' = \frac{Y - np_0}{\sqrt{np_0(1 - p_0)}} \tag{7-30}$$

to test hypotheses (7-18), (7-19), and (7-20). According to the central limit theorem and Eq. (6-16), the random variable $Z' = (Y - n\mu)/\sigma \sqrt{n}$ has approximately a standard normal distribution. Here, the random variable X of the theorem has probability function (7-17) with $\mu = p$, $\sigma = \sqrt{p(1 - p)}$. Further, each of the three standard hypotheses specifies that $p = p_0$. Hence, in each case (7-30) has approximately a standard normal distribution. When $n > 1,000$ we no longer need to concern ourselves with the difference between an aimed-at significance level α_0 and an actual significance level α. For all practical purposes a desired level can now be achieved.

When testing (7-18) we found that only small values of y support H_1. Thus small y, and hence small z', lead to rejection of H_0. If the significance level is α, then the critical region is

$$z' = \frac{y - np_0}{\sqrt{np_0(1 - p_0)}} \leq z_\alpha \tag{7-31}$$

where z_α is defined by Eq. (6-5). Similarly, only large y and large z' support the H_1 of (7-19). A test with significance level α is obtained by rejecting H_0 when

$$z' \geq z_{1-\alpha} \tag{7-32}$$

Finally, when (7-20) is tested, both small and large values of y, and hence z', provide support for H_1. Thus, it is reasonable to reject H_0 with both small and large values of z', and the critical region usually chosen to yield a test with significance level α is

$$z' \leq z_{\alpha/2} \quad \text{and} \quad z' \geq z_{1-\alpha/2} \tag{7-33}$$

EXAMPLE 7-19

Redo Example 7-7 for a random sample of 2,500 bolts that produce 277 requiring rework.

Solution

The only differences are that another statistic is used and the critical region is found from Appendix B5. Since $\alpha_0 = .05$, the critical region is

$z' \geqq z_{.95} = 1.645$.   We calculate

$$z' = \frac{270 - 250}{\sqrt{2,500(.10)(.90)}} = \frac{27}{\sqrt{2,500}\sqrt{.09}} = \frac{27}{50(.3)} = \frac{27}{15} = 1.8$$

Since the observed value of the statistic is in the critical region, again the hypothesis is rejected and we do not purchase the machine.

Formulas that yield approximate power for the large sample tests of (7-18), (7-19), and (7-20) are obtained with a little algebraic manipulation. For example, the power of the test for (7-18) is

$$\Pr\left[\frac{Y - np_0}{\sqrt{np_0(1 - p_0)}} \leqq z_\alpha\right] \qquad (7\text{-}34)$$

If $p = p_1$, then in the right-hand side of (7-30), p_0 should be replaced by p_1 if Z' is to have a distribution that is approximately standard normal. Thus, to evaluate the probability (7-34), the left-hand side of the inequality has to be converted to $(Y - np_1)/\sqrt{np_1(1 - p_1)}$.   This is achieved by multiplying each side of the inequality by $\sqrt{np_0(1 - p_0)}$, adding np_0 to each side, subtracting np_1 from each side, and dividing each side by $\sqrt{np_1(1 - p_1)}$.   We can then write (7-34) as

$$\Pr\left[\frac{Y - np_1}{\sqrt{np_1(1 - p_1)}} \leqq \frac{z_\alpha \sqrt{p_0(1 - p_0)} + \sqrt{n}\,(p_0 - p_1)}{\sqrt{p_1(1 - p_1)}}\right]$$

$$\cong \Pr\left[Z \leqq \frac{z_\alpha \sqrt{p_0(1 - p_0)} + \sqrt{n}\,(p_0 - p_1)}{\sqrt{p_1(1 - p_1)}}\right] \quad (7\text{-}35)$$

from which power can be approximated by using the standard normal table.   Similar calculations for the test of (7-19) show that the expression

$$\Pr\left[Z \leqq \frac{z_\alpha \sqrt{p_0(1 - p_0)} + \sqrt{n}\,|p_0 - p_1|}{\sqrt{p_1(1 - p_1)}}\right] \qquad (7\text{-}36)$$

yields approximate power for both (7-18) and (7-19) if p_1 is a value of p in the range of the alternative designated by H_1.   For approximate power in the two-sided case (7-20), replace α by $\alpha/2$ in (7-36).   To get the latter result we need to observe that either $\Pr(Z' \leqq z_{\alpha/2})$ or $\Pr(Z' \geqq z_{1-\alpha/2})$ is very small ($< \alpha/2$) and can be neglected.

If a power of at least $1 - \beta$ is required for some p_1 in the alternative H_1 of (7-18), then (7-35) must be $\geqq 1 - \beta$.   The condition will be fulfilled when

$$\frac{z_\alpha \sqrt{p_0(1 - p_0)} + \sqrt{n}\,(p_0 - p_1)}{\sqrt{p_1(1 - p_1)}} \geqq z_{1-\beta} \qquad (7\text{-}37)$$

254

Using the fact that $z_{1-\beta} = -z_\beta$, and solving inequality (7-37) for n, yields

$$n \geq \left[\frac{z_\alpha \sqrt{p_0(1 - p_0)} + z_\beta \sqrt{p_1(1 - p_1)}}{p_0 - p_1} \right]^2 \qquad (7\text{-}38)$$

a result that holds for (7-19) as well as (7-18). For the two-sided case of (7-20), formula (7-38) applies after α is replaced by $\alpha/2$.

EXAMPLE 7-20

Find the power of the test used in Example 7-19 if $p = .12$.

Solution

The hypothesis and alternative are a special case of (7-19) and the power is given by (7-36). Thus, since $\alpha = .05$, $n = 2,500$, $p_0 = .10$, $p_1 = .12$, we get

$$\text{Power} \cong \Pr\left(Z \leq \frac{-1.645 \sqrt{(.10)(.90)} + \sqrt{2,500} \, |.10 - .12|}{\sqrt{(.12)(.88)}} \right)$$

$$= \Pr\left(Z \leq \frac{-1.645(.3) + .50|-.02|}{.325} \right)$$

$$= \Pr\left(Z \leq 1.56 \right) = .94$$

EXAMPLE 7-21

Find the n required to yield a power of .90 when $p = .12$ for the test of Example 7-19.

Solution

We use (7-38) with $\alpha = .05$, $\beta = .10$, $p_0 = .10$, $p_1 = .12$. We get

$$n \geq \left(\frac{-1.645 \sqrt{(.10)(.90)} - 1.282 \sqrt{(.12)(.88)}}{.10 - .12} \right)^2$$

$n \geq (45.5)^2$ or $n = 2,070$.

To get a large-sample equal-tail confidence interval for p, we start with

$$\Pr\left(-z_{1-\alpha/2} \leq \frac{Y - np}{\sqrt{np(1 - p)}} \leq z_{1-\alpha/2} \right) \cong 1 - \alpha \qquad (7\text{-}39)$$

We then attempt to reduce (7-39) to the form

$$\Pr[u_1(Y) \leq p \leq u_2(Y)] \cong 1 - \alpha$$

by working with the two inequalities that appear on the left-hand side of $\cong$. This procedure requires that we solve a quadratic inequality. If certain small terms are then discarded, we get

$$u_1(Y) = \frac{Y}{n} - z_{1-\alpha/2} \sqrt{\frac{\frac{Y}{n}\left(1 - \frac{Y}{n}\right)}{n}} \qquad u_2(Y) = \frac{Y}{n} + z_{1-\alpha/2} \sqrt{\frac{\frac{Y}{n}\left(1 - \frac{Y}{n}\right)}{n}}$$

$$(7\text{-}40)$$

Then $[u_1(y), u_2(y)]$ is a confidence interval for p with confidence coefficient approximately $1 - \alpha$. (For details of the derivation, see Mood and Graybill [7], p. 263.)

EXAMPLE 7-22

Find a confidence interval for p with approximate confidence coefficient .95 using the information of Example 7-19.

Solution

We have $n = 2,500$, $y = 277$, $y/n = .1108$, $\alpha = .05$, $1 - \alpha/2 = .975$, $z_{.975} = 1.96$. Thus

$$z_{1-\alpha/2} \sqrt{\frac{\frac{y}{n}\left(1 - \frac{y}{n}\right)}{n}} = \frac{1.96 \sqrt{(.1108)(.8892)}}{50} = .012$$

and the interval is $(.111 - .012, .111 + .012)$, which reduces to $(.099, .123)$.

Large sample results for Poisson μ are similar to those for binomial p. If $n\mu \leq 100$, then we can proceed as in Sec. 7-7 and use the Molina tables [6] described in Chap. 4. If $n\mu > 100$, then the statistic

$$Z' = \frac{Y - n\mu_0}{\sqrt{n\mu_0}} \qquad\qquad (7\text{-}41)$$

can be used to test hypotheses (7-25), (7-26), and (7-27). Because of the central limit theorem, the random variable Z' has approximately a standard normal distribution when $X_1, X_2, \ldots, X_n$ is a random sample from a distribution whose probability function is (7-28) with $\mu = \mu_0$.

Critical regions are selected in the same manner as in the binomial case. For (7-25), reject if

$$z' \leq z_\alpha \qquad\qquad (7\text{-}42)$$

256

For (7-26), reject if

$$z' \geqq z_{1-\alpha} \tag{7-43}$$

Finally for (7-27) the critical region is

$$z' \leqq z_{\alpha/2} \quad \text{and} \quad z' \geqq z_{1-\alpha/2} \tag{7-44}$$

The counterparts of (7-36), (7-38), and (7-40) are

$$\Pr\left(Z \leqq \frac{z_\alpha \sqrt{\mu_0} + \sqrt{n}\,|\mu_0 - \mu_1|}{\sqrt{\mu_1}}\right) \tag{7-45}$$

$$n \geqq \left[\frac{z_\alpha \sqrt{\mu_0} + z_\beta \sqrt{\mu_1}}{\mu_0 - \mu_1}\right]^2 \tag{7-46}$$

and

$$u_1(Y) = \frac{Y}{n} - z_{1-\alpha/2}\sqrt{\frac{Y/n}{n}} \qquad u_2(Y) = \frac{Y}{n} + z_{1-\alpha/2}\sqrt{\frac{Y/n}{n}} \tag{7-47}$$

EXAMPLE 7-23

Suppose that the individual of Example 7-14 selects 200 days at random during which he received a total of 364 telephone calls. Find the new critical region and make the appropriate decision. How large does n have to be to achieve a power of .95 if $\mu = 1.8$?

Solution

The new critical region is given by (7-42) with $\alpha = .05$, $n = 200$, $\mu_0 = 2$. Hence, we now reject if

$$\frac{y - 400}{\sqrt{400}} \leqq -1.645$$

Since $y = 364$, the statistic has value $(364 - 400)/20 = -1.8$ and H_0: $\mu = 2$ is rejected.

To achieve a power of .95 when $\mu = 1.8$, we must have

$$n \geqq \left[\frac{-1.645\sqrt{2} - 1.645\sqrt{1.8}}{2 - 1.8}\right]^2 \cong 513$$

EXERCISES

7-37 Suppose that in Exercise 7-19 the seed company's claim is investigated by using 1,600 seeds, 184 of which fail to germinate. Test the appropriate hypothesis using $\alpha = .05$.

7-38 Find the power of the test used in Exercise 7-37 if $p = .88$. How large would n have to be to raise this to .90?

7-39 Find an equal-tail confidence interval with coefficient approximately .90 for the value of p in Exercise 7-37.

7-40 According to a genetic theory, 25 percent of a species have a certain characteristic. A random sample of 1,200 of the species contains 324 having the characteristic. If $\alpha = .05$ is used, is the theory contradicted?

7-41 Find the power of the test used in Exercise 7-40 if 23 percent have the characteristic. If 27 percent have the characteristic.

7-42 Find an equal-tail confidence interval with coefficient approximately .95 for the value of p in Exercise 7-40.

7-43 Suppose that the executive in Exercise 7-30 looks at a random sample of 225 pages of the secretary's work and finds 252 errors. Recommend the appropriate decision using $\alpha = .05$.

7-44 Find the power of the test in Exercise 7-43 if $\mu = 1.21$. How large would n have to be to raise this to .95?

7-45 Find an equal-tail confidence interval with coefficient approximately .95 for the value of μ in Exercise 7-43.

REFERENCES

1 BLYTH, COLIN R., and DAVID W. HUTCHINSON: Table of Neyman—Shortest Unbiased Confidence Intervals for the Binomial Parameter, *Biometrika*, vol. 47, pp. 381–391, 1960.

2 BLYTH, COLIN R., and DAVID W. HUTCHINSON: Table of Neyman—Shortest Unbiased Confidence Intervals for the Poisson Parameter, *Biometrika*, vol. 48, pp. 191–194, 1961.

3 CROW, EDWIN L., and ROBERT S. GARDNER: Confidence Intervals for the Expectation of a Poisson Variable, *Biometrika*, vol. 46, pp. 441–453, 1959.

4 GUENTHER, WILLIAM C.: "Concepts of Statistical Inference," McGraw-Hill Book Company, New York, 1965.

5 HARVARD UNIVERSITY COMPUTATION LABORATORY: "Tables of the Cumulative Binomial Probability Distribution," Harvard University Press, Cambridge, Mass., 1955.

6 MOLINA, E. C.: "Poisson's Exponential Binomial Limit," D. Van Nostrand Company, Inc., Princeton, N.J., 1949.

7 MOOD, ALEXANDER M., and FRANKLIN A. GRAYBILL: "Introduction to the Theory of Statistics," 2d ed., McGraw-Hill Book Company, New York, 1963.

8 OWEN, D. B.: "Handbook of Statistical Tables," Addison-Wesley Publishing Company, Inc., Reading, Mass., 1962.

9 PATIL, G. P.: On the Equivalence of Binomial and Inverse Binomial Acceptance Sampling Plans and an Acknowledgement, *Technometrics*, vol. 5, pp. 119–121, 1963.

10 U.S. ARMY ORDNANCE CORPS: Tables of Cumulative Binomial Probabilities, *Ordnance Corps Pamphlet ORDP*20-1, September, 1952.

7-1 A parameter is a constant which appears in a probability function. A hypothesis is an assumption concerning the parameters of a probability function. A test is a rule indicating when to reject a hypothesis. Statistics is concerned with drawing inferences about unknown parameters of a probability distribution. A random variable that is observed in a statistical problem is called a statistic. The probability distribution of a statistic is called a sampling distribution.

7-2 The observed value t of a random variable T yields one value that can serve as a point estimate of a parameter θ. However, for a good estimate we desire that $E(T) = \theta$ and $\text{Var}(T)$ be a minimum. An estimator T with these properties is called "best." If X_1, X_2, $\ldots$, X_n is a random sample from a distribution that has probability function

(a) $f(x;p) = p^x(1 - p)^{1-x} \qquad x = 0, 1$
$$0 < p < 1$$

then $\bar{X}$ is the best estimator of p.

(b) $f(x;p) = p(1 - p)^{x-1} \qquad x = 1, 2, 3, \ldots$
$$0 < p < 1$$

then $(n - 1)/(Y - 1)$, where $Y = \sum_{i=1}^{n} X_i$ is the best estimator of p.

(c) $f(x;\mu) = \dfrac{e^{-\mu}\mu^x}{x!} \qquad x = 0, 1, 2, \ldots$

then $\bar{X}$ is the best estimator of p.

7-3 An interval $[u_1(t), u_2(t)]$ is called a confidence interval for the parameter θ with coefficient $1 - \alpha$ if

$$\Pr[u_1(T) < \theta < u_2(T)] = 1 - \alpha$$

When θ is a parameter in a discrete distribution, determination of a confidence interval is usually difficult without specially prepared tables or graphs.

7-4 The rejection of a true hypothesis is called a Type I error. The probability of committing a Type I error is called the level of significance. The acceptance of a false hypothesis is called a Type II error. The probability of rejecting the hypothesis is called the power of the test. The critical region is the set of outcomes which leads to rejection of the hypothesis.

259

7-6 Let $X_1, X_2, \ldots, X_n$ be a random sample from a distribution with probability function

$$f(x;p) = p^x(1 - p)^{1-x} \qquad \begin{aligned} & x = 0, 1 \\ & 0 < p < 1 \end{aligned}$$

and $Y = X_1 + X_2 + \cdots + X_n$, which has a binomial distribution with parameter n, p. To test

$$H_0\text{:} \quad p = p_0 \qquad \text{against} \qquad H_1\text{:} \quad p < p_0 \tag{7-18}$$

the hypothesis is rejected when y is small. The critical region usually selected is $y \leqq y_1$, where y_1 is the largest value of y such that $\Pr(Y \leqq y) \leqq \alpha_0$, where α_0 is a prechosen aimed-at significance level. To test

$$H_0\text{:} \quad p = p_0 \qquad \text{against} \qquad H_1\text{:} \quad p > p_0 \tag{7-19}$$

the hypothesis is rejected when y is large. The critical region usually selected is $y \geqq y_1$, where y_1 is the smallest y such that $\Pr(Y \geqq y_1) \leqq \alpha_0$. To test

$$H_0\text{:} \quad p = p_0 \qquad \text{against} \qquad H_1\text{:} \quad p \neq p_0 \tag{7-20}$$

the hypothesis is rejected if y is too small or too large. The critical region usually selected is $y \leqq y_1$ and $y \geqq y_2$, where y_1 is the largest y such that $\Pr(Y \leqq y) \leqq \alpha_0/2$ and y_2 is the smallest y such that $\Pr(Y \geqq y) \leqq \alpha_0/2$. All critical regions are found with $p = p_0$. The powers of the tests of (7-18), (7-19), and (7-20) are respectively

$$\sum_{y=0}^{y_1} b(y;n,p) \qquad \sum_{y=y_1}^{n} b(y;n,p)$$

and

$$\sum_{y=0}^{y_1} b(y;n,p) + \sum_{y=y_2}^{n} b(y;n,p)$$

With good binomial tables, the minimum n can be found such that

Power $\leqq \alpha_0$ when $p = p_0$
Power $\geqq 1 - \beta$ when $p = p_1$

7-7 Let $X_1, X_2, \ldots, X_n$ be a random sample from a distribution with probability function

$$f(x;\mu) = \frac{e^{-\mu}\mu^x}{x!} \qquad x = 0, 1, 2, \ldots$$

and $Y = X_1 + X_2 + \cdots + X_n$ which has a Poisson distribution with mean $n\mu$. To test

$$H_0: \quad \mu = \mu_0 \quad \text{against} \quad H_1: \quad \mu < \mu_0 \tag{7-25}$$

the hypothesis is rejected when y is small. The critical region usually selected is $y \leq y_1$, where y_1 is the largest value of y such that $\Pr(Y \leq y) \leq \alpha_0$. To test

$$H_0: \quad \mu = \mu_0 \quad \text{against} \quad H_1: \quad \mu > \mu_0 \tag{7-26}$$

the hypothesis is rejected when y is large. The critical region usually selected is $y \geq y_1$, where y_1 is the smallest value of y such that $\Pr(Y \geq y) \leq \alpha_0$. To test

$$H_0: \quad \mu = \mu_0 \quad \text{against} \quad H_1: \quad \mu \neq \mu_0 \tag{7-27}$$

the hypothesis is rejected when y is too small or too large. The critical region usually selected is $y \leq y_1$ and $y \geq y_2$, where y_1 is the largest y such that $\Pr(Y \leq y) \leq \alpha_0/2$ and y_2 is the smallest y such that $\Pr(Y \geq y) \leq \alpha_0/2$. All critical regions are found with $\mu = n\mu_0$. The powers of the tests of (7-25), (7-26), and (7-27) are respectively

$$\sum_{y=0}^{y_1} p(y;n\mu) \qquad \sum_{y=y_1}^{\infty} p(y;n\mu)$$

and

$$\sum_{y=0}^{y_1} p(y;n\mu) + \sum_{y=y_2}^{\infty} p(y;n\mu)$$

where

$$p(y;n\mu) = \frac{e^{-n\mu}(n\mu)^y}{y!} \qquad y = 0, 1, 2, \ldots$$

With good Poisson tables the minimum n can be found such that

Power $\leq \alpha_0$ when $\mu = \mu_0$
Power $\geq 1 - \beta$ when $\mu = \mu_1$

7-8   When n is large, then

$$Z' = \frac{Y - np_0}{\sqrt{np_0(1 - p_0)}} \tag{7-30}$$

is used to test (7-18), (7-19), and (7-20). Critical regions usually selected are respectively $z' < z_\alpha$, $z' > z_{1-\alpha}$, $z' < z_{\alpha/2}$, and $z' > z_{1-\alpha/2}$. A large-sample confidence interval for binomial p is $[u_1(y), u_2(y)]$

261

where

$$u_1(y) = \frac{y}{n} - z_{1-\alpha/2} \sqrt{\frac{\frac{y}{n}\left(1 - \frac{y}{n}\right)}{n}}$$

(7-40)

$$u_2(y) = \frac{y}{n} + z_{1-\alpha/2} \sqrt{\frac{\frac{y}{n}\left(1 - \frac{y}{n}\right)}{n}}$$

When n is large,

$$Z' = \frac{Y - n\mu_0}{\sqrt{n\mu_0}}$$

(7-41)

is used to test (7-25), (7-26), and (7-27). Critical regions usually selected are respectively $z' < z_\alpha$, $z' > z_{1-\alpha}$, $z' < z_{\alpha/2}$, and $z' > z_{1-\alpha/2}$. A large-sample confidence interval for Poisson μ is $[u_1(y), u_2(y)]$, where

$$u_1(y) = \frac{y}{n} - z_{1-\alpha/2} \sqrt{\frac{y/n}{n}} \qquad u_2(y) = \frac{y}{n} + z_{1-\alpha/2} \sqrt{\frac{y/n}{n}} \qquad (7\text{-}47)$$

CHAPTER EIGHT

Sampling

8-1 INTRODUCTION

On many occasions in the previous chapters we have encountered the concept of sampling. In Sec. 4-4 we learned how to use a table of random numbers to draw a random sample from a finite population. In order to derive the probability functions for the hypergeometric and generalized hypergeometric distributions, formulas (4-24) and (5-69), we had to assume that a sample is drawn randomly from a finite population of N objects. In Sec. 5-4 with the assistance of formula (5-34) we defined a random sample from a distribution (sometimes called a random sample from an infinite population) and we used this concept throughout Chap. 7 to estimate parameters and to test hypotheses. If we check conditions (b) and (c) of (4-1) used to derive the binomial probability function, conditions (b) and (c) of (4-8) used to derive the negative binomial probability function, and conditions (b) and (c) of (5-55) used to derive the multinomial probability function, we shall observe that these conditions are

263

another way of stating that these models require a random sample from a distribution (or infinite population). Random samples were required on many occasions in Chap. 5 in the discussion of derived random variables.

In this chapter we shall be concerned with *survey sampling*. By this we mean that the sampled population consists of N units (as in the situation encountered in the derivation of the hypergeometric probability function). With each unit we shall associate a characteristic, denoting these numbers by $v_1, v_2, \ldots, v_N$. As an example the N units may be all of the families in the city of Laramie and the v's could be any characteristic associated with a family such as the number of children attending public school or total family income per year. One of the major objectives of survey sampling is to obtain a good estimate of the total

$$\hat{V} = \sum_{i=1}^{N} v_i = v_1 + v_2 + \cdots + v_N \tag{8-1}$$

or the average

$$\bar{V} = \frac{\sum_{i=1}^{N} v_i}{N} = \frac{v_1 + v_2 + \cdots + v_N}{N} \tag{8-2}$$

and a good estimate of the variance of $\hat{V}$ or $\bar{V}$, whichever is used. If the characteristic in question is the number of children attending public school, it is quite likely that the school board would be interested in an estimate of $\hat{V}$, the total. If the characteristic were income, we might be interested in either the total or the average, depending upon our objectives. At any rate, if we know N an estimate of one yields an estimate of the other, since the two quantities differ only by a factor of N.

When N is small and the characteristic of each unit is readily available, then there is no point in obtaining estimates. However, in situations where survey sampling is usually employed N is relatively large and measurement of each unit may be prohibitively costly. In addition some units may not be available for measurement or N may not be known. In situations like these it is obviously advantageous to use a sample to estimate the average or the total, particularly if the estimators have relatively small variability.

8-2 SIMPLE RANDOM SAMPLING

When a sample of size n is selected from N units in such a way that each of the $\binom{N}{n}$ possible samples has probability $1 \bigg/ \binom{N}{n}$ of being chosen, then the procedure is called *simple random sampling*. In Sec. 4-4 we have already observed the relationship between the uniform distribution and simple random sampling and discussed the use of a random number table in selecting such a sample.

Suppose that from N values $v_1, v_2, \ldots, v_N$ we draw a simple random sample. Let Y_1 be the value associated with the first unit drawn, Y_2 be the value associated with the second unit drawn, $\ldots$, Y_n be the value associated with the nth unit drawn. Then we might expect that

$$\bar{Y} = \frac{Y_1 + Y_2 + \cdots + Y_n}{n} \tag{8-3}$$

would be a good estimator of $\bar{V}$. We next seek the mean and variance of $\bar{Y}$.

Let us define N random variables $X_1, X_2, \ldots, X_N$. The random variable X_i will be associated with v_i, $i = 1, 2, \ldots, N$ and has the value 1 if v_i is in the sample of size n and the value 0 if v_i is not in the sample. Hence in terms of the X_i we can rewrite (8-3) as

$$\bar{Y} = \frac{X_1 v_1 + X_2 v_2 + \cdots + X_N v_N}{n} = \frac{\sum\limits_{i=1}^{N} v_i X_i}{n} \tag{8-4}$$

Since only n of the X_i's can be 1 the numerator of (8-4) contains only n terms. For example suppose $N = 10$, $n = 3$, and the table of random numbers selects v_2, v_4, and v_9 for inclusion in the sample. Then

$$X_2 = X_4 = X_9 = 1$$

all other X_i are 0, and

$$\bar{y} = \frac{\begin{aligned}v_1(0) + v_2(1) + v_3(0) + v_4(1) + v_5(0) + v_6(0) + v_7(0) + v_8(0) \\ + v_9(1) + v_{10}(0)\end{aligned}}{3}$$

$$= \frac{v_2 + v_4 + v_9}{3}$$

From (5-45) we have

$$E\left(\frac{v_1}{n} X_1 + \frac{v_2}{n} X_2 + \cdots + \frac{v_N}{n} X_N\right) = \frac{v_1}{n} E(X_1) + \frac{v_2}{n} E(X_2) + \cdots$$
$$+ \frac{v_N}{n} E(X_N)$$

or

$$E(\bar{Y}) = \frac{1}{n} \sum_{i=1}^{N} v_i E(X_i) \tag{8-5}$$

and from (5-47) we get

$$\mathrm{Var}(\bar{Y}) = \sum_{i=1}^{N} \left(\frac{v_i}{n}\right)^2 \mathrm{Var}(X_i) + 2 \sum_{\substack{i,j=1 \\ i<j}}^{N} \left(\frac{v_i}{n}\right)\left(\frac{v_j}{n}\right) \mathrm{Cov}(X_i, X_j)$$

or

$$\text{Var}(\bar{Y}) = \frac{1}{n^2}\left[\sum_{i=1}^{N} v_i^2 \, \text{Var}(X_i) + 2 \sum_{\substack{i,j=1 \\ i<j}}^{N} v_i v_j \, \text{Cov}(X_i,X_j)\right] \tag{8-6}$$

Hence we need $E(X_i)$, $\text{Var}(X_i)$, and $\text{Cov}(X_i,X_j)$. Observe that conditions (5-68) for the generalized hypergeometric distribution are satisfied with $r = N$ and each $k_i = 1$. Hence the joint probability function of $X_1, X_2, \ldots, X_N$ is a special case of (5-69), being

$$f(x_1, x_2, \ldots, x_N) = \frac{\binom{1}{x_1}\binom{1}{x_2}\cdots\binom{1}{x_N}}{\binom{N}{n}} \tag{8-7}$$

where $x_i = 0, 1, i = 1, 2, \ldots, N$ subject to the condition $x_1 + x_2 + \cdots + x_N = n$. We have already studied the marginal distribution for the hypergeometric in Sec. 5-6 and for (8-7) specifically in Exercise 5-59. Hence the marginal probability function of X_i is

$$f_i(x_i) = \frac{N-n}{N} \qquad x_i = 0$$

$$= \frac{n}{N} \qquad x_i = 1 \tag{8-8}$$

from which we readily obtain

$$E(X_i) = \frac{n}{N} \tag{8-9}$$

$$\text{Var}(X_i) = \frac{n}{N} - \left(\frac{n}{N}\right)^2 = \frac{n(N-n)}{N^2} \tag{8-10}$$

The joint probability function of X_i and X_j appears in Exercise 5-59, and formula (5-22) yields

$$\text{Cov}(X_i,X_j) = \frac{n(n-1)}{N(N-1)} - \left(\frac{n}{N}\right)\left(\frac{n}{N}\right) = -\frac{n(N-n)}{N^2(N-1)} \tag{8-11}$$

Substituting the result of (8-9) in (8-5) yields

$$E(\bar{Y}) = \frac{1}{n}\sum_{i=1}^{N} v_i \frac{n}{N} = \sum_{i=1}^{N} \frac{v_i}{N} = \bar{V}$$

Hence $\bar{Y}$ is an unbiased estimator of $\bar{V}$. Also

$$E\left(\frac{N}{n}\sum_{i=1}^{N} v_i X_i\right) = \frac{N}{n}\sum_{i=1}^{N} v_i E(X_i) = \frac{N}{n}\sum_{i=1}^{N} v_i \frac{n}{N} = \sum_{i=1}^{N} v_i = \hat{V}$$

266

so that

$$\hat{Y} = \frac{N}{n} \sum_{i=1}^{N} v_i X_i = \frac{N}{n} (Y_1 + Y_2 + \cdots + Y_n) = N\bar{Y} \qquad (8\text{-}12)$$

is an unbiased estimator of the total $\hat{V}$. Next, using the results of (8-10) and (8-11) in (8-6) we get

$$\begin{aligned}
\text{Var}(\bar{Y}) &= \frac{1}{n^2} \left[\sum_{i=1}^{N} v_i^2 \frac{n(N-n)}{N^2} - 2 \sum_{\substack{i,j=1 \\ i<j}}^{N} v_i v_j \frac{n(N-n)}{N^2(N-1)} \right] \\
&= \frac{1}{n^2} \frac{n(N-n)}{N^2} \left[\sum_{i=1}^{N} v_i^2 - \frac{2}{N-1} \sum_{\substack{i,j=1 \\ i<j}}^{N} v_i v_j \right]
\end{aligned}$$

But since

$$(v_1 + v_2 + \cdots + v_N)^2 = \sum_{i=1}^{N} v_i^2 + 2 \sum_{\substack{i,j=1 \\ i<j}}^{N} v_i v_j$$

we have

$$2 \sum_{\substack{i,j=1 \\ i<j}}^{N} v_i v_j = \left(\sum_{i=1}^{N} v_i \right)^2 - \sum_{i=1}^{N} v_i^2 \qquad (8\text{-}13)$$

Hence

$$\begin{aligned}
\text{Var}(\bar{Y}) &= \frac{N-n}{nN^2} \left[\sum_{i=1}^{N} v_i^2 - \frac{1}{N-1} \left(\hat{V}^2 - \sum_{i=1}^{N} v_i^2 \right) \right] \\
&= \frac{N-n}{nN^2} \left[\frac{N}{N-1} \sum_{i=1}^{N} v_i^2 - \frac{\hat{V}^2}{N-1} \right] \\
&= \frac{N-n}{nN^2(N-1)} \left[N \sum_{i=1}^{N} v_i^2 - (N\bar{V})^2 \right] \\
&= \frac{N-n}{nN^2(N-1)} \left[N \left(\sum_{i=1}^{N} v_i^2 - N\bar{V}^2 \right) \right]
\end{aligned}$$

By squaring the terms $(v_i - \bar{V})$ and forming three sums we easily show that

$$\sum_{i=1}^{N} (v_i - \bar{V})^2 = \sum_{i=1}^{N} v_i^2 - N\bar{V}^2$$

Hence

$$\sum_{i=1}^{N} v_i^2 - \frac{2}{N-1} \sum_{\substack{i,j=1 \\ i<j}}^{N} v_i v_j = \frac{N}{N-1} \sum_{i=1}^{N} (v_i - \bar{V})^2 \tag{8-14}$$

Now if we let

$$\sigma'^2 = \frac{\sum_{i=1}^{N} (v_i - \bar{V})^2}{N-1} = \frac{N}{N-1} \frac{\sum_{i=1}^{N} (v_i - \bar{V})^2}{N} = \frac{N\sigma^2}{N-1} \tag{8-15}$$

we can write

$$\text{Var}(\bar{Y}) = \frac{N-n}{nN} \sigma'^2 \tag{8-16}$$

It is perhaps more natural to use σ^2 than σ'^2. Both can be regarded as measures of variability but it is more convenient to use σ'^2 in the derivations found later in this section. To get the variance of $\hat{Y}$ observe that $\hat{Y} = N\bar{Y}$ and use formula (5-47) with $a_1 = N$, all other a's zero. We get

$$\text{Var}(\hat{Y}) = N^2 \,\text{Var}(\bar{Y})$$

or

$$\text{Var}(\hat{Y}) = \frac{N(N-n)\sigma'^2}{n} \tag{8-17}$$

Although we may have a reasonable estimate of σ'^2 based upon past experience, usually that quantity will be unknown. The usefulness of (8-16) as a measure of variability will be limited by our ability to produce a good estimate of σ'^2. A reasonable choice for an estimator is

$$S^2 = \frac{\sum_{i=1}^{n} (Y_i - \bar{Y})^2}{n-1} \tag{8-18}$$

Recalling the definition of the X_i, we can also write

$$S^2 = \frac{\sum_{i=1}^{N} (v_i - \bar{Y})^2 X_i}{n-1} \tag{8-19}$$

a form more useful for showing $E(S^2) = \sigma'^2$, a fact we shall examine very shortly. If S^2 is an unbiased estimator of σ'^2, then an unbiased estimator of $\text{Var}(\bar{Y})$ is

$$\text{var}(\bar{Y}) = \frac{N-n}{nN} S^2 \tag{8-20}$$

268

and an unbiased estimator of $\mathrm{Var}(\hat{Y})$ is

$$\mathrm{var}(\hat{Y}) = \frac{N(N-n)}{n} S^2 \tag{8-21}$$

results which follow immediately from formula (3-12). To denote observed values of the random variables (8-20) and (8-21) we shall use the notation $\mathrm{var}(\bar{y})$ and $\mathrm{var}(\hat{y})$. To show S^2 is unbiased rewrite (8-19) as

$$
\begin{aligned}
(n-1)S^2 &= \sum_{i=1}^{N} [(v_i - \bar{V}) - (\bar{Y} - \bar{V})]^2 X_i \\
&= \sum_{i=1}^{N} (v_i - \bar{V})^2 X_i - 2 \sum_{i=1}^{N} (v_i - \bar{V})(\bar{Y} - \bar{V}) X_i \\
&\qquad\qquad\qquad\qquad\qquad + \sum_{i=1}^{N} (\bar{Y} - \bar{V})^2 X_i \\
&= \sum_{i=1}^{N} (v_i - \bar{V})^2 X_i - 2(\bar{Y} - \bar{V}) \Big[\sum_{i=1}^{N} v_i X_i - \sum_{i=1}^{N} \bar{V} X_i \Big] \\
&\qquad\qquad\qquad\qquad\qquad + n(\bar{Y} - \bar{V})^2 \\
&= \sum_{i=1}^{N} (v_i - \bar{V})^2 X_i - 2(\bar{Y} - \bar{V})[n\bar{Y} - n\bar{V}] + n(\bar{Y} - \bar{V})^2 \\
&= \sum_{i=1}^{N} (v_i - \bar{V})^2 X_i - n(\bar{Y} - \bar{V})^2
\end{aligned}
$$

since the last two terms combine. Next we have

$$
\begin{aligned}
E(S^2) &= E\left[\frac{1}{n-1} \sum_{i=1}^{N} (v_i - \bar{V})^2 X_i - \frac{n}{n-1} (\bar{Y} - \bar{V})^2 \right] \\
&= \frac{1}{n-1} \sum_{i=1}^{N} (v_i - \bar{V})^2 E(X_i) - \frac{n}{n-1} E[(\bar{Y} - \bar{V})^2] \tag{8-22}
\end{aligned}
$$

But $E(X_i)$ and $E[(\bar{Y} - \bar{V})^2]$ are given by (8-9) and (8-14). Using these results yields

$$
\begin{aligned}
E(S^2) &= \frac{1}{n-1} \sum_{i=1}^{N} (v_i - \bar{V})^2 \frac{n}{N} - \frac{n}{n-1} \frac{N-n}{nN} \sigma'^2 \\
&= \frac{1}{n-1} \frac{n}{N} (N-1)\sigma'^2 - \frac{1}{n-1} \frac{N-n}{N} \sigma'^2 \\
&= \frac{nN - n - N + n}{N(n-1)} \sigma'^2 = \sigma'^2
\end{aligned}
$$

as was to be proved.   Hence (8-20) and (8-21) are unbiased estimators of (8-16) and (8-17), respectively.

EXAMPLE 8-1

Suppose that there are 200 families in the city of Pinedale.   With the aid of a table of random numbers 20 families were selected.   The names were forwarded to the Bureau of Internal Revenue and the Bureau released the following information based upon a random sample drawn without replacement:

Observed sample mean $\bar{y} = \$5,000$ per year
Observed sample variance $s^2 = 1,000,000$

Estimate $\hat{V}$, $\text{Var}(\bar{Y})$, and $\text{Var}(\hat{Y})$.

Solution

From (8-12) $\hat{Y} = N\bar{Y}$ is an unbiased estimator of $\hat{V}$.   Here

$$\hat{y} = 200(\$5,000) = \$1,000,000$$

our estimate of the total yearly income for the people of Pinedale.   An unbiased estimate of $\text{Var}(\bar{Y})$ is obtained from (8-20) and is

$$\frac{N-n}{nN}s^2 = \frac{200-20}{20(200)}(1,000,000) = 45,000$$

Similarly, an unbiased estimate of $\text{Var}(\hat{Y})$ is obtained from (8-21) and is

$$\frac{N(N-n)}{n}s^2 = \frac{200(200-20)}{20}(1,000,000) = 1,800,000,000$$

To make probability statements about $\bar{Y}$ and $\hat{Y}$ we may attempt to apply the central limit theorem and use the random variable Z' as given by (6-15) or (6-16).   Two difficulties arise.   First, $Y_1, Y_2, \ldots, Y_n$ is not a random sample from a distribution.   [Recall Eq. (5-34) and accompanying definition.]   It can be shown that

$$\Pr(Y_j = v_i) = \frac{1}{N} \qquad i = 1, 2, \ldots, N \qquad (8\text{-}23)$$

$$j = 1, 2, \ldots, n$$

so that each Y_j has the same marginal distribution.   However, we can also show that the Y_j's are not independent.   Hence, it would appear that the central limit theorem does not apply.   Second, usually we do not know μ and σ^2, parameters required to evaluate probabilities involving Z'.   In practice $\bar{y}$ and $s \sqrt{N-n}/\sqrt{nN}$ have been substituted for

μ and $\sigma/\sqrt{n}$ (since $\sigma/\sqrt{n}$ is the standard deviation of $\bar{Y}$ when we have a random sample from a distribution) and

$$Z' = \frac{\bar{Y} - \bar{y}}{s\sqrt{N-n}/\sqrt{nN}} \tag{8-24}$$

$$= \frac{\hat{Y} - \hat{y}}{s\sqrt{N(N-n)}/\sqrt{n}} \tag{8-25}$$

is treated as a random variable having approximately a standard normal distribution. Experimental evidence seems to indicate that probability statements so obtained are good enough for practical purposes if n is not too small. It is, however, fairly difficult to give a satisfactory rule for specifying how large n must be. For further discussion on the validity of the normal approximation see Cochran [1], page 38.

Let us next consider the problem of obtaining a confidence interval for $\bar{V}$. If Z' has approximately a standard normal distribution, then we can make probability statements such as

$$\Pr\left(-1.96 < \frac{\bar{Y} - \bar{V}}{\sigma/\sqrt{n}} < 1.96\right) \cong .95 \tag{8-26}$$

or more generally

$$\Pr\left(-z_{1-\alpha/2} < \frac{\bar{Y} - \bar{V}}{\sigma/\sqrt{n}} < z_{1-\alpha/2}\right) \cong 1 - \alpha \tag{8-27}$$

Consider the two inequalities in (8-26). One is $-1.96 < (\bar{Y} - \bar{V})/(\sigma/\sqrt{n})$. Since inequalities, like equations, can be multiplied by a positive number we can write $-1.96\sigma/\sqrt{n} < \bar{Y} - \bar{V}$. The same number can be added to or subtracted from each side of an inequality. Hence $-1.96\sigma/\sqrt{n} < \bar{Y} - \bar{V}$ and $-1.96\sigma/\sqrt{n} - \bar{Y} < -\bar{V}$ are equivalent. Finally, if both sides of an inequality are multiplied by a negative number, the sense is reversed. Thus if $-1.96\sigma/\sqrt{n} - \bar{Y} < -\bar{V}$, then $1.96\sigma/\sqrt{n} + \bar{Y} > \bar{V}$ or $\bar{V} < \bar{Y} + 1.96\sigma/\sqrt{n}$. The second inequality $(\bar{Y} - \bar{V})/(\sigma/\sqrt{n}) < 1.96$ can be written $\bar{Y} - 1.96\sigma/\sqrt{n} < \bar{V}$. Thus

$$-1.96 < \frac{\bar{Y} - \bar{V}}{\sigma/\sqrt{n}} < 1.96 \quad \text{and} \quad \bar{Y} - 1.96\frac{\sigma}{\sqrt{n}} < \bar{V} < \bar{Y} + 1.96\frac{\sigma}{\sqrt{n}}$$

are equivalent inequalities. That is, if the former is true, then so is the latter. Thus (8-26) and (8-27) could be rewritten as

$$\Pr\left(\bar{Y} - 1.96\frac{\sigma}{\sqrt{n}} < \bar{V} < \bar{Y} + 1.96\frac{\sigma}{\sqrt{n}}\right) \cong .95 \tag{8-28}$$

and

$$\Pr\left(\bar{Y} - z_{1-\alpha/2}\frac{\sigma}{\sqrt{n}} < \bar{V} < \bar{Y} + z_{1-\alpha/2}\frac{\sigma}{\sqrt{n}}\right) \cong 1 - \alpha \qquad (8\text{-}29)$$

If σ were known and a sample of size n yields $\bar{Y} = \bar{y}$, then we can compute

$$\left(\bar{y} - z_{1-\alpha/2}\frac{\sigma}{\sqrt{n}}, \quad \bar{y} + z_{1-\alpha/2}\frac{\sigma}{\sqrt{n}}\right) \qquad (8\text{-}30)$$

The interval (8-30) is called a confidence interval for $\bar{V}$ with coefficient $1 - \alpha$ (approximately $1 - \alpha$). According to (8-29) intervals so computed will contain $\bar{V}$ approximately $1 - \alpha$ of the time. The unfortunate feature of the interval (8-30) is that it depends on σ, which may be unknown. A common procedure in such instances is to replace $\sigma/\sqrt{n}$ by $s\sqrt{N-n}/\sqrt{nN}$, obtaining

$$\left(\bar{y} - z_{1-\alpha/2}\frac{s\sqrt{N-n}}{\sqrt{nN}}, \quad \bar{y} + z_{1-\alpha/2}\frac{s\sqrt{N-n}}{\sqrt{nN}}\right) \qquad (8\text{-}31)$$

This substitution involves a further approximation and we can usually expect that the confidence coefficient will be slightly lower than $1 - \alpha$ for (8-31). The corresponding intervals for $\hat{V}$ are obtained by multiplying the end points of (8-30) and (8-31) by N. This gives for the counterpart of (8-31)

$$\left(\hat{y} - z_{1-\alpha/2}\frac{s\sqrt{N(N-n)}}{\sqrt{n}}, \quad \hat{y} + z_{1-\alpha/2}\frac{s\sqrt{N(N-n)}}{\sqrt{n}}\right) \qquad (8\text{-}32)$$

EXAMPLE 8-2

Using the results of Example 8-1 estimate $\Pr(\bar{Y} > \$5,200)$ and find confidence intervals for $\bar{V}$ and $\hat{V}$ with coefficients approximately .90.

Solution

We can write

$$\Pr(\bar{Y} > 5,200) = \Pr\left(\frac{\bar{Y} - 5,000}{\sqrt{45,000}} > \frac{5,200 - 5,000}{\sqrt{45,000}}\right)$$

$$\cong \Pr(Z > .943)$$
$$= 1 - \Pr(Z < .943)$$
$$= 1 - .83 = .17$$

To find the confidence interval we observe that $\alpha = .10$, $\alpha/2 = .05$, $z_{1-\alpha/2} = z_{.95} = 1.645$ from Appendix B5. Hence (8-31) becomes $[5,000 -$

$1.645\sqrt{45,000}$, $5,000 + 1.645\sqrt{45,000}$] which reduces to ($4,651, $5,349). Multiplying the end points of the interval by $N = 200$ gives ($930,200, $1,069,800) as the confidence interval for $\hat{V}$.

In Example 8-1 s^2 was given, but usually this quantity will have to be calculated from $y_1, y_2, \ldots, y_n$. According to (8-18) we would evaluate

$$s^2 = \frac{\sum_{i=1}^{n} (y_i - \bar{y})^2}{n - 1} \tag{8-33}$$

A form of s^2 more adaptable to hand calculation or use with a desk calculator is obtained by squaring out all the terms in the numerator. We get

$$\sum_{i=1}^{n} (y_i - \bar{y})^2 = \sum_{i=1}^{n} (y_i^2 - 2\bar{y}y_i + \bar{y}^2)$$

$$= \sum_{i=1}^{n} y_i^2 - 2\bar{y} \sum_{i=1}^{n} y_i + n\bar{y}^2$$

$$= \sum_{i=1}^{n} y_i^2 - 2\bar{y}n\bar{y} + n\bar{y}^2$$

$$= \sum_{i=1}^{n} y_i^2 - n\bar{y}^2$$

$$= \sum_{i=1}^{n} y_i^2 - n \left(\frac{\sum_{i=1}^{n} y_i}{n} \right)^2$$

$$= \frac{n \sum_{i=1}^{n} y_i^2 - \left(\sum_{i=1}^{n} y_i \right)^2}{n}$$

Hence we can write

$$s^2 = \frac{n \sum_{i=1}^{n} y_i^2 - \left(\sum_{i=1}^{n} y_i \right)^2}{n(n - 1)} \tag{8-34}$$

EXAMPLE 8-3

Evaluate s^2 by using (8-34) if $y_1 = 6$, $y_2 = 4$, $y_3 = 4$, $y_4 = 5$, $y_5 = 2$, $y_6 = 5$, $y_7 = 3$, $y_8 = 4$, $y_9 = 6$, $y_{10} = 1$.

273

Solution

We need $\sum_{i=1}^{10} y_i = 6 + 4 + 4 + 5 + 2 + 5 + 3 + 4 + 6 + 1 = 40$ and

$\sum_{i=1}^{10} y_i^2 = 6^2 + 4^2 + 4^2 + 5^2 + 2^2 + 5^2 + 3^2 + 4^2 + 6^2 + 1^2 = 184$. Thus

$$s^2 = \frac{10(184) - 40^2}{10(9)} = \frac{1,840 - 1,600}{90} = 2.67$$

EXERCISES

8-1 Suppose $N = 5$ and $v_1 = 7$, $v_2 = 8$, $v_3 = 5$, $v_4 = 16$, $v_5 = 14$. Calculate $\bar{y}$ for every (ordered) sample of size $n = 2$ drawn at random without replacement and verify that $E(\bar{Y}) = \bar{V}$, $\text{Var}(\bar{Y}) = (N - n)\sigma'^2/nN$.

8-2 Using the finite population given in Exercise 8-1, find s^2 for every (ordered) sample of size $n = 2$ drawn at random without replacement and verify that $E(S^2) = \sigma'^2$.

8-3 A city contains 4,000 families. From these a simple random sample of 20 families was selected and these families had the following numbers of children attending public school:

2, 0, 2, 1, 2, 0, 0, 3, 1, 5, 2, 2, 0, 4, 1, 1, 1, 4, 3, 0

Estimate the total number of children attending public school. Find a confidence interval for the total with confidence coefficient approximately .95. If there are actually 7,000 children in the city attending public school, estimate the probability that $\hat{Y}$ is less than the observed $\hat{y}$.

8-4 A large firm has 1,000 accounts receivable as of July 1. A random sample of 100 accounts yields $\bar{y} = \$50$, $s^2 = 400$. Find estimates of $\text{Var}(\bar{Y})$, $\text{Var}(\hat{Y})$ and obtain confidence intervals with coefficient approximately .90 for $\bar{V}$ and $\hat{V}$.

8-5 Consider a population of 4 items with values v_1, v_2, v_3, v_4. Write down all 24 (ordered) samples of size 3 and observe that under random sampling (so that each sample has probability $\frac{1}{24}$ of being selected)

$$\begin{aligned}
\text{Pr}(v_1 \text{ is selected first}) &= \text{Pr}(v_2 \text{ is selected first}) \\
&= \text{Pr}(v_3 \text{ is selected first}) \\
&= \text{Pr}(v_4 \text{ is selected first}) \\
&= \frac{6}{24} = \frac{1}{4}
\end{aligned}$$

Also observe that these probabilities are all $\frac{1}{4} = 1/N$ if the word "first" is replaced by "second," "third," or "fourth" in each of the four probabilities.

8-6 Generalize the result of Exercise 8-5 to show that the probability function of each Y_i is

$$f(y_i) = 1/N \qquad y_i = v_1, v_2, \ldots, v_N \qquad i = 1, 2, \ldots, n$$

Also show that the joint probability function of Y_i, Y_j, $i \neq j$, is

$$g(y_i, y_j) = 1/N(N-1) \qquad y_i, y_j = v_1, v_2, \ldots, v_N$$

where y_i and y_j cannot both assume the same value of v. Hence show that Y_i and Y_j are not independent. *Hint:* For the first part take, for example, Y_5 and v_7. There are $N(N-1) \cdots (N - n + 1)$ ordered samples of size n, all having the same probability of selection if the sampling is random. Since we want $Y_5 = v_7$ there are $(N-1)(N-2)(N-3)(N-4) \cdot 1 \cdot (N-5) \cdots (N - n + 1)$ ordered samples so constructed. Hence $\Pr(Y_5 = v_7) = 1/N$, the same argument holding for any other value of v. For Y_i (instead of Y_5) the 1 occurs in the ith place (instead of the 5th) in the latter product. To obtain the joint probability function of Y_i, Y_j consider, for example, Y_3, Y_6 and calculate $\Pr(Y_3 = v_5, Y_6 = v_8) = 1/N(N-1)$. Observe that the number of ordered samples such that $Y_3 = v_5$, $Y_6 = v_8$ is the product $(N-2)(N-3) \cdot 1 \cdot (N-4)(N-5) \cdot 1 \cdot (N-6) \cdots (N - n + 1)$. The answer is the same for all $N(N-1)$ choices for the v's. A similar argument also holds for Y_i, Y_j (instead of Y_3, Y_6) with 1's occuring in the ith and jth positions (instead of the 3rd and 6th) in the latter product.

8-7 Given that the probability functions $f(y_i)$ and $g(y_i, y_j)$ are correct as defined in Exercise 8-6 show that

(a) $E(Y_i) = \bar{V}$

(b) $\operatorname{Var}(Y_i) = \dfrac{N-1}{N} \sigma'^2$

(c) $\operatorname{Cov}(Y_i, Y_j) = \dfrac{\displaystyle\sum_{\substack{i,j=1 \\ i \neq j}}^{N} (v_i - \bar{V})(v_j - \bar{V})}{N(N-1)}$

$$= \dfrac{\left[\displaystyle\sum_{i=1}^{N} (v_i - \bar{V}) \right]^2 - \displaystyle\sum_{i=1}^{N} (v_i - \bar{V})^2}{N(N-1)} = -\dfrac{\sigma'^2}{N}$$

(d) $E(\bar{Y}) = \bar{V}$

(e) $\operatorname{Var}(\bar{Y}) = \dfrac{\sigma'^2(N-n)}{nN}$

by using (b), (c), and (5-47).

Next suppose that a sample of size n is to be chosen at random from N items by sampling with replacement. This can also be accomplished by using a table of random numbers exactly the same way it is used in sampling without replacement except that now repeated values are included in the sample.

Now $Y_1, Y_2, \ldots, Y_n$ is a random sample from the distribution whose probability function is

$$f(y) = \frac{1}{N} \qquad y = v_1, v_2, \ldots, v_N \tag{8-35}$$

with

$$E(Y) = v_1 \frac{1}{N} + v_2 \frac{1}{N} + \cdots + v_N \frac{1}{N} = \frac{v_1 + v_2 + \cdots + v_N}{N} = \bar{V}$$
$$\tag{8-36}$$

$$\text{Var}(Y) = (v_1 - \bar{V})^2 \frac{1}{N} + (v_2 - \bar{V})^2 \frac{1}{N} + \cdots + (v_N - \bar{V})^2 \frac{1}{N}$$
$$= \frac{(v_1 - \bar{V})^2 + (v_2 - \bar{V})^2 + \cdots + (v_N - \bar{V})^2}{N} \tag{8-37}$$

Now define

$$\sigma^2 = \frac{\sum\limits_{i=1}^{N} (v_i - \bar{V})^2}{N} = \text{Var}(Y) \tag{8-38}$$

which is different from (8-15), having N instead of $N - 1$ in the denominator. We already know from (5-53) and (5-54) that

$$E(\bar{Y}) = E(Y) = \bar{V} \tag{8-39}$$

$$\text{Var}(\bar{Y}) = \frac{\text{Var}(Y)}{n} = \frac{\sigma^2}{n} \tag{8-40}$$

Hence $\bar{Y}$ is an unbiased estimator of $\bar{V}$ (and $\hat{Y}$ is an unbiased estimator of $\hat{V}$). S^2, as previously defined by (8-18), is unbiased. To show this write

$$E(S^2) = E\left[\frac{\sum\limits_{i=1}^{n} (Y_i - \bar{Y})^2}{n - 1} \right]$$

$$= \frac{1}{n - 1} E\left[\sum\limits_{i=1}^{n} (Y_i - \bar{Y})^2 \right]$$

$$= \frac{1}{n-1} E \left\{ \sum_{i=1}^{n} [(Y_i - \bar{V}) - (\bar{Y} - \bar{V})]^2 \right\}$$

$$= \frac{1}{n-1} E \left\{ \sum_{i=1}^{n} (Y_i - \bar{V})^2 \right.$$

$$\left. - 2(\bar{Y} - \bar{V}) \sum_{i=1}^{n} (Y_i - \bar{V}) + \sum_{i=1}^{n} (\bar{Y} - \bar{V})^2 \right\}$$

$$= \frac{1}{n-1} E \left\{ \sum_{i=1}^{n} (Y_i - \bar{V})^2 - 2(\bar{Y} - \bar{V})n(\bar{Y} - \bar{V}) + n(\bar{Y} - \bar{V})^2 \right\}$$

$$= \frac{1}{n-1} E \left\{ \sum_{i=1}^{n} (Y_i - \bar{V})^2 - n(\bar{Y} - \bar{V})^2 \right\}$$

$$= \frac{1}{n-1} E \left\{ \left[\sum_{i=1}^{n} (Y_i - \bar{V})^2 \right] - nE[(\bar{Y} - \bar{V})^2] \right\}$$

$$= \frac{1}{n-1} \left\{ \sum_{i=1}^{n} E(Y_i - \bar{V})^2 - n \operatorname{Var}(\bar{Y}) \right\}$$

$$= \frac{1}{n-1} \left\{ n\sigma^2 - \frac{n\sigma^2}{n} \right\} = \sigma^2$$

Thus an unbiased estimator of $\operatorname{Var}(\bar{Y})$ is

$$\operatorname{var}(\bar{Y}) = \frac{S^2}{n} \qquad (8\text{-}41)$$

and an unbiased estimator of $\operatorname{Var}(\hat{Y})$ is

$$\operatorname{var}(\hat{Y}) = N^2 \frac{S^2}{n} \qquad (8\text{-}42)$$

EXAMPLE 8-4

In Example 8-1 suppose that the sample of 20 was drawn with replacement yielding $\bar{y} = \$5,000$, $s^2 = 1,000,000$. Now estimate $\hat{V}$, $\operatorname{Var}(\bar{Y})$, $\operatorname{Var}(\hat{Y})$.

Solution

Again $\bar{y} = \$5,000$ is an unbiased estimate of $\bar{V}$ and $\hat{y} = \$1,000,000$ is an unbiased estimate of $\hat{V}$. According to (8-41) an unbiased estimate of $\operatorname{Var}(\bar{Y})$ is

$$\operatorname{var}(\bar{y}) = \frac{s^2}{n} = \frac{1,000,000}{20} = 50,000$$

Similarly, because of (8-42) an unbiased estimate of $\text{Var}(\hat{Y})$ is

$$\text{var}(\hat{y}) = \frac{N^2 s^2}{n} = \frac{(200)^2 (1,000,000)}{20} = 2,000,000,000$$

The variance estimates obtained in Example 8-4 turned out to be larger than those obtained by simple random sampling in Example 8-1. It is easy to see that this is always the case since the estimate of $\text{Var}(\bar{Y})$ without replacement can be written $(s^2/n)(1 - n/N)$ which is always smaller than s^2/n unless $n = N$.

Intuitively it is obvious that if N is large by comparison with n, random sampling without replacement and random sampling with replacement are almost equivalent. Hence, if n/N is small, we would expect that $\text{Var}(\bar{Y})$ as given by (8-16) would be approximately equal to $\text{Var}(\bar{Y})$ as given by (8-40). To show this we can rewrite the former expression:

$$\text{Var}(\bar{Y}) = \left(1 - \frac{n}{N}\right)\frac{\sigma'^2}{n}$$

$$= \left(1 - \frac{n}{N}\right)\left(\frac{N}{N-1}\right)\frac{\sigma^2}{n}$$

$$\cong \frac{\sigma^2}{n}$$

Thus, when n/N is small, we might just as well use the simpler expression for the variance (8-40) even if sampling is without replacement.

When sampling with replacement, it is not necessary for us to use equal probabilities. In fact, we might suspect that if we could choose a set of p's, $p_1, p_2, \ldots, p_N$ $\left(\sum_{i=1}^{N} p_i = 1\right)$ such that each p_i is approximately proportional to the corresponding v_i, we could get an estimator of $\bar{V}$ with smaller variance. Of course, we cannot do this exactly since the v's are unknown before sampling. However, we may have available another variable which gives some indication as to the size of the v's. For example, suppose that from 100 farms in a county we wish to estimate average income for the year from a random sample of 20 farms. Since income may be closely related to size, we could select a given farm with probability proportional to acreage (which we shall assume is known). The procedure for drawing such a sample will be illustrated later. Having determined the p's let $Y_1, Y_2, \ldots, Y_n$ be a random sample from the distribution with probability function

$$f(y) = p_i \qquad y = \frac{v_1}{Np_1}, \frac{v_2}{Np_2}, \ldots, \frac{v_N}{Np_N} \qquad (8\text{-}43)$$

If $p_i = 1/N$, then (8-43) reduces to (8-35). We have immediately

$$E(Y) = \frac{v_1}{Np_1}(p_1) + \frac{v_2}{Np_2}(p_2) + \cdots + \frac{v_N}{Np_N}(p_N)$$

$$= \frac{v_1 + v_2 + \cdots + v_n}{N} = \bar{V}$$

and

$$\text{Var}(Y) = \sum_{i=1}^{N} \left(\frac{v_i}{Np_i} - \bar{V} \right)^2 p_i = \sigma^2 \tag{8-44}$$

Again (8-39) and (8-40) hold true. Thus $\bar{Y}$ is still unbiased and

$$\text{Var}(\bar{Y}) = \frac{\text{Var}(Y)}{n} = \frac{1}{n} \sum_{i=1}^{N} \left(\frac{v_i}{Np_i} - \bar{V} \right)^2 p_i \tag{8-45}$$

The proof of the fact that $E(S^2) = \sigma^2$ still holds, as it does for any random sample from a distribution. Hence an unbiased estimate of $\text{Var}(\bar{Y})$ is

$$\text{var}(\bar{y}) = \frac{s^2}{n} = \left(\frac{1}{n} \right) \frac{1}{n-1} \sum_{i=1}^{n} (y_i - \bar{y})^2 \tag{8-46}$$

As usual, to get $\text{Var}(\hat{Y})$ and $\text{var}(\hat{y})$ multiply (8-45) and (8-46) by N^2.

EXAMPLE 8-5

From 5 farms whose sizes in acres are 200, 360, 180, 120, 140 draw a random sample of two farms using probabilities proportional to the size.

Solution

Since $200 + 360 + 180 + 120 + 140 = 1,000$ we can use as probabilities $p_1 = .200, p_2 = .360, p_3 = .180, p_4 = .120, p_5 = .140$. Divide the three-digit numbers 1, 2, . . . , 1,000 as follows:

Let 1, 2, . . . , 200 correspond to the first farm
201, . . . , 560 correspond to the second farm
561, . . . , 740 correspond to the third farm
741, . . . , 860 correspond to the fourth farm
861, . . . , 1,000 correspond to the fifth farm

Now enter the table of random numbers by some method such as those suggested in Sec. 4-4. Suppose we start in row 22, column 3. Then 104 is the first three-digit number. If we proceed down the column of three digits formed by columns 3, 4, and 5 the next number is 550. We see that 104 and 550 correspond to the first and second farms, and our sample includes the 200-acre and 360-acre farms. Since the table of random numbers is so constructed such that the probability of drawing any three-digit number is 1/1,000, the five farms have the desired probabilities of selection.

EXAMPLE 8-6 ──

Suppose we determine that the income of the 200-acre farm in Example 8-5 is $10,000 and the income of the 360-acre farm is $14,400. Estimate $\bar{V}$, $\hat{V}$, $\mathrm{Var}(\bar{Y})$, $\mathrm{Var}(\hat{Y})$ based upon sampling with unequal probabilities.

Solution

We have $N = 5$, $p_1 = .200$, $p_2 = .360$, $v_1 = \$10,000$, $v_2 = \$14,400$. Hence

$$y_1 = \frac{v_1}{Np_1} = \frac{\$10,000}{5(.200)} = \$10,000$$

$$y_2 = \frac{v_2}{Np_2} = \frac{\$14,400}{5(.360)} = \$8,000$$

Hence $\bar{y} = \$9,000$, $\hat{y} = \$45,000$,

$$\mathrm{var}(\bar{y}) = \left(\frac{1}{n}\right)\frac{1}{n-1}\sum_{i=1}^{2}(y_i - \bar{y})^2$$

$$= \left(\frac{1}{2}\right)\left(\frac{1}{1}\right)[(10,000 - 9,000)^2 + (8,000 - 9,000)^2]$$

$$= 1,000,000$$
$$\mathrm{var}(\hat{y}) = 5^2(1,000,000) = 25,000,000$$

If we had used sampling with equal probabilities, then $\bar{y} = \$12,200$,

$$\mathrm{var}(\bar{y}) = \left(\frac{1}{2}\right)\left(\frac{1}{1}\right)[(10,000 - 12,200)^2 + (14,400 - 12,200)^2]$$

$$= 4,840,000$$

$\hat{y} = \$61,000$ $\mathrm{var}(\hat{y}) = 25(4,840,000) = 121,000,000$

──

Example 8-6 demonstrates that if the p's are chosen approximately proportional to the v's, the variance of the estimators will be smaller. If it were possible to choose $p_i = v_i/N\bar{V}$, then it is easy to see from (8-45) that $\mathrm{Var}(\bar{Y}) = 0$. In addition $Y_i = v_i/N(v_i/N\bar{V}) = \bar{V}$, $\bar{Y} = \bar{V}$, and $\mathrm{var}(\bar{y}) = 0$.

The counterparts of (8-24) and (8-25) are

$$Z' = \frac{\bar{Y} - \bar{y}}{s/\sqrt{n}} \tag{8-47}$$

$$= \frac{\hat{Y} - \hat{y}}{sN/\sqrt{n}} \tag{8-48}$$

Now $Y_1, Y_2, \ldots, Y_n$ are independent and form a random sample from the distribution whose probability function is (8-35) when sampling with equal probabilities, or (8-43) when sampling with unequal probabilities. Hence it might be expected that the normal approximation would be a little better than when sampling without replacement.

The confidence interval (8-30) still applies, but (8-31) should be replaced by

$$\left(\bar{y} - z_{1-\alpha/2} \frac{s}{\sqrt{n}}, \quad \bar{y} + z_{1-\alpha/2} \frac{s}{\sqrt{n}} \right) \tag{8-49}$$

and (8-32) by

$$\left(\hat{y} - z_{1-\alpha/2} \frac{sN}{\sqrt{n}}, \quad \hat{y} + z_{1-\alpha/2} \frac{sN}{\sqrt{n}} \right) \tag{8-50}$$

EXAMPLE 8-7

Using the sample results of Example 8-1 but assuming sampling with replacement, estimate $\Pr(\bar{Y} > \$5,200)$ and find confidence intervals for $\bar{V}$ and $\hat{V}$ with coefficient approximately .90.

Solution

In Example 8-4 we found $\bar{y} = 5,000$ and $\mathrm{var}(\bar{y}) = 50,000$. Hence, using (8-47), we can write

$$\Pr(\bar{Y} > 5,200) = \Pr\left(\frac{\bar{Y} - 5,000}{\sqrt{50,000}} > \frac{5,200 - 5,000}{\sqrt{50,000}} \right)$$

$$\cong \Pr(Z > .894)$$
$$= 1 - \Pr(Z < .894)$$
$$= 1 - .80 = .20$$

The confidence interval for $\bar{V}$ is $(5,000 - 1.645\sqrt{50,000}, 5,000 + 1.645\sqrt{50,000})$ which reduces to (\$4,632, \$5,368), and multiplying the end points by 200 gives (\$926,400, \$1,073,600) as the confidence interval for $\hat{V}$.

EXAMPLE 8-8

Using the sample results of Example 8-6, estimate $\Pr(\bar{Y} > \$10,000)$ and find a confidence interval for $\bar{V}$ with coefficient approximately .95.

Solution

Since $n = 2$ we should have little faith in the results. However, we shall give the calculations to demonstrate the procedures. We had $\bar{y} = 9,000$, $\mathrm{var}(\bar{y}) = 1,000,000 = s^2/n$ so that $s/\sqrt{n} = 1,000$. Hence

$$\Pr(\bar{Y} > 10,000) = \Pr\left(\frac{\bar{Y} - 9,000}{1,000} > \frac{10,000 - 9,000}{1,000}\right)$$

$$\cong \Pr(Z > 1)$$

$$= .16$$

The confidence interval for $\bar{V}$ is given by (8-49) with $\alpha = .05$, $1 - \alpha/2 = .975$, $z_{.975} = 1.96$. Hence we get $[9,000 - 1.96(1,000),\ 9,000 + 1.96(1,000)]$, which reduces to ($\$7,040, \$10,960$).

EXERCISES

8-8 Suppose $N = 5$ and $v_1 = 7$, $v_2 = 8$, $v_3 = 5$, $v_4 = 16$, $v_5 = 14$. Calculate $\bar{y}$ for every (ordered) sample of size $n = 2$ drawn at random with replacement and equal probabilities and verify $E(\bar{Y}) = \bar{V}$, $\mathrm{Var}(\bar{Y}) = \sigma^2/n$.

8-9 Using the finite population given in Exercise 8-8, find s^2 for every (ordered) sample of size $n = 2$ drawn at random with replacement and equal probabilities and verify $E(S^2) = \sigma^2$.

8-10 Repeat Exercise 8-3 under the assumption that sampling has been performed at random with replacement and equal probabilities.

8-11 Repeat Exercise 8-4 under the assumption that sampling has been performed at random with replacement and equal probabilities.

8-12 Suppose a population consists of four items with values $v_1 = 20$, $v_2 = 48$, $v_3 = 84$, $v_4 = 96$. Find $\mathrm{Var}(\bar{Y})$ where $\bar{Y}$ is based upon samples of size $n = 2$ drawn with replacement and equal probabilities. Next suppose the items are drawn with replacement and probabilities $p_1 = .1$, $p_2 = .2$, $p_3 = .3$, $p_4 = .4$ and find $\mathrm{Var}(\bar{Y})$.

8-13 A county has 100 lakes with total area 1,000 acres. To estimate the total number of fish caught in the county over a given weekend, game wardens are stationed at five lakes and count the number of fish caught by requiring that fishermen report to them before leaving the lake. Suppose that the five wardens report totals of 5, 60, 80, 15, and 40 fish. If lakes were selected at random with replacement and equal probabilities find $\hat{y}$ and $\mathrm{var}(\hat{y})$. Suppose that the areas of the lakes were respectively 1, 15, 20, 5, and 10 acres and that sampling was performed with replacement and probabilities proportional to the areas (that is, $p_1 = .001$, $p_2 = .015$, $p_3 = .020$, $p_4 = .005$, $p_5 = .010$ for the five selected lakes). Now what are $\hat{y}$ and $\text{var}(\hat{y})$?

8-14 Assuming that n is large enough in Exercise 8-13 (which it probably is not) find confidence intervals for $\hat{Y}$ with coefficient approximately .95 using both methods of sampling.

8-15 Another method of sampling might be called variable-sample-size sampling. Under this method each item in the population is selected with probability p (which can be accomplished by spinning a balanced pointer N times or drawing N numbers from a table of random numbers). Define a random variable X with probability function

$$f(x) = p \qquad x = 1$$
$$ = 1 - p \qquad x = 0$$

Let $X_1, X_2, \ldots, X_N$ be a random sample from the distribution where $X_i = 1$ if v_i is to be included in the sample, $X_i = 0$ if v_i is not included in the sample ($X_i, X_2, \ldots, X_n$ will be independent if a balanced spinner or a table of random numbers is used to determine whether or not to include each value of v). Show that

$$\bar{Y} = \frac{1}{Np} \sum_{i=1}^{N} v_i X_i = \frac{Y_1 + Y_2 + \cdots + Y_n}{Np} = \frac{1}{Np} \sum_{i=1}^{n} Y_i$$

is an unbiased estimator of $\bar{V}$ where $y_1, y_2, \ldots, y_n$ are the values of the v's that are included in the sample. Show also that

$$\text{Var}(\bar{Y}) = \frac{1-p}{pN^2} \sum_{i=1}^{N} v_i^2 \qquad \text{and} \qquad \text{var}(\bar{Y}) = \frac{1-p}{p^2 N^2} \sum_{i=1}^{N} v_i^2 X_i$$

is an unbiased estimator of $\text{Var}(\bar{Y})$. What is the distribution of the sample size and the expected sample size?

8-16 In Exercise 8-15 show that

$$\text{Var}(\bar{Y}) = \frac{1-p}{pN} (\sigma^2 + \bar{V}^2)$$

Observe that $\bar{Y}$ has the property (usually undesirable) that its variance depends upon the mean of the population, $\bar{V}$.

8-17 Let $v_1 = -1$, $v_2 = 0$, $v_3 = 1$, $p = \frac{1}{3}$. If variable-sample-size sampling is used, find the probability distribution of $\bar{Y}$. Show that $E(\bar{Y}) = \bar{V}$. Find $\text{Var}(\bar{Y})$ and demonstrate that the formula

$\text{Var}(\bar{Y}) = [(1 - p)/pN^2] \sum\limits_{i=1}^{N} v_i^2$ is correct. Observe that the expected sample size is $Np = 1$ and that $\text{Var}(\bar{Y})$ is less than the variance of the estimator of $\bar{V}$ one gets with a fixed sample size $n = 1$. Hence there exist problems in which variable-sample-size sampling produces an estimator of $\bar{V}$ with smaller variance than the fixed-sample-size procedure where n of the fixed-sample-size procedure equals the expected number of observations under variable-sample-size sampling.

8-18 Suppose $p = .25$, $N = 20$, and variable-sample-size sampling as described in Exercise 8-15 is used. If we start in row 1, column 1 of the table of random numbers and include the value of v if a two-digit number in the range 1, 2, . . . , 25 is obtained, which values of v should be included? Proceed down the column of two-digit numbers.

8-19 It is interesting to note that except for minor changes the same method of derivation used in Sec. 8-2 can be used to derive the results of this section. The procedure is more complicated than the one we used since it requires the use of dependent (rather than independent) random variables. Define

$$\bar{Y} = \frac{1}{n} \sum_{i=1}^{N} \frac{v_i X_i}{Np_i} \tag{8-51}$$

where X_i is the number of times that v_i appears in the sample. It is easy to see that this is the same as $\bar{Y}$ based upon a random sample from (8-43). Verify that conditions (5-55) are satisfied with N categories and probabilities $p_1, p_2, \ldots, p_N$ so that the joint probability function of $X_1, X_2, \ldots, X_N$ is given by (5-57) with k replaced by N. Hence, by (5-60), the marginal probability distribution of each X_i is a binomial with parameters n and p_i and $E(X_i) = np_i$, $\text{Var}(X_i) = np_i(1 - p_i)$. The joint probability function of X_i and X_j is given by (5-62), and from (5-67) we have $\text{Cov}(X_i, X_j) = -np_ip_j$. Show that $\bar{Y}$ as defined by (8-51) is an unbiased estimator of $\bar{V}$. Use (5-47) to get $\text{Var}(\bar{Y})$ in terms of $\text{Var}(X_i)$ and $\text{Cov}(X_i, X_j)$. This should give

$$\text{Var}(\bar{Y}) = \frac{1}{nN^2} \left[\sum_{i=1}^{N} v_i^2 \left(\frac{1 - p_i}{p_i} \right) - 2 \sum_{\substack{i,j=1 \\ i<j}}^{N} v_i v_j \right]$$

Now, using (8-13), show that this reduces to

$$\text{Var}(\bar{Y}) = \frac{1}{n} \left[\sum_{i=1}^{N} \frac{v_i^2}{N^2 p_i} - \bar{V}^2 \right]$$

which can be rewritten

$$\text{Var}(\bar{Y}) = \frac{1}{n} \sum_{i=1}^{N} \left(\frac{v_i}{Np_i} - \bar{V} \right)^2 p_i$$

which agrees with (8-45). The sample variance S^2 can now be written

$$S^2 = \frac{1}{n-1} \sum_{i=1}^{N} \left(\frac{v_i}{Np_i} - \bar{Y} \right)^2 X_i \qquad (8\text{-}52)$$

which is the same as (8-19) with v_i replaced by v_i/Np_i. If this replacement is made all the algebra leading up to (8-22) is correct. Using $E(X_i) = np_i$, $\text{Var}(\bar{Y}) = \sigma^2/n$ where σ^2 is defined by (8-44), show $E(S^2) = \sigma^2$. Hence an unbiased estimator of $\text{Var}(\bar{Y})$ is $\text{var}(\bar{Y}) = S^2/n$ with S^2 defined by (8-52).

8-4   STRATIFIED SAMPLING

In Exercise 2-30 we have already encountered the concept of stratification. In that problem we had three nonoverlapping subgroups or subpopulations. Town A had $N_1 = 100$ people, town B had $N_2 = 200$ people, and $N_3 = 300$ people lived in rural areas. These subpopulations are known as *strata*. Obviously another way to divide the population of $N = 600$ into strata is to assign the $N_1 = 320$ women to one subgroup and the $N_2 = 280$ men to a second subgroup.

The procedure in stratified sampling consists of first dividing the population into strata and then drawing a random sample from each stratum. The objectives are the same as those of Secs. 8-2 and 8-3. That is, we wish to estimate $\bar{V}$, the population mean, and $\hat{V}$, the population total, to obtain the variances of the estimators, to estimate these variances, and perhaps to obtain confidence intervals and make probability statements.

The main reason for stratification is to reduce the variability of the estimators. For example, suppose we wish to estimate the average income of all individuals in the county mentioned in Exercise 2-30. If it were known that every man had the same income and every woman had the same income, then a sample of 1 from each of these strata would produce a perfect estimator of $\bar{V}$, an estimator with variance 0. Suppose the man who is sampled reports an income of \$5,000 for last year while the woman who is sampled reports \$2,000. Then

$$\bar{V} = \frac{320(\$2,000) + 280(\$5,000)}{600} = \$3,400$$

is the exact population average. If simple random sampling is used, the estimator will have positive variance unless $n = N$.

The above example suggests that strata should be formed by selecting sampling units for the same strata that are as nearly alike as possible with respect to the variable being measured. Besides sex, other factors that might be used for stratification include geographic area, age, educational background, and income bracket.

In general there will be L strata with $N_1, N_2, \ldots, N_L$ units respectively. The total number of units in the population will be $N_1 + N_2 + \cdots + N_L = N$. The sample sizes taken will be $n_1, n_2, \ldots, n_L$ respectively with total sample size $n_1 + n_2 + \cdots + n_L = n$. Let $\bar{V}$ be the population average and $\bar{V}_1, \bar{V}_2, \ldots, \bar{V}_L$ be the strata averages. Then the population average is the total of all units divided by N, or

$$\bar{V} = \frac{N_1 \bar{V}_1 + N_2 \bar{V}_2 + \cdots + N_L \bar{V}_L}{N} \tag{8-53}$$

$$= \frac{N_1}{N} \bar{V}_1 + \frac{N_2}{N} \bar{V}_2 + \cdots + \frac{N_L}{N} \bar{V}_L = \sum_{i=1}^{L} \frac{N_i}{N} \bar{V}_i$$

If we let the sample means of the strata be denoted by $\bar{Y}_1, \bar{Y}_2, \ldots, \bar{Y}_L$, then by (5-45)

$$\bar{Y} = \frac{N_1}{N} \bar{Y}_1 + \frac{N_2}{N} \bar{Y}_2 + \cdots + \frac{N_L}{N} \bar{Y}_L = \sum_{i=1}^{L} \frac{N_i}{N} \bar{Y}_i \tag{8-54}$$

is an unbiased estimator of $\bar{V}$ for any method of sampling that produces unbiased estimates of the strata means. That is,

$$E(\bar{Y}) = \frac{N_1}{N} E(\bar{Y}_1) + \frac{N_2}{N} E(\bar{Y}_2) + \cdots + \frac{N_L}{N} E(\bar{Y}_L)$$

$$= \frac{N_1}{N} \bar{V}_1 + \frac{N_2}{N} \bar{V}_2 + \cdots + \frac{N_L}{N} \bar{V}_L = \bar{V}$$

If sampling in each stratum is performed independently of sampling in any other stratum, then $\bar{Y}_1, \bar{Y}_2, \ldots, \bar{Y}_L$ would be independent and by (5-49) we have

$$\text{Var}(\bar{Y}) = \sum_{i=1}^{L} \frac{N_i^2}{N^2} \text{Var}(\bar{Y}_i) \tag{8-55}$$

If simple random sampling discussed in Sec. 8-2 is used, then by using (8-16) we get

$$\text{Var}(\bar{Y}) = \sum_{i=1}^{L} \frac{N_i^2}{N^2} \left(\frac{N_i - n_i}{n_i N_i} \right) \sigma_i'^2 = \frac{1}{N^2} \sum_{i=1}^{L} N_i (N_i - n_i) \frac{\sigma_i'^2}{n_i} \tag{8-56}$$

where each $\sigma_i'^2$ is the variance of the ith stratum defined by (8-15). Obviously, an unbiased estimator of this variance is

$$\text{var}(\bar{Y}) = \frac{1}{N^2} \sum_{i=1}^{L} N_i(N_i - n_i) \frac{S_i^2}{n_i} \tag{8-57}$$

where S_i^2 is the sample variance in the ith stratum. If sampling with replacement and equal probabilities is used, then (8-55) becomes

$$\text{Var}(\bar{Y}) = \sum_{i=1}^{L} \frac{N_i^2}{N^2} \frac{\sigma_i^2}{n_i} \tag{8-58}$$

with an unbiased estimator being given by

$$\text{var}(\bar{Y}) = \sum_{i=1}^{L} \frac{N_i^2}{N^2} \frac{S_i^2}{n_i} \tag{8-59}$$

Here each σ_i^2 is defined by (8-38). Formulas (8-58) and (8-59) hold for sampling with replacement and unequal probabilities, except that now each σ_i^2 has to be computed from (8-44) and the total number of p's required is N (N_i in each stratum and summing to 1 in each stratum). Similarly, we could write down $\text{Var}(\bar{Y})$ and $\text{var}(\bar{Y})$ if the variable-sample-size sampling procedure of Exercise 8-15 is used.

Since the variance formulas (8-56) and (8-58) both have n_i appearing in the denominator, it is obvious that the variance can be decreased by increasing the n_i. Sometimes the total sample size n is fixed and, subject to this condition, we may wish to select the n_i in some reasonable manner. One method of determining the n_i is called proportional allocation. This means that the sample sizes in each stratum are chosen in proportion to the number of units in the stratum. Using this procedure we select each n_i so that

$$\frac{n_i}{N_i} = \frac{n}{N} \qquad i = 1, 2, \ldots, L$$

or

$$n_i = \frac{n}{N} N_i \tag{8-60}$$

Thus the fraction of each stratum that is sampled is equal to the fraction of the total population sampled. It is easy to verify that

$$\sum_{i=1}^{L} n_i = \sum_{i=1}^{L} \frac{n}{N} N_i = \frac{n}{N} \sum_{i=1}^{L} N_i = \frac{n}{N} N = n$$

as should be. If the n_i as given by (8-60) are not integers, the usual procedure is to round them off to one of the nearest integer values, keeping the total sample size equal to n. If the values of the n_i given by

(8-60) are substituted in (8-58) and (8-59) we get

$$\text{Var}(\bar{Y}) = \sum_{i=1}^{L} \frac{N_i \sigma_i^2}{Nn} \qquad (8\text{-}61)$$

and

$$\text{var}(\bar{Y}) = \sum_{i=1}^{L} \frac{N_i S_i^2}{Nn} \qquad (8\text{-}62)$$

for proportional allocation. It can be shown (see Exercise 8-24) that (8-61) can never exceed the variance of sampling with replacement but without stratification.

EXAMPLE 8-9

Suppose we wish to estimate the average income for residents of the county mentioned at the beginning of the section. As strata we shall use residents of town A, residents of town B, and rural residents, with $N_1 = 100$, $N_2 = 200$, $N_3 = 300$. If we plan to use a sample of size 60 and proportional allocation, what are n_1, n_2, n_3? Using these n_i we then use random sampling with replacement in each stratum and find $\bar{y}_1 = \$3,600$, $\bar{y}_2 = \$3,300$, $\bar{y}_3 = \$3,000$, $s_1 = \$100$, $s_2 = \$200$, $s_3 = \$300$. Find the estimate of the average income and estimate the variance of the estimator.

Solution

Using (8-60) we get

$$n_1 = \frac{60}{600}(100) = 10$$

$$n_2 = \frac{60}{600}(200) = 20$$

$$n_3 = \frac{60}{600}(300) = 30$$

Then, using (8-54) we get

$$\bar{y} = \frac{100}{600}(\$3,600) + \frac{200}{600}(\$3,300) + \frac{300}{600}(\$3,000) = \$3,200$$

The estimate of the variance can be computed from (8-62) and we get for the observed value of $\text{var}(\bar{Y})$

$$\text{var}(\bar{y}) = \frac{100(100)^2 + 200(200)^2 + 300(300)^2}{600(60)}$$

$$= (100)^2 \left[\frac{1 + 8 + 27}{6(60)} \right] = 1,000$$

EXAMPLE 8-10

Suppose that in Example 8-9 we plan to use $n = 70$ with proportional allocation. Find n_1, n_2, n_3.

Solution

We have

$$n_1 = \frac{70}{600} (100) = 11.67$$

$$n_2 = \frac{70}{600} (200) = 23.33$$

$$n_3 = \frac{70}{600} (300) = 35$$

We can verify that $11.67 + 23.33 + 35 = 70$. We would use $n_1 = 12$, $n_2 = 23$, $n_3 = 35$. Now the conditions for the use of (8-62) are not exactly satisfied, and perhaps (8-59) should be used to estimate the variance of $\bar{Y}$. However, one would expect the two estimates to differ by very little and no great error can be made by using (8-62).

Another reasonable method of allocation for fixed total sample size n is based upon selecting the n_i so that (8-58) is a minimum. By a straightforward application of the minimization techniques of calculus it can be shown that this leads to

$$n_i = n \frac{N_i \sigma_i}{\sum\limits_{i=1}^{L} N_i \sigma_i} \qquad i = 1, 2, \ldots, L \tag{8-63}$$

If the n_i are determined by (8-63), this statification procedure is called optimum allocation. Even though more advanced mathematics may be required to find the n_i given by (8-63), once the result is given it is fairly easy to show that these values of n_i actually minimize the variance (8-58) (see Exercise 8-23). Two facts concerning optimum allocation are readily apparent. First, the formula for n_i contains $\sigma_1, \sigma_2, \ldots, \sigma_L$, quantities that are usually unknown. Hence, unless we knew the σ_i or had some reasonable advance estimates for them, (8-63) would be useless. Second, if $\sigma_1 = \sigma_2 = \cdots = \sigma_L = \sigma$, then (8-63) and (8-60) are the same. Consequently, unless we feel that the σ_i are radically different in magnitude, we might just as well use proportional allocation. Small differences in the σ_i are not going to make an appreciable difference in the variances obtained by the two methods of allocation.

If the n_i of (8-63) are substituted in (8-58) we get

$$\text{Var}(\bar{Y}) = \frac{1}{nN^2} \left(\sum_{i=1}^{L} N_i \sigma_i \right)^2 \tag{8-64}$$

As we have implied above, since optimum allocation minimizes $\text{Var}(\bar{Y})$, the variance of $\bar{Y}$ obtained by proportional allocation must be at least as large as the variance of $\bar{Y}$ obtained by optimum allocation with equality occurring if all σ_i are equal.

As previously

$$Z' = \frac{\bar{Y} - \bar{y}}{\sqrt{\text{var}(\bar{y})}} \tag{8-65}$$

is treated as a random variable having approximately a standard normal distribution, and a confidence interval with coefficient approximately $1 - \alpha$ for $\bar{V}$ is given by

$$[\bar{y} - z_{1-\alpha/2} \sqrt{\text{var}(\bar{y})}, \quad \bar{y} + z_{1-\alpha/2} \sqrt{\text{var}(\bar{y})}] \tag{8-66}$$

We should, of course, use the correct estimates of $\text{Var}(\bar{Y})$ as given in this section.

EXAMPLE 8-11

Using the information given in Example 8-9, estimate $\text{Pr}(\bar{Y} > \$3,250)$. Find a confidence interval for $\bar{V}$ with coefficient approximately equal to .95.

Solution

We have

$$\text{Pr}(\bar{Y} > 3,250) = \text{Pr}\left(\frac{\bar{Y} - 3,200}{\sqrt{1,000}} > \frac{3,250 - 3,200}{\sqrt{1,000}} \right)$$

$$\cong \text{Pr}(Z > 1.58)$$

$$= .057$$

The interval is

$(3,200 - 1.96 \sqrt{1,000}, \ 3,200 + 1.96 \sqrt{1,000})$

which reduces to ($\$3,138$, $\$3,262$).

EXERCISES

8-20   In Example 8-9 suppose we had known in advance from past surveys that the standard deviation of incomes for people living in rural areas has been three times as large as for people living in town.

Use this fact and optimum allocation to determine the sample sizes so that $n_1 + n_2 + n_3 = 60$. Assuming that the estimates given in Example 8-9 were determined with these sample sizes, estimate $\bar{V}$ and $\text{Var}(\bar{Y})$.

8-21 Suppose we wish to estimate the average number of dates per week for girls attending the University. Of 3,000 girls 600 live in sororities, 1,500 live in dormitories, and 900 live in town (in privately owned dwellings). Determine the sample size for each of these strata if a total sample size of 100 is used with proportional allocation. With these sample sizes sampling is performed at random with equal probabilities and with replacement. The results are:

Sororities: sample mean 4.5, sample variance 4.4
Dormitories: sample mean 2.8, sample variance 7.5
Town: sample mean 2.0, sample variance 5.4

Find $\bar{y}$ and $\text{var}(\bar{y})$. Estimate $\text{Pr}(\bar{Y} > 3.0)$. Find a confidence interval for $\bar{V}$ with coefficient approximately equal to .95.

8-22 In Exercise 8-21 suppose that similar surveys conducted in other institutions showed that the standard deviation for sorority girls is $\frac{2}{3}$, for dormitory girls is 1.9, and for town girls is 1.5. Use these figures to determine the sample sizes for optimum allocation. Assuming these variances are correct, find $\text{Var}(\bar{Y})$.

8-23 The inequality

$$\sum_{i=1}^{L} \left(\frac{N_i \sigma_i}{N n_i} - \frac{D}{N n} \right)^2 n_i \geq 0$$

where $D = \sum_{i=1}^{L} N_i \sigma_i$ is obviously true since all terms on the left side of the inequality sign are nonnegative. Square the binomial term to get three separate sums and use the result to show that (8-64) is the minimum value of (8-58).

8-24 We have stated that $\text{Var}(\bar{Y})$ given by (8-61), which was achieved by proportional allocation, can never exceed the variance of sampling with replacement but without stratification. To prove this we first note that

$$\sigma^2 = \frac{1}{N} \sum_{i=1}^{L} \sum_{j=1}^{N_i} (v_{ij} - \bar{V})^2$$

where v_{ij} is the j value in the ith primary. (See Appendix A1 for more on double summation.) That is, according to (8-38) we subtract $\bar{V}$ from every value of v, square all the results, and add. The double sum means that we first evaluate

$$\sum_{j=1}^{N_i} (v_{ij} - \bar{V})^2$$

in strata i, $i = 1, 2, \ldots, L$ and then sum these results for all strata. Now, in each stratum subtract and add the strata mean $\bar{V}_i$ getting

$$\sigma^2 = \frac{1}{N} \sum_{i=1}^{L} \sum_{j=1}^{N_i} [(v_{ij} - \bar{V}_i) + (\bar{V}_i - \bar{V})]^2$$

After squaring and separating into three terms this becomes

$$\sigma^2 = \frac{1}{N} \sum_{i=1}^{L} \sum_{j=1}^{N_i} (v_{ij} - \bar{V}_i)^2 + \frac{2}{N} \sum_{i=1}^{L} \sum_{j=1}^{N_i} (v_{ij} - \bar{V}_i)(\bar{V}_i - \bar{V})$$

$$+ \frac{1}{N} \sum_{i=1}^{L} \sum_{j=1}^{N_i} (\bar{V}_i - \bar{V})^2$$

Note that

$$\sum_{j=1}^{N_i} (v_{ij} - \bar{V}_i)(\bar{V}_i - \bar{V}) = (\bar{V}_i - \bar{V}) \sum_{j=1}^{N_i} (v_{ij} - \bar{V}_i)$$

$$= (\bar{V}_i - \bar{V})(N_i \bar{V}_i - N_i \bar{V}_i)$$

$$= 0$$

Hence, the middle sum is equal to 0 and we can write

$$\frac{\sigma^2}{n} = \frac{1}{Nn} \sum_{i=1}^{L} \sum_{j=1}^{N_i} (v_{ij} - \bar{V}_i)^2 + \frac{1}{Nn} \sum_{i=1}^{L} \sum_{j=1}^{N_i} (\bar{V}_i - \bar{V})^2$$

But, by definition we have

$$\sigma_i^2 = \frac{1}{N_i} \sum_{j=1}^{N_i} (v_{ij} - \bar{V}_i)^2$$

Using this definition in the previous equation, show that the variance of $\bar{Y}$ with random sampling with replacement is equal to the variance of $\bar{Y}$ using proportional allocation and sampling with replacement plus a nonnegative term. Hence, proportional allocation yields a smaller variance (unless every stratum has the same mean).

REFERENCE

COCHRAN, WILLIAM G.: "Sampling Techniques," 2d ed, John Wiley & Sons, Inc., New York, 1963.

8-1 With each of N units is associated a number $v_1, v_2, \ldots, v_N$. One of the major objectives of this chapter is to obtain good estimates of

$$\hat{V} = v_1 + v_2 + \cdots + v_N \qquad (8\text{-}1)$$

and

$$\bar{V} = \frac{\hat{V}}{N} \qquad (8\text{-}2)$$

by using a sample of size $n < N$.

8-2 When a sample of size n is selected from N units in such a way that each sample of the same size has the same probability of being chosen, the procedure is called simple random sampling. This implies that sampling is performed without replacement. If the observed values of a sample are $y_1, y_2, \ldots, y_n$, then

$$\bar{y} = \frac{y_1 + y_2 + \cdots + y_n}{n}$$

is an unbiased estimate of $\bar{V}$. That is, $E(\bar{Y}) = \bar{V}$. Further,

$$\text{Var}(\bar{Y}) = \frac{N-n}{nN} \frac{\sum_{i=1}^{N}(v_i - \bar{V})^2}{N-1} \qquad (8\text{-}16)$$

for which

$$\text{var}(\bar{y}) = \frac{N-n}{nN} \frac{\sum_{i=1}^{n}(y_i - \bar{y})^2}{n-1} = \frac{N-n}{nN} s^2$$

is an unbiased estimate. An unbiased estimate of $\hat{V}$ is

$$\hat{y} = N\bar{y}$$

$$\text{Var}(\hat{Y}) = \frac{N(N-n)}{n} \frac{\sum_{i=1}^{n}(v_i - \bar{V})^2}{N-1} \qquad (8\text{-}17)$$

for which

$$\text{var}(\hat{y}) = \frac{N(N-n)}{n} \frac{\sum_{i=1}^{n}(y_i - \bar{y})^2}{n-1}$$

is an unbiased estimate.

To make probability statements about $\bar{Y}$ and $\hat{Y}$, we use the fact that

$$Z' = \frac{\bar{Y} - \bar{y}}{s\sqrt{N - n}/\sqrt{nN}} \tag{8-24}$$

$$= \frac{Y - \hat{y}}{s\sqrt{N(N - n)}/\sqrt{n}} \tag{8-25}$$

has approximately a standard normal distribution. Confidence intervals for $\bar{V}$ and $\hat{V}$ with confidence coefficients approximately equal to $1 - \alpha$ are, respectively,

$$\left(\bar{y} - z_{1-\alpha/2}\frac{s\sqrt{N - n}}{\sqrt{nN}}, \quad \bar{y} + z_{1-\alpha/2}\frac{s\sqrt{N - n}}{\sqrt{nN}}\right) \tag{8-31}$$

and

$$\left(\hat{y} - z_{1-\alpha/2}\frac{s\sqrt{N(N - n)}}{\sqrt{n}}, \quad \hat{y} + z_{1-\alpha/2}\frac{s\sqrt{N(N - n)}}{\sqrt{n}}\right) \tag{8-32}$$

The computing formula for s^2 is

$$s^2 = \frac{n\sum\limits_{i=1}^{n} y_i^2 - (\sum\limits_{i=1}^{n} y_i)^2}{n(n - 1)} \tag{8-34}$$

8-3 In random sampling with replacement from a finite number of units, $Y_1, Y_2, \ldots, Y_n$ can be regarded as a random sample from the distribution whose probability function is

$$f(y) = \frac{1}{N} \qquad y = v_1, v_2, \ldots, v_N \tag{8-35}$$

An unbiased estimator of $\bar{V}$ is $\bar{Y}$, whose variance is

$$\text{Var}(\bar{Y}) = \frac{\text{Var}(Y)}{n} \tag{8-40}$$

An unbiased estimate of $\text{Var}(\bar{Y})$ is

$$\text{var}(\bar{y}) = \frac{s^2}{n}$$

An unbiased estimator of $\hat{V}$ is $\hat{Y} = N\bar{Y}$, whose variance

$$\text{Var}(\hat{Y}) = \frac{N^2\,\text{Var}(Y)}{n}$$

is estimated by

$$\text{var}(\hat{y}) = \frac{N^2 s^2}{n}$$

If n/N is small, there is little difference between the estimates obtained by sampling with and without replacement. If v_i is drawn independently with probability p_i and

$$f(y) = p_i \qquad y = \frac{v_i}{Np_i} \qquad i = 1, 2, \ldots, N$$

then $\bar{Y}$ is an unbiased estimator of $\bar{V}$ with

$$\text{Var}(\bar{Y}) = \frac{1}{n} \sum_{i=1}^{N} \left(\frac{v_i}{Np_i} - \bar{V} \right)^2 p_i \tag{8-45}$$

for which an unbiased estimate is

$$\text{var}(\bar{y}) = \frac{1}{n} \frac{\sum_{i=1}^{n} (y_i - \bar{y})^2}{n - 1} = \frac{s^2}{n} \tag{8-46}$$

8-4 Nonoverlapping subpopulations are called strata. In general one considers L strata with $N_1, N_2, \ldots, N_L$ units from which L random samples of size $n_1, n_2, \ldots, n_L$ are drawn. If the strata sample means are $\bar{Y}_1, \bar{Y}_2, \ldots, \bar{Y}_L$, then

$$\bar{Y} = \sum_{i=1}^{L} \frac{N_i}{N} \bar{Y}_i \tag{8-54}$$

is an unbiased estimator of $\bar{V}$. Here $N = \sum_{i=1}^{L} N_i$, with

$$\text{Var}(\bar{Y}) = \sum_{i=1}^{L} \frac{N_i^2}{N^2} \text{Var}(\bar{Y}_i) \tag{8-55}$$

being estimated by

$$\text{var}(\bar{y}) = \sum_{i=1}^{L} \frac{N_i^2}{N^2} \text{var}(\bar{y}_i)$$

where $\text{var}(\bar{y}_i)$ is computed from the appropriate formula, depending upon which kind of sampling is used in the strata. If the n_i are chosen so that

$$n_i = \frac{n}{N} N_i \tag{8-60}$$

where $n = \sum_{i=1}^{L} n_i$ is fixed, the method of determining the n_i is called proportional allocation. With proportional allocation

$$\text{var}(\bar{Y}) = \sum_{i=1}^{L} \frac{N_i S_i^2}{Nn} \tag{8-62}$$

If n is fixed and the n_i are chosen to minimize $\mathrm{Var}(\bar{Y})$, then

$$n_i = n \frac{N_i \sigma_i}{\sum\limits_{i=1}^{L} N_i \sigma_i} \tag{8-63}$$

and the method of obtaining the n_i is called optimum allocation. With the n_i so selected,

$$\mathrm{Var}(\bar{Y}) = \frac{1}{nN^2} \left(\sum\limits_{i=1}^{L} N_i \sigma_i \right)^2 \tag{8-64}$$

As in previous situations, a confidence interval with confidence coefficient approximately equal to $1 - \alpha$ is

$$[\bar{y} - z_{1-\alpha/2} \sqrt{\mathrm{var}(\bar{y})}, \quad \bar{y} + z_{1-\alpha/2} \sqrt{\mathrm{var}(\bar{y})}] \tag{8-66}$$

and

$$Z' = \frac{\bar{Y} - \bar{y}}{\sqrt{\mathrm{var}(\bar{y})}} \tag{8-65}$$

is treated as a random variable that is approximately standard normal.

CHAPTER NINE

Markov Chains

9-1 STOCHASTIC PROCESSES

We have already considered a number of situations involving a series (or sequence) of random experiments.   As the word random implies, probabilities were associated with the various outcomes of each experiment. As a simple illustration consider a series of three binomial experiments, each consisting of the throwing of a six-sided symmetric die.   If we consider the rolling of a 6 a success, then the probability of a success is $\frac{1}{6}$ and the various possible outcomes that may be obtained, together with the associated probabilities, can be quickly listed.   For convenience in displaying this information we shall use a tree diagram (see Fig. 9-1).   We have, of course, studied a number of interesting properties for the binomial sequence of experiments.   As another example, suppose that three

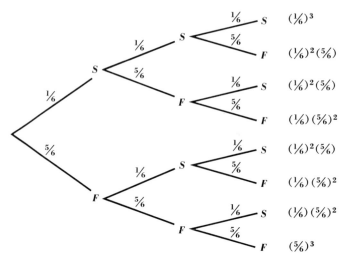

Figure 9-1 *Tree diagram for sequence of three binomial experiments with* $p = \frac{1}{6}$. *S is success, F is failure.*

cards are drawn from a well-shuffled deck, one at a time without replacement. Let the drawing of a spade be a success. The various possibilities together with probabilities are given in Fig. 9-2. Some of the properties of this type of sequence were investigated during our study of the hypergeometric distribution. Any such sequence of random experiments, such as the two we have discussed, is often referred to as a *stochastic process*.

Figure 9-2 *Tree diagram for drawing three cards from deck.*

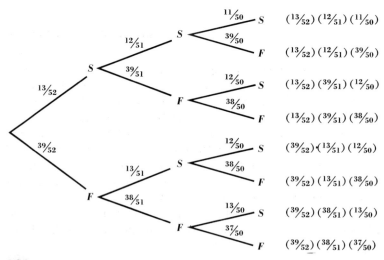

In comparing the two sequences of the previous paragraph we observe one contrasting difference. In the first case the outcome of each experiment was independent of the outcome of *any* previous experiment, while in the second case each outcome depended upon the results of *all* previous experiments. For some sequences the result of each experiment will depend at most upon the result of the immediately preceding experiment. A stochastic process of this type is called a *Markov chain*. As an illustration, suppose that a machine produces bolts. If a good bolt is turned out, the probability that the next bolt is also good is .9. However, if a defective bolt is produced, then the probability that the next one is good is .8. Thus, the result at each stage depends upon the immediately preceding result and no others.

A Markov chain is, of course, just another model that can be used to characterize a series of experiments. As in the case of previous models, we first hypothesize a set of conditions. Then, assuming that the conditions hold, we can derive interesting properties of the model, which can be worked out once and for all. Later, we can attempt to fit the model to real-life situations and use the derived properties to make probability statements about the future behavior of the process. Such a procedure is not new to us as we have done this repeatedly in previous chapters.

9-2 THE MARKOV CHAIN MODEL AND BASIC TERMINOLOGY

Consider a sequence of experiments with the following properties:

(a) *The result of each experiment can be classified into one of k categories, say, $C_1, C_2, \ldots, C_k$.*

(b) *The probabilities of falling into these categories for any experiment depends at most upon the outcome of the immediately preceding experiment and upon no other previous experiment.*   (9-1)

(c) *The probability that category j occurs given that category i has occurred on the immediately preceding experiment remains constant throughout the sequence of experiments. (This probability will be denoted by p_{ij}. Since i and j can each assume any of the values 1, 2, \ldots, k, there are k^2 such probabilities.)*

To illustrate let us again consider the bolt machine mentioned in the previous section. We could let C_1 correspond to the result "good bolt" and C_2 correspond to the result "bad bolt." The probabilities associated with the process are

p_{11} = .9, the probability that a good bolt follows a good bolt
p_{12} = .1, the probability that a bad bolt follows a good bolt
p_{21} = .8, the probability that a good bolt follows a bad bolt
p_{22} = .2, the probability that a bad bolt follows a bad bolt

In attempting to justify the use of this model with a real-life machine, one may have real doubts that the (b) and (c) conditions of (9-1) are satisfied. However, if these conditions are fulfilled, then the sequence of experiments forms a Markov chain.

The categories C_1, C_2, . . . , C_k are called *states* and when the result C_i occurs the process is said to be in state C_i. The probabilities p_{ij} are called transition probabilities and can be conveniently exhibited in a square matrix P where

$$P = \begin{bmatrix} p_{11} & p_{12} & \cdot & \cdot & p_{1k} \\ p_{21} & p_{22} & \cdot & \cdot & p_{2k} \\ \cdot & \cdot & \cdot & \cdot & \cdot & \cdot & \cdot & \cdot & \cdot \\ p_{k1} & p_{k2} & \cdot & \cdot & p_{kk} \end{bmatrix} \tag{9-2}$$

For the bolt machine example we have

$$P = \begin{bmatrix} .9 & .1 \\ .8 & .2 \end{bmatrix}$$

It is easy to see that the sum of the p_{ij} in each row must be 1, that is,

$$\sum_{j=1}^{k} p_{ij} = 1 \qquad i = 1, 2, \ldots, k \tag{9-3}$$

since one of the k states must follow state C_i if the next experiment in the sequence is performed. A matrix P of the type (9-2) satisfying Eq.(9-3) is called a stochastic matrix or transition matrix. Mathematicians have devoted considerable effort to the investigation of interesting properties of stochastic matrices.

Some information which we may wish to have for a Markov chain might include:

1 The probability that a process which starts in state C_i will be in state C_j after the sequence of experiments has been performed n times. Denote this by $p_{ij}^{(n)}$.
2 The number which $p_{ij}^{(n)}$ approaches as n increases if such a number exists.
3 The average number of repetitions required for a process to go from state C_i to state C_j.

300

4 The probability that x_1 outcomes are in C_1, x_2 outcomes are in C_2, . . . , x_k outcomes are in C_k if the series starts in a given state and is repeated n times. We have, of course, already derived this result for the case in which successive experiments are independent of any previous experiment when considering the multinomial model (binomial when $k = 2$). (Observe that the multinomial model is a special case of a Markov chain.)

These are some of the things which we shall investigate in this chapter.

EXAMPLE 9-1

Suppose that a six-sided symmetric die is being rolled. Let the categories or states be

C_1: the result 1 or 2
C_2: the result 3, 4, or 5
C_3: the result 6

When one of six sides shows, that result is not allowed on the next experiment. Hence, if a 5 is rolled, then to determine the next state, the die is rolled until something other than a 5 shows. Find the matrix of transition probabilites.

Solution

If the last result was state C_1, then of the five possible results on the next experiment, one is in C_1, three are in C_2, and one is in C_3. Hence $p_{11} = \frac{1}{5} = .2, p_{12} = \frac{3}{5} = .6, p_{13} = \frac{1}{5} = .2$. If the last result was state C_2, then of the five possible results on the next experiment two are in C_1, two are in C_2, and one is in C_3. This yields $p_{21} = \frac{2}{5} = .4, p_{22} = \frac{2}{5} = .4$, $p_{23} = \frac{1}{5} = .2$. If the last result was state C_3, then of the five possible results on the next experiment two are in C_1, three are in C_2. Thus $p_{31} = \frac{2}{5} = .4$, $p_{32} = \frac{3}{5} = 6$, $p_{33} = 0$. The matrix of transition probabilities is

$$P = \begin{bmatrix} .2 & .6 & .2 \\ .4 & .4 & .2 \\ .4 & .6 & 0 \end{bmatrix}$$

EXAMPLE 9-2

Suppose we have two urns containing red balls and white balls. Urn 1 has four red and two white balls while urn 2 has three red balls and 1 white ball. The experiment consists of drawing a ball at random from an urn, observing its color, and replacing it in the same urn. If the ball is red, the next draw is made from the other urn. If it is white, the next draw is made from the same urn. Let the states be

C_1: the last draw was from urn 1
C_2: the last draw was from urn 2

Find the matrix of transition probabilities.

Solution

If the last draw was made from urn 1, then $p_{11} = \frac{1}{3}$, $p_{12} = \frac{2}{3}$. If the last draw was made from urn 2, then $p_{21} = \frac{3}{4}$, $p_{22} = \frac{1}{4}$. Hence

$$P = \begin{bmatrix} \frac{1}{3} & \frac{2}{3} \\ \frac{3}{4} & \frac{1}{4} \end{bmatrix}$$

EXAMPLE 9-3

Two individuals, say, A and B, have three pennies between them. They each toss a coin with A winning a penny from his opponent if both show heads or both show tails. If the pennies do not match, A loses a penny. If one individual has all the pennies, the game ceases. Let the states be

C_1: A has 0 pennies
C_2: A has 1 penny
C_3: A has 2 pennies
C_4: A has 3 pennies

Find the matrix of transition probabilities.

Solution

If A has 0 pennies, he cannot leave state C_1. Hence $p_{11} = 1$, $p_{12} = p_{13} = p_{14} = 0$. Since A's probability of winning is $\frac{1}{2}$ on each toss, he can go from state C_2 to C_1 or C_3 with probability $\frac{1}{2}$. Hence, he cannot remain in C_2 or jump to C_4 in one toss. Thus $p_{21} = \frac{1}{2}$, $p_{22} = 0$, $p_{23} = \frac{1}{2}$, $p_{24} = 0$. Similarly, if he has two pennies, he must have either 1 or 3 after the next toss and $p_{31} = 0$, $p_{32} = \frac{1}{2}$, $p_{33} = 0$, $p_{34} = \frac{1}{2}$. Finally, if he has all three pennies he cannot leave state C_4 and $p_{41} = p_{42} = p_{43} = 0$, $p_{44} = 1$. The matrix of transition probabilities is

$$P = \begin{bmatrix} 1 & 0 & 0 & 0 \\ \frac{1}{2} & 0 & \frac{1}{2} & 0 \\ 0 & \frac{1}{2} & 0 & \frac{1}{2} \\ 0 & 0 & 0 & 1 \end{bmatrix}$$

302

EXAMPLE 9-4

A student is taking too many courses to do justice to all of them. Consequently, he decides how to prepare his assignments in his probability course by rolling a six-sided symmetric die, regarding a 1, 2, 3, or 4 as a success, and using the following procedure:

1 If, for the last assignment, he worked all of the problems, he will work all the problems for the next assignment if the die produces a success; otherwise, he will work half of the problems.
2 If, for the last assignment, he worked half of the problems, he will work all of the problems if the die produces a success; otherwise he will work no problems.
3 If, for the last assignment, he worked no problems, he will work half the problems if the die produces a success; otherwise, he will work no problems.

Find the matrix of transition probabilities.

Solution

Let the states be

C_1: all problems are worked on the last assignment
C_2: half the problems are worked on the last assignment
C_3: none of the problems are worked on the last assignment

Then, since the probability of a success is $\frac{2}{3}$ we find from

1 that $p_{11} = \frac{2}{3}$, $p_{12} = \frac{1}{3}$, $p_{13} = 0$
2 that $p_{21} = \frac{2}{3}$, $p_{22} = 0$, $p_{23} = \frac{1}{3}$
3 that $p_{31} = 0$, $p_{32} = \frac{2}{3}$, $p_{33} = \frac{1}{3}$

and

$$P = \begin{bmatrix} \frac{2}{3} & \frac{1}{3} & 0 \\ \frac{2}{3} & 0 & \frac{1}{3} \\ 0 & \frac{2}{3} & \frac{1}{3} \end{bmatrix}$$

EXERCISES

9-1 Suppose that two urns, say, urn 1 and urn 2, each contain two balls. Two of the four balls are black and two are white. The experiment consists of drawing one ball from each urn, placing the ball drawn from urn 1 into urn 2, and placing the ball drawn from urn 2 into urn 1. Let the states be

C_1: 0 white balls in urn 1
C_2: 1 white ball in urn 1
C_3: 2 white balls in urn 1

Find the matrix of transition probabilities.

9-2 Suppose that two urns, say, urn 1 and urn 2, contain a total of three balls numbered 1, 2, and 3. The experiment consists of selecting one of three numbers at random (probably with a table of random numbers) and moving the ball with that number to the other urn. Let the states be

C_1: urn 1 contains 0 balls
C_2: urn 1 contains 1 ball
C_3: urn 1 contains 2 balls

Find the matrix of transition probabilities.

9-3 Suppose that a four-sided symmetric die is being rolled. Let the states C_j, $j = 1, 2, 3, 4$, be the highest number that has appeared on the down side of the die. Find the matrix of transition probabilities.

9-4 Four rooms are connected by doorways as indicated in Fig. 9-3. An individual moves from room to room at regular intervals, choosing an exit at random from those available. Let the state be the number of the room in which the individual finds himself. Find the matrix of transition probabilities.

9-5 A gambler is playing with two slot machines. It is known that machine 1 pays off $\frac{1}{4}$ of the time while machine 2 pays off $\frac{1}{3}$ of the time but the gambler does not know which machine is machine 2. If he wins he plays the same machine again while if he loses he changes and plays the other machine. Let the state be the number of machine being played. Find the matrix of transition probabilities.

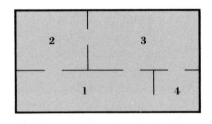

Figure 9-3   Rooms connected by doorways.

9-6   A symmetric six-sided die is being rolled with the appearance of a 1 or 2 being regarded as a success.   Let the states be

C_1: two successive rolls produce successes
C_2: a success is followed by a failure
C_3: a failure is followed by a success
C_4: two successive rolls produce failures

Find the matrix of transition probabilities.

9-7   Suppose that in Example 9-3 the individuals do not quit when one of them has all the pennies.   The one with 0 pennies is allowed to play free (by being loaned a penny for the purpose of tossing). Hence, if A has 0 pennies before tossing, he has either 0 or 1 after the toss.   Using the same states as in the example, find the matrix of transition probabilities.

9-8   A housewife uses three brands of soap—A, B, C—in her washing machine.   When she finishes a package of brand A soap the probability is .8 that she buys brand A again.   When she finishes a box of brand B the probability is .6 that she buys brand B again. Finally, when she finishes a package of brand C soap the probability is .4 that she buys brand C again.   When she switches to another brand, she does so with equal probability for the other two brands. Let the states be

C_1: brand A is being used
C_2: brand B is being used
C_3: brand C is being used

Find the matrix of transition probabilities.

9-3   PROBABILITIES AFTER n REPETITIONS

Let us again consider the bolt machine that makes good and defective bolts.   We have already observed that the transition matrix is

$$P = \begin{bmatrix} .9 & .1 \\ .8 & .2 \end{bmatrix} \tag{9-4}$$

Suppose the machine begins by producing a good bolt.   That is, the initial state of the process is "good."   Now with the assistance of tree diagrams we can trace the process through several steps or repetitions.   After two

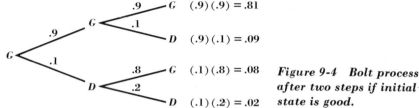

Figure 9-4 Bolt process after two steps if initial state is good.

steps we get the possibilities exhibited in Fig. 9-4. (G is good, D is defective.)

From Fig. 9-4 we observe that

$$p_{11}^{(2)} = (.9)(.9) + (.1)(.8) = .81 + .08 = .89$$
$$p_{12}^{(2)} = (.9)(.1) + (.1)(.2) = .09 + .02 = .11$$

Thus, if the process starts in state C_1, after two steps the probability of being in state C_1 is .89 and the probability of being in C_2 is .11. Similar analysis for an initial state "defective" yields Fig. 9-5 and probabilities

$$p_{21}^{(2)} = (.8)(.9) + (.2)(.8) = .72 + .16 = .88$$
$$p_{22}^{(2)} = (.8)(.1) + (.2)(.2) = .08 + .04 = .12$$

Now let us write the four probabilities in the matrix form

$$P^{(2)} = \begin{bmatrix} p_{11}^{(2)} & p_{12}^{(2)} \\ p_{21}^{(2)} & p_{22}^{(2)} \end{bmatrix} = \begin{bmatrix} .89 & .11 \\ .88 & .12 \end{bmatrix}$$

It is easy to verify that $P^{(2)} = PP = P^2$. In other words $p_{ij}^{(2)}$ is the element in the ith row and jth column of the P^2 matrix.

Next let us investigate the results obtained by following the process through the third step. If the process starts in state G, we could summarize the possible results and corresponding probabilities by adding to Fig. 9-4 another column consisting of eight branches. However, we can simplify the calculations a little if we use the results already obtained. That is, we already know that if the process starts in state G, after two steps it will be in state G with probability .88, state D with probability .12.

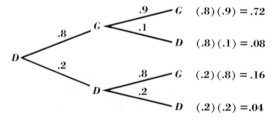

Figure 9-5 Bolt process after two steps if initial state is defective.

306

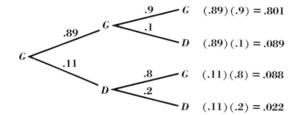

Hence, the first set of branches in Fig. 9-6 combines the results of the first two steps. The corresponding diagram for an initial state D is given in Fig. 9-7. Hence, we see

$$p_{11}^{(3)} = (.89)(.9) + (.11)(.8) = .889$$
$$p_{12}^{(3)} = (.89)(.1) + (.11)(.2) = .111$$
$$p_{21}^{(3)} = (.88)(.9) + (.12)(.8) = .888$$
$$p_{22}^{(3)} = (.88)(.1) + (.12)(.2) = .112$$

These results can be written in the matrix form

$$P^{(3)} = \begin{bmatrix} .889 & .111 \\ .888 & .112 \end{bmatrix}$$

It is easy to verify that $P^{(3)} = P^2 P = P^3$.
Now let us define

$$P^{(n)} = \begin{bmatrix} p_{11}^{(n)} & p_{12}^{(n)} & \cdots & p_{1k}^{(n)} \\ p_{21}^{(n)} & p_{22}^{(n)} & \cdots & p_{2k}^{(n)} \\ \cdots & \cdots & \cdots & \cdots \\ p_{k1}^{(n)} & p_{k2}^{(n)} & \cdots & p_{kk}^{(n)} \end{bmatrix} \tag{9-5}$$

where, as previously mentioned, $p_{ij}^{(n)}$ is the probability that a process that starts in state C_i will be in state C_j after n steps. The computations with the bolt example suggest that

$$P^{(n)} = P^n \tag{9-6}$$

By generalizing the bolt discussion to the case with k categories (instead of 2) and unspecified p_{ij} it can be shown (again tree diagrams are helpful) that (9-6) is true.

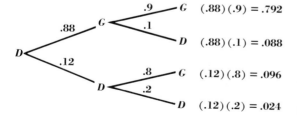

In order to compute the probabilities $p_{ij}^{(n)}$ we had to assume that the initial state of the process was C_i. Suppose that the initial state C_i is chosen with probability $p_i^{(0)}$. These probabilities can be represented as a row vector

$$p^{(0)} = [p_1^{(0)}, p_2^{(0)}, \ldots, p_k^{(0)}] \qquad (9\text{-}7)$$

For the bolt example a sensible choice might be $p^{(0)} = [.9,.1]$. In Example 9-1, if the process is started by the roll of a die, then $p^{(0)} = [\frac{2}{6},\frac{3}{6},\frac{1}{6}]$. For Example 9-2 we might pick an urn at random in which case $p^{(0)} = [.5,.5]$. In Example 9-3 we must have $p_1^{(0)} = p_4^{(0)} = 0$ and, if A owns one penny when the game is started, then $p^{(0)} = [0,1,0,0]$. If the student of Example 9-4 selects his initial state at random, then $p^{(0)} = [\frac{1}{3},\frac{1}{3},\frac{1}{3}]$. We shall call $p^{(0)}$ the *vector of initial probabilities*.

Any row vector $a = [a_1, a_2, \ldots, a_k]$ such that each $a_i \geq 0$ and $\sum_{i=1}^{k} a_i = 1$ is called a *probability vector*.

Suppose that for some unexplained reason the probability that the bolt machine produces a good bolt on the initial trial is .6. Then, the vector of initial probabilities is $[.6,.4]$. Assuming that the matrix of transition probabilities is still given by (9-4), we see from Fig. 9-8 that the probabilities of being in states G and D after 1 step are respectively

$$p_1^{(1)} = (.6)(.9) + (.4)(.8) = .86$$
$$p_2^{(1)} = (.6)(.1) + (.4)(.2) = .14$$

These results can be written in the vector form $p^{(1)} = [.86,.14]$. It is easy to see that $p^{(1)} = p^{(0)}P$.

Let us follow the process through a second step. We could extend Fig. 9-8, but it is easier to consider the process as one with a vector of initial probabilities $[.86,.14]$. Hence, from Fig. 9-9 we see that

$$p_1^{(2)} = (.86)(.9) + (.14)(.8) = .886$$
$$p_2^{(2)} = (.86)(.1) + (.14)(.2) = .114$$

If we let $p^{(2)} = [.886,.114]$, then obviously $p^{(2)} = p^{(1)}P$ and, since $p^{(1)} = p^{(0)}P$, we have $p^{(2)} = p^{(0)}PP = p^{(0)}P^2$.

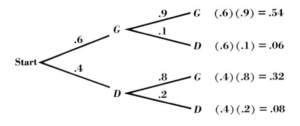

Figure 9-8 Bolt process after one step if initial probabilities are .6, .4.

308

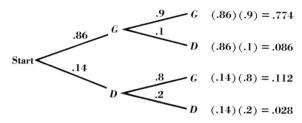

Figure 9-9   Bolt process after two steps if ini-
tial probabilities are .6, .4 or after one step if
initial probabilities are .86, .14.

If we let $p_i^{(n)}$ be the probability that the process will be in state C_i after n repetitions and

$$p^{(n)} = [p_1^{(n)}, p_2^{(n)}, \ldots, p_k^{(n)}] \qquad (9\text{-}8)$$

then the bolt example suggests

$$p^{(n)} = p^{(0)}P^n \qquad (9\text{-}9)$$

a result that can be proved by generalizing the above discussion.

EXAMPLE 9-5

For the urn problem of Example 9-2 find the matrix of the $p_{ij}^{(2)}$. If the process is started by selecting an urn at random, find $p^{(2)}$.

Solution

We know from (9-6) that the elements of P^2 are the $p_{ij}^{(2)}$. Hence

$$P^{(2)} = \begin{bmatrix} \frac{1}{3} & \frac{2}{3} \\ \frac{3}{4} & \frac{1}{4} \end{bmatrix}\begin{bmatrix} \frac{1}{3} & \frac{2}{3} \\ \frac{3}{4} & \frac{1}{4} \end{bmatrix} = \begin{bmatrix} (\frac{1}{9}) + (\frac{6}{12}) & (\frac{2}{9}) + (\frac{2}{12}) \\ (\frac{3}{12}) + (\frac{3}{16}) & (\frac{6}{12}) + (\frac{1}{16}) \end{bmatrix}$$

$$= \begin{bmatrix} \frac{11}{18} & \frac{7}{18} \\ \frac{7}{16} & \frac{9}{16} \end{bmatrix}$$

If the process is started by selecting an urn at random, then $p^{(0)} = [\frac{1}{2}, \frac{1}{2}]$ and

$$p^{(2)} = [\frac{1}{2}, \frac{1}{2}]\begin{bmatrix} \frac{11}{18} & \frac{7}{18} \\ \frac{7}{16} & \frac{9}{16} \end{bmatrix} = \left[\frac{11}{36} + \frac{7}{32}, \frac{7}{36} + \frac{9}{32}\right]$$

$$= [\frac{151}{288}, \frac{137}{288}]$$

Thus, after two draws, the probability that the next draw is to be made from urn 1 is $\frac{151}{288}$ while the probability that the next draw is to be made from urn 2 is $\frac{137}{288}$.

EXAMPLE 9-6

For the penny-matching problem of Example 9-3, find $P^{(2)}$ and $P^{(3)}$.

Solution

$$P^{(2)} = \begin{bmatrix} 1 & 0 & 0 & 0 \\ \frac{1}{2} & 0 & \frac{1}{2} & 0 \\ 0 & \frac{1}{2} & 0 & \frac{1}{2} \\ 0 & 0 & 0 & 1 \end{bmatrix} \begin{bmatrix} 1 & 0 & 0 & 0 \\ \frac{1}{2} & 0 & \frac{1}{2} & 0 \\ 0 & \frac{1}{2} & 0 & \frac{1}{2} \\ 0 & 0 & 0 & 1 \end{bmatrix} = \begin{bmatrix} 1 & 0 & 0 & 0 \\ \frac{1}{2} & \frac{1}{4} & 0 & \frac{1}{4} \\ \frac{1}{4} & 0 & \frac{1}{4} & \frac{1}{2} \\ 0 & 0 & 0 & 1 \end{bmatrix}$$

$$P^{(3)} = \begin{bmatrix} 1 & 0 & 0 & 0 \\ \frac{1}{2} & 0 & \frac{1}{2} & 0 \\ 0 & \frac{1}{2} & 0 & \frac{1}{2} \\ 0 & 0 & 0 & 1 \end{bmatrix} \begin{bmatrix} 1 & 0 & 0 & 0 \\ \frac{1}{2} & \frac{1}{4} & 0 & \frac{1}{4} \\ \frac{1}{4} & 0 & \frac{1}{4} & \frac{1}{2} \\ 0 & 0 & 0 & 1 \end{bmatrix} = \begin{bmatrix} 1 & 0 & 0 & 0 \\ \frac{5}{8} & 0 & \frac{1}{8} & \frac{2}{8} \\ \frac{2}{8} & \frac{1}{8} & 0 & \frac{5}{8} \\ 0 & 0 & 0 & 1 \end{bmatrix}$$

Hence, for example, if A starts with one penny, after three plays the probability is $\frac{5}{8}$ that he is broke, $\frac{1}{8}$ that he has two pennies, and $\frac{2}{8}$ that he has all three. He cannot have one penny.

EXAMPLE 9-7

For the urn problem of Example 9-2, suppose that the process is started by selecting urn 1 with probability $\frac{9}{17}$ and urn 2 with probability $\frac{8}{17}$. Find $p^{(1)}$ and $p^{(2)}$.

Solution

Now $p^{(0)} = [\frac{9}{17}, \frac{8}{17}]$ and

$$p^{(1)} = p^{(0)}P = [\frac{9}{17}, \frac{8}{17}] \begin{bmatrix} \frac{1}{3} & \frac{2}{3} \\ \frac{3}{4} & \frac{1}{4} \end{bmatrix} = [\frac{9}{17}, \frac{8}{17}] = p^{(0)}$$

and

$$p^{(2)} = p^{(0)}P^2 = [p^{(0)}P]P = p^{(1)}P = p^{(0)}P = p^{(0)}$$

Thus, the probability that the next draw is to be made from urn 1 is $\frac{9}{17}$ after either 1 or 2 draws. It is obvious $p^{(n)} = p^{(0)}$ for any n since

$$p^{(n)} = p^{(0)}PP^{n-1} = p^{(0)}P^{n-1} = p^{(0)}PP^{n-2}$$
$$= \cdots = p^{(0)}P = p^{(0)}$$

Hence, at any step, the probability is $\frac{9}{17}$ that the next draw is to be made from urn 1.

EXAMPLE 9-8

Show that if the process described in Example 9-1 is started by rolling a die with the initial state being C_1, C_2, C_3, depending upon which of the three results is obtained, then the probabilities of being in the various states are the same for every step.

Solution

We have $p^{(0)} = [\frac{1}{3}, \frac{1}{2}, \frac{1}{6}]$. It is easy to verify that $p^{(1)} = p^{(0)}P = p^{(0)}$. That is

$$[\frac{1}{3}, \frac{1}{2}, \frac{1}{6}] \begin{bmatrix} .2 & .6 & .2 \\ .4 & .4 & .2 \\ .4 & .6 & 0 \end{bmatrix} = [\frac{1}{3}, \frac{1}{2}, \frac{1}{6}]$$

Hence $p^{(2)} = p^{(0)}PP = p^{(0)}P = p^{(0)}$. Similarly, $p^{(3)} = p^{(0)}, \ldots, p^{(n)} = p^{(0)}$ as successive multiplications by P still yield $p^{(0)}$.

In Examples 9-7 and 9-8 the vector of initial probabilities satisfied the equation $p^{(0)}P = p^{(0)}$. A probability vector $a = [a_1, a_2, \ldots, a_k]$ is called a stationary probability vector if

$$aP = a \qquad\qquad\qquad (9\text{-}10)$$

Let us attempt to find a stationary probability vector for the transition matrix of the bolt process. We must have

$$[a_1, a_2] \begin{bmatrix} .9 & .1 \\ .8 & .2 \end{bmatrix} = [a_1, a_2] \qquad\qquad\qquad (9\text{-}11)$$

where $a_2 = 1 - a_1$. After the vector on the left-hand side of (9-11) is multiplied by the matrix, we get

$$[.9a_1 + .8a_2, .1a_1 + .2a_2] = [a_1, a_2]$$

which means that

$$.9a_1 + .8(1 - a_1) = a_1$$
$$.1a_1 + .2(1 - a_1) = 1 - a_1$$

Both equations yield $a_1 = \frac{8}{9}$. Hence $a_2 = \frac{1}{9}$ and $[\frac{8}{9}, \frac{1}{9}]$ is the only stationary probability vector for the Markov chain associated with the bolt process.

Suppose that the matrix of transition probabilities is

$$P = \begin{bmatrix} 1 & 0 \\ 0 & 1 \end{bmatrix}$$

Then (9-10) becomes

$$[a_1, a_2] \begin{bmatrix} 1 & 0 \\ 0 & 1 \end{bmatrix} = [a_1, a_2]$$

or

$$[a_1, a_2]P = [a_1, a_2]$$

and any probability vector is stationary for this chain. We might wonder what condition will guarantee the existence of a unique stationary probability vector. In this connection let us make the following definition:

If some power of a transition matrix has only positive entries, then it is called a regular matrix.

It can be shown that if a transition matrix is regular, then a unique stationary probability vector exists.

Some nonregular transition matrices also have a unique stationary probability vector. For example, consider

$$P = \begin{bmatrix} 0 & 1 \\ 1 & 0 \end{bmatrix}$$

Then, $aP = a$ yields $a_1 = a_2$ so that $[\frac{1}{2}, \frac{1}{2}]$ is the only solution for a. But P is not regular since

$$P^n = P \qquad \text{if } n \text{ is odd}$$

$$= \begin{bmatrix} 1 & 0 \\ 0 & 1 \end{bmatrix} \qquad \text{if } n \text{ is even}$$

and no power of P has all positive entries.

It is known that a $k \times k$ transition matrix P is regular if and only if P^{k^2-2k+2} has no zero entries and that a similar statement cannot be made for any smaller power of P. Thus, in checking for regularity, a decision can be made as soon as the exponent on P equals $k^2 - 2k + 2$ (and possibly sooner). Hence, if $k = 2, 3, 4$, then P^2, P^5, P^{10} respectively are the maximum powers of P that need be considered.

EXAMPLE 9-9

Show that

$$P = \begin{bmatrix} 0 & 1 \\ \frac{1}{3} & \frac{2}{3} \end{bmatrix}$$

is a regular transition matrix.

Solution

$$P^2 = \begin{bmatrix} 0 & 1 \\ \frac{1}{3} & \frac{2}{3} \end{bmatrix} \begin{bmatrix} 0 & 1 \\ \frac{1}{3} & \frac{2}{3} \end{bmatrix} = \begin{bmatrix} \frac{1}{3} & \frac{2}{3} \\ \frac{2}{9} & \frac{7}{9} \end{bmatrix}$$

All entries in P^2 are positive so that P is a regular transition matrix.

At the beginning of the section in which we considered the bolt process we found that

$$P^{(2)} = P^2 = \begin{bmatrix} .89 & .11 \\ .88 & .12 \end{bmatrix} \qquad P^{(3)} = P^3 = \begin{bmatrix} .889 & .111 \\ .888 & .112 \end{bmatrix}$$

Similarly we can find

$$P^{(4)} = P^4 = \begin{bmatrix} .8889 & .1111 \\ .8888 & .1112 \end{bmatrix}$$

The calculations suggest not only that each row is approaching the same row vector but that this vector is $[\frac{8}{9}, \frac{1}{9}]$, the unique stationary probability vector of P. The following results can be proved: If a Markov chain has a regular transition matrix P, then

(a) *As n increases, each row of P^n approaches the same row vector a.*

(b) *Each entry a_i, $i = 1, 2, \ldots, k$ of a is positive.*

(c) *The vector a is the unique stationary probability vector of P.*

(d) *If $p^{(0)}$ is any vector of initial probabilities, then as n increases the entries of $p^{(n)} = p^{(0)}P^n$ approach the corresponding elements of a. That is, $p_i^{(n)}$ approaches a_i, $i = 1, 2, \ldots, k$.*

(9-12)

The implication of part (d), (9-12), is very important from a practical point of view. This result implies that if the matrix of transition probabilities is regular, then, regardless of how the initial probabilities are selected, the probability that the process is in state C_i is very close to a_i after the series of experiments is performed a sufficiently large number of times. In the bolt example, we found that the rows of P^4 were very close to $[\frac{8}{9}, \frac{1}{9}]$. Thus only four repetitions are required to practically stabilize the process. Here the elements of the two rows were not too different, being $[.9, .1]$ in the first row and $[.8, .2]$ in the second. If the transition matrix were

$$P = \begin{bmatrix} .9 & .1 \\ .1 & .9 \end{bmatrix}$$

then we would expect that more repetitions of the experiment would be required to achieve a similar degree of closeness.

If the Markov chain has only two states and transition matrix

$$P = \begin{bmatrix} p_{11} & p_{12} \\ p_{21} & p_{22} \end{bmatrix}$$

with p_{12} and p_{21} not both 0, then it is easy to show (see Exercise 9-20) that the unique stationary probability vector is

$$a = \left[\frac{p_{21}}{p_{12} + p_{21}}, \frac{p_{12}}{p_{12} + p_{21}} \right] \tag{9-13}$$

EXAMPLE 9-10

If a certain baseball team wins a game, then the probability that they win the next game also is $\frac{2}{3}$. If they lose a game, then their probability of winning the next game is only $\frac{1}{2}$. In the long run, what fraction of their games could the team expect to win?

Solution

Let the states be

C_1: if the team won the last game
C_2: if the team lost the last game

Then

$$P = \begin{bmatrix} \frac{2}{3} & \frac{1}{3} \\ \frac{1}{2} & \frac{1}{2} \end{bmatrix}$$

According to (9-13), the stationary probability vector is

$$a = \left[\frac{\frac{1}{2}}{(\frac{1}{3}) + (\frac{1}{2})}, \frac{\frac{1}{3}}{(\frac{1}{3}) + (\frac{1}{2})} \right] = \left[\frac{3}{5}, \frac{2}{5} \right]$$

Hence, in the long run the team would expect to win $\frac{3}{5}$ of their games or, in baseball terminology, play .600 ball.

EXAMPLE 9-11

Show that the transition matrix of Example 9-4 is regular and find the unique stationary probability vector.

Solution

$$P^2 = \begin{bmatrix} \frac{2}{3} & \frac{1}{3} & 0 \\ \frac{2}{3} & 0 & \frac{1}{3} \\ 0 & \frac{2}{3} & \frac{1}{3} \end{bmatrix} \begin{bmatrix} \frac{2}{3} & \frac{1}{3} & 0 \\ \frac{2}{3} & 0 & \frac{1}{3} \\ 0 & \frac{2}{3} & \frac{1}{3} \end{bmatrix} = \begin{bmatrix} \frac{6}{9} & \frac{2}{9} & \frac{1}{9} \\ \frac{4}{9} & \frac{4}{9} & \frac{1}{9} \\ \frac{4}{9} & \frac{2}{9} & \frac{3}{9} \end{bmatrix}$$

has all positive entries.
From

$$[a_1, a_2, a_3] \begin{bmatrix} \frac{2}{3} & \frac{1}{3} & 0 \\ \frac{2}{3} & 0 & \frac{1}{3} \\ 0 & \frac{2}{3} & \frac{1}{3} \end{bmatrix} = [a_1, a_2, a_3]$$

314

we get

$$\tfrac{2}{3}a_1 + \tfrac{2}{3}a_2 = a_1$$
$$\tfrac{1}{3}a_1 + \tfrac{2}{3}a_3 = a_2$$
$$\tfrac{1}{3}a_2 + \tfrac{1}{3}a_3 = a_3$$

The first and third equations reduce to

$$2a_2 = a_1 \qquad a_2 = 2a_3$$

Hence $a_1 = 4a_3$. Since $a_1 + a_2 + a_3 = 1$ we get

$$4a_3 + 2a_3 + a_3 = 1$$
$$a_3 = \tfrac{1}{7}$$
$$a_2 = \tfrac{2}{7}$$
$$a_1 = \tfrac{4}{7}$$
$$a = [\tfrac{4}{7}, \tfrac{2}{7}, \tfrac{1}{7}]$$

It is easy to verify that these results also satisfy the middle equation. Thus, in the long run the student works all the problems about $\tfrac{4}{7}$ of the time, half of the problems about $\tfrac{2}{7}$ of the time, and no problems about $\tfrac{1}{7}$ of the time.

EXERCISES

9-9   Show that the transition matrix found in Exercise 9-1 is regular and find the unique stationary probability vector.   Interpret the result.

9-10   Show that the transition matrix found in Exercise 9-2 is not regular.

9-11   For the transition matrix found in Exercise 9-3 find $P^{(2)}$ and $P^{(4)}$. Is the matrix P regular?   Is there a unique stationary probability vector?   What matrix do you think P^n approaches as n increases?

9-12   Show that the transition matrix found in Exercise 9-4 is regular. Find the unique stationary probability vector and interpret the result.

9-13   Find the unique stationary probability vector for the transition matrix of Exercise 9-5.

9-14   Show that the transition matrix found in Exercise 9-6 is regular. Find the unique stationary probability vector and interpret the result.

9-15   Find $P^{(2)}$ and $P^{(3)}$ for the transition matrix found in Exercise 9-7. Is the matrix regular?   Find the unique stationary probability vector and interpret the result.

9-16   Find the unique stationary probability vector for the transition matrix of Exercise 9-8.

9-17 Generalize Example 9-4 by replacing $\frac{2}{3}$ by p and $\frac{1}{3}$ by q. Find the stationary probability vector. Check your result for the case $p = \frac{2}{3}$, $q = \frac{1}{3}$. (See Example 9-11.)

9-18 Generalize Exercise 9-6 by replacing $\frac{1}{3}$ by p and $\frac{2}{3}$ by q. Find the stationary probability vector. Check your result for the case $p = \frac{1}{3}$, $q = \frac{2}{3}$. (See Exercise 9-14.) *Hint:* The sum of the first two equations yields $a_1 + a_3 = p$ and the sum of the last two equations yields $a_2 + a_4 = q$. From this the result follows easily.

9-19 Generalize Exercise 9-7 by assuming that A's probability of winning is p instead of $\frac{1}{2}$. Find the stationary probability vector. Check your result for the case $p = \frac{1}{2}$. (See Exercise 9-15.) *Hint:* Starting with the fourth equation, express a_3 in terms of a_4. Then the third equation yields a_2 in terms of a_4 and the second equation yields a_1 in terms of a_4.

9-20 Prove that Eq. (9-13) is correct.

9-21 Suppose that a Markov chain has the transition matrix

$$P = \begin{bmatrix} 1 & 0 & 0 \\ p & 0 & q \\ 0 & 0 & 1 \end{bmatrix}$$

where $p + q = 1$. Show that P is not regular. Show that any vector of the type $a = [a_1, 0, a_2]$ is a stationary probability vector.

9-4 AVERAGE NUMBER OF STEPS REQUIRED TO GO FROM ONE STATE TO ANOTHER IF TRANSITION MATRIX IS REGULAR

Let us assume that a Markov chain has a regular transition matrix. Then, for some n, $P^{(n)} = P^n$ has all nonzero entries so that each $p_{ij}^{(n)} > 0$. In other words, it is possible to go from any state to any other state in n steps. In this section we seek the average number of steps required to go from one state to another when the Markov chain has a regular transition matrix.

If we let X_{ij} be the number of steps required to go from state C_i to state C_j, then we seek $E(X_{ij}) = n_{ij}$. Obviously, potential values for X_{ij} are $1, 2, 3, \ldots$. If we could calculate $f_1 = \Pr(X_{ij} = 1)$, $f_2 = \Pr(X_{ij} = 2)$, $f_3 = \Pr(X_{ij} = 3)$, $\ldots$, then by definition of expected value we have

$$E(X_{ij}) = (1)f_1 + (2)f_2 + (3)f_3 + \cdots \tag{9-14}$$

First let us consider a chain with two states and regular transition matrix

$$P = \begin{bmatrix} p_{11} & p_{12} \\ p_{21} & p_{22} \end{bmatrix} \tag{9-15}$$

In this case the probabilities f_1, f_2, f_3, . . . can be found quickly. Consider X_{12}. Then $X_{12} = 1$ if the process goes from state C_1 to state C_2 and $f_1 = p_{12}$. In order to use two steps so that $X_{12} = 2$ the process must stay in state C_1 on the first step and go to state C_2 on the second. Hence $f_2 = p_{11}p_{12}$. To have $X_{12} = 3$ the process must stay in state C_1 on the first two steps and then move to state C_2 on the third. Thus $f_3 = p_{11}^2 p_{12}$. To require r steps the process must stay in state C_1 on the first $r - 1$ steps and move to state C_2 on the rth step. This gives $f_r = p_{11}^{r-1} p_{12}$. We recognize that f_r is a special case of the negative binomial probability function, given by formula (4-9) with $c = 1$, $p_{12} = p$, $p_{11} = q$, $r = n$. Hence, by formula (4-12), $E(X_{12}) = c/p = 1/p_{12}$. By interchanging the roles of subscripts 1 and 2 we have immediately $E(X_{21}) = 1/p_{21}$. Next consider X_{11}. If $X_{11} = 1$, the process stays in state C_1 and $f_1 = p_{11}$. To use two steps so that $X_{11} = 2$ the process must go to state C_2 on the first step and return to state C_1 on the second. Thus $f_2 = p_{12}p_{21}$. To have $X_{11} = 3$ the process must move to state C_2 on the first step, stay in C_2 on the second, and move to C_1 on the third. Hence, $f_3 = p_{12}p_{22}p_{21}$. Similarly $f_4 = p_{12}p_{22}^2 p_{21}$, $f_r = p_{12}p_{22}^{r-2}p_{21}$, $r \geq 2$. Then

$$E(X_{11}) = (1)p_{11} + (2)p_{12}p_{21} + (3)p_{12}p_{22}p_{21} + (4)p_{12}p_{22}^2 p_{21} + \cdots$$

$$= p_{11} + p_{12}[2p_{21} + 3p_{22}p_{21} + 4p_{22}^2 p_{21} + \cdots]$$

$$= p_{11} + \frac{p_{12}}{p_{22}}[2p_{21}p_{22} + 3p_{21}p_{22}^2 + 4p_{21}p_{22}^3 + \cdots]$$

$$= p_{11} + \frac{p_{12}}{p_{22}}[(p_{21} + 2p_{21}p_{22} + 3p_{21}p_{22}^2 + \cdots) - p_{21}]$$

The sum in the parenthesis is the expected value of a negative binomial random variable with $c = 1$, $p = p_{21}$, $q = p_{22}$ and hence equal to $1/p_{21}$. With this simplification we get

$$E(X_{11}) = p_{11} + \frac{p_{12}}{p_{22}}\left(\frac{1}{p_{21}} - p_{21}\right) = p_{11} + \frac{p_{12}}{p_{22}}\left(\frac{1 - p_{21}^2}{p_{21}}\right)$$

$$= p_{11} + \frac{p_{12}}{p_{22}}\frac{(1 - p_{21})(1 + p_{21})}{p_{21}} = p_{11} + \frac{p_{12}(1 + p_{21})}{p_{21}}$$

$$= 1 - p_{12} + \frac{p_{12}(1 + p_{21})}{p_{21}}$$

Finally, the latter result reduces to

$$E(X_{11}) = \frac{p_{12} + p_{21}}{p_{21}} = \frac{1}{a_1}$$

where a_1 is the first component in the stationary probability vector. By interchanging the subscripts 1 and 2 we have immediately

$$E(X_{22}) = \frac{p_{21} + p_{12}}{p_{12}} = \frac{1}{a_2}$$

where a_2 is the second entry in the stationary probability vector. Summarizing our results in matrix form, we have

$$N = \begin{bmatrix} n_{11} & n_{12} \\ n_{21} & n_{22} \end{bmatrix} = \begin{bmatrix} 1/a_1 & 1/p_{12} \\ 1/p_{21} & 1/a_2 \end{bmatrix} \tag{9-16}$$

For a Markov chain with only two states it is easy to evaluate probabilities for X_{ij}. We found that the probability function for X_{12} is

$$f_r = p_{12} p_{11}^{r-1} \qquad r = 1, 2, \ldots$$

where $r = x_{12}$. As we have already observed X_{12} has a negative binomial distribution $c = 1$, $p = p_{12}$ and

$$\Pr(X_{12} \leq s) = \sum_{r=1}^{s} b^*(r;1,p_{12}) = \sum_{y=1}^{s} b(y;s,p_{12}) \tag{9-17}$$

where the b and b^* symbols are defined by (4-2) and (4-9), respectively. Similarly

$$\Pr(X_{21} \leq s) = \sum_{y=1}^{s} b(y;s,p_{21}) \tag{9-18}$$

For X_{11} the probability function is

$$\begin{aligned} f_r &= p_{11} & r &= 1 \\ &= p_{12} p_{22}^{r-2} p_{21} & r &= 2, 3, 4, \ldots \end{aligned}$$

where $x_{11} = r$. Hence, if $s \geq 2$,

$$\Pr(X_{11} \leq s) = p_{11} + p_{12}[p_{21} + p_{21}p_{22} + p_{21}p_{22}^2 + \cdots + p_{21}p_{22}^{s-2}]$$

$$= p_{11} + p_{12} \sum_{r=1}^{s-1} b^*(r;1,p_{21})$$

or

$$\Pr(X_{11} \leq s) = p_{11} + p_{12} \sum_{y=1}^{s-1} b(y; s-1, p_{21}) \qquad s \geq 2 \tag{9-19}$$

$$= p_{11} \qquad s = 1$$

Similarly (or by interchanging subscripts 1 and 2) we get

$$\Pr(X_{22} \leqq s) = p_{22} + p_{21} \sum_{y=1}^{s-1} b(y; s-1, p_{12}) \qquad s \geqq 2$$

$$= p_{22} \qquad s = 1$$

(9-20)

EXAMPLE 9-12

For the Markov chain of Example 9-2 find the n_{ij}. Find the probability that the number of steps required to go from state C_2 to state C_1 is five or less. Find the probability that the number of steps required to go from state C_1 back to state C_1 is six or less.

Solution

The transition matrix was

$$P = \begin{bmatrix} \frac{1}{3} & \frac{2}{3} \\ \frac{3}{4} & \frac{1}{4} \end{bmatrix}$$

and in Example 9-7 it was verified that the stationary probability vector was $(\%_{17}, \%_{17})$. Hence $n_{11} = 1/a_1 = {}^{17}\!/_9$, $n_{22} = 1/a_2 = {}^{17}\!/_8$, $n_{12} = 1/p_{12} = \frac{3}{2}$, $n_{21} = 1/p_{21} = \frac{4}{3}$. Thus, for example, it takes on the average 1.5 steps to go from state C_1 to state C_2.

The probability that the number of steps required to go from state C_2 to state C_1 is five or less is [from (9-18)]

$$\Pr(X_{21} \leqq 5) = \sum_{y=1}^{5} b(y;5,.75)$$

$$= \sum_{y=0}^{4} b(y;5,.25) \qquad \textit{by method used to get (4-4)}$$

$$= .99902 \qquad \textit{from Appendix B1}$$

The probability that the number of steps required to go from state C_1 back to state C_1 is six or less is [from (9-19)]

$$\Pr(X_{11} \leqq 6) = \frac{1}{3} + \frac{2}{3} \sum_{y=1}^{5} b(y;5,.75)$$

$$= \frac{1}{3} + \frac{2}{3}(.99902)$$

$$= .99934$$

When a Markov chain has three or more states the direct method which we have used for two-state chains becomes more tedious. We shall make no attempt to exhibit the probability distribution of X_{ij}. (Feller [1],

p. 352, expresses these probabilities in terms of recursive formulas.) An indirect method will be used to obtain expressions which will yield the n_{ij}.

Suppose that a three-state Markov chain starts in state C_2 and we seek $E(X_{21}) = n_{21}$, the average number of steps required to go to state C_1. Let Y be a random variable that assumes value 1, 2, or 3 depending upon whether the process goes to C_1, C_2, or C_3, respectively, on the first step. Then Y will assume the value

1   1 with probability p_{21} (in which case $X_{21} = 1$)
2   2 with probability p_{22}
3   3 with probability p_{23}

We easily calculate

$$
\begin{aligned}
\phi(y) = E(X_{21}|y) &= 1 & y &= 1 \\
&= 1 + n_{21} & y &= 2 \\
&= 1 + n_{31} & y &= 3
\end{aligned}
$$

This follows by observing that in case *1* only one step is required, in case *2* one step is required to stay in state C_2 and then, on the average, n_{21} further steps are required to go to state C_1, and in case *3* one step is required to go to state C_3 and then, on the average, n_{31} further steps are required to go to state C_1. Now recall Exercise 5-42 and make the identification

$$ X_1 = X_{21} \qquad X_2 = Y \qquad w(X_1, X_2) = w(X_{21}, Y) = X_{21} $$

According to the theorem of that exercise we have $E[\phi(Y)] = E(X_{21}) = n_{21}$. Thus we get

$$
\begin{aligned}
E[\phi(Y)] &= (1)p_{21} + (1 + n_{21})p_{22} + (1 + n_{31})p_{23} \\
&= n_{21}
\end{aligned}
$$

which, after rearrangement, is

$$ n_{21} = (p_{21} + p_{22} + p_{23}) + n_{21}p_{22} + n_{31}p_{23} $$

or

$$ n_{21} = 1 + p_{22}n_{21} + p_{23}n_{31} $$

Similarly, we can get

$$
\begin{aligned}
n_{11} &= 1 + p_{12}n_{21} + p_{13}n_{31} \\
n_{12} &= 1 + p_{11}n_{12} + p_{13}n_{32} \\
n_{13} &= 1 + p_{11}n_{13} + p_{12}n_{23} \\
n_{22} &= 1 + p_{21}n_{12} + p_{23}n_{32} \\
n_{23} &= 1 + p_{21}n_{13} + p_{22}n_{23} \\
n_{31} &= 1 + p_{32}n_{21} + p_{33}n_{31} \\
n_{32} &= 1 + p_{31}n_{12} + p_{33}n_{32} \\
n_{33} &= 1 + p_{31}n_{13} + p_{32}n_{23}
\end{aligned}
\tag{9-21}
$$

The solution of these equations yields the n_{ij}.

It is easy to check that the equations can be written in matrix form as

$$
\begin{bmatrix} n_{11} & n_{12} & n_{13} \\ n_{21} & n_{22} & n_{23} \\ n_{31} & n_{32} & n_{33} \end{bmatrix} = \begin{bmatrix} p_{11} & p_{12} & p_{13} \\ p_{21} & p_{22} & p_{23} \\ p_{31} & p_{32} & p_{33} \end{bmatrix} \begin{bmatrix} 0 & n_{12} & n_{13} \\ n_{21} & 0 & n_{23} \\ n_{31} & n_{32} & 0 \end{bmatrix} + \begin{bmatrix} 1 & 1 & 1 \\ 1 & 1 & 1 \\ 1 & 1 & 1 \end{bmatrix}
$$

or

$$ N = PN_d + J \tag{9-22} $$

where N_d is obtained from the N matrix by replacing the diagonal elements by 0's and J is a square matrix with all entries equal to 1. By generalizing the above argument, it can be shown that the matrix equation (9-22) holds for a Markov chain with k states, $k \geq 2$.

One interesting result is immediately obtainable from (9-22). If we multiply both sides of (9-22) by a, the stationary probability vector of P, we get

$$ aN = aPN_d + aJ $$

But since $aP = a$ and $aJ = [1, 1, \ldots, 1]$ this becomes

$$ aN = aN_d + [1, 1, \ldots, 1] $$

or

$$ a(N - N_d) = [1, 1, \ldots, 1] \tag{9-23} $$

Since $N - N_d$ is a matrix in which all entries are 0 except for diagonal terms n_{ii}, (9-23) yields equations of the type $a_i n_{ii} = 1$ or

$$ n_{ii} = \frac{1}{a_i} \tag{9-24} $$

EXAMPLE 9-13

For the Markov chain of Example 9-1, whose transition matrix is regular, find the n_{ij}.

Solution

We had

$$ P = \begin{bmatrix} .2 & .6 & .2 \\ .4 & .4 & .2 \\ .4 & .6 & 0 \end{bmatrix} \qquad a = [\tfrac{1}{3}, \tfrac{1}{2}, \tfrac{1}{6}] $$

having verified that a is the stationary probability vector in Example 9-8. From (9-24) we get $n_{11} = 3$, $n_{22} = 2$, $n_{33} = 6$. The equations (9-21) become

$$3 = 1 + .6n_{21} + .2n_{31}$$
$$n_{12} = 1 + .2n_{12} + .2n_{32}$$
$$n_{13} = 1 + .2n_{13} + .6n_{23}$$
$$n_{21} = 1 + .4n_{21} + .2n_{31}$$
$$2 = 1 + .4n_{12} + .2n_{32}$$
$$n_{23} = 1 + .4n_{13} + .4n_{23}$$
$$n_{31} = 1 + .6n_{21}$$
$$n_{32} = 1 + .4n_{12}$$
$$6 = 1 + .4n_{13} + .6n_{23}$$

Using the value of n_{31} given by the seventh equation in the first equation yields

$$3 = 1 + .6n_{21} + .2(1 + .6n_{21}) \qquad \text{or} \qquad n_{21} = \tfrac{5}{2}$$

Hence $n_{31} = 1 + .6(\tfrac{5}{2}) = \tfrac{5}{2}$. Similarly, using the value of n_{32} given by the eighth equation in the second equation gives

$$n_{12} = 1 + .2n_{12} + .2(1 + .4n_{12}) \qquad \text{or} \qquad n_{12} = \tfrac{5}{3}$$

Hence $n_{32} = 1 + .4(\tfrac{5}{3}) = \tfrac{5}{3}$. Equations three and six involve only n_{13} and n_{23} and can be written

$$.8n_{13} - .6n_{23} = 1$$
$$-.4n_{13} + .6n_{23} = 1$$

Thus $n_{13} = 5$, $n_{23} = 5$ and the matrix N is

$$N = \begin{bmatrix} 3 & \tfrac{5}{3} & 5 \\ \tfrac{5}{2} & 2 & 5 \\ \tfrac{5}{2} & \tfrac{5}{3} & 6 \end{bmatrix}$$

Thus, for example, it takes on the average five steps to go from C_1 to C_3 and six steps to go from C_3 back to C_3.

EXERCISES

9-22 Find the N matrix for the Markov chain of Exercise 9-5. The stationary probability vector was found in Exercise 9-13.

9-23 Find the N matrix for the Markov chain of Exercise 9-1. The stationary probability vector was found in Exercise 9-9. Interpret the values of n_{12} and n_{32}. If urn 1 contains one white ball, on the average how many steps are required before it will contain two white balls?

9-24   Suppose that the matrix of transition probabilities for a Markov chain is

$$P = \begin{bmatrix} .9 & .1 \\ .8 & .2 \end{bmatrix}$$

(as in the bolt example). We have already observed that the stationary probability vector is $[\frac{8}{9}, \frac{1}{9}]$. Find the N matrix. Find the probability that the number of steps required to go from state C_1 to state C_2 is 10 or less. Find the probability that the number of steps required to go from state C_2 back to C_2 is 6 or less; 2 or less (correct entry is not in table so use original formula).

9-25   Find the N matrix for the Markov chain of Exercise 9-8. The stationary probability vector was found in Exercise 9-16. Suppose that the housewife last purchased brand A. On the average how many purchases does she have to make before she buys brand C?

9-26   Find the N matrix for the Markov chain of Example 9-4. The stationary probability vector was found in Example 9-11. If the student worked no problems for his last assignment, on the average how many assignments are made before he again works no problems?

9-27   Find the N matrix for the Markov chain of Exercise 9-7. The stationary probability vector was found in Exercise 9-15. Suppose that A starts with two pennies. On the average how many tosses of the coin are required before A finds himself penniless?

9-28   Find the N matrix for the Markov chain of Exercise 9-6. The stationary probability vector was found in Exercise 9-14. Suppose that the last two rolls of the die produced a success followed by a failure. On the average how many rolls are required to obtain two successive successes?

9-29   Find the N matrix for the Markov chain of Exercise 9-4. The stationary probability vector was found in Exercise 9-12. On the average how many steps are required to go from room 4 to room 2?

9-30   In Exercise 9-17 we generalized Example 9-4 by replacing $\frac{2}{3}$ by p and $\frac{1}{3}$ by q and then found the stationary probability vector. Find the N matrix in terms of p and q and check your answer with the results of Exercise 9-26.

9-31   In Exercise 9-18 we generalized Exercise 9-6 by replacing $\frac{1}{3}$ by p and $\frac{2}{3}$ by q and then found the stationary probability vector. Find the N matrix in terms of p and q and check your answer with the results of Exercise 9-28.

9-32   In Exercise 9-19 we generalized Exercise 9-7 by assuming that A's probability of winning was p instead of $\frac{1}{2}$ and then found the stationary probability vector. Find the N matrix in terms of p and q and check your answer with the results of Exercise 9-27.

We have investigated some interesting properties of Markov chains, restricting the discussion almost entirely to chains with a regular transition matrix. There are, of course, other important types of Markov chains which we could study but we shall not do so in this book. We shall, however, give a brief discussion of one of these types.

Suppose that the matrix of transition probabilities for a Markov chain is

$$P = \begin{bmatrix} 1 & 0 \\ \frac{2}{3} & \frac{1}{3} \end{bmatrix} \tag{9-25}$$

Obviously, if the process ever enters state C_1 it can never leave since $p_{11} = 1$ and $p_{12} = 0$. Any state that is impossible to leave is called an *absorbing* state. We observe that state C_i is absorbing if $p_{ii} = 1$ (and hence, $p_{ij} = 0$, $i \neq j$). A Markov chain with one or more absorbing states is called an *absorbing Markov chain* provided it is possible to go to an absorbing state from every nonabsorbing state. Thus, the chain with transition matrix (9-25) is an absorbing chain as is the chain with transition matrix (9-26).

$$P = \begin{bmatrix} 1 & 0 & 0 \\ \frac{1}{6} & \frac{2}{6} & \frac{3}{6} \\ 0 & 0 & 1 \end{bmatrix} \tag{9-26}$$

However, the chain with transition matrix (9-27)

$$P = \begin{bmatrix} 1 & 0 & 0 \\ 0 & \frac{1}{3} & \frac{2}{3} \\ 0 & \frac{1}{2} & \frac{1}{2} \end{bmatrix} \tag{9-27}$$

does not fit the definition since the process cannot reach state C_1 from either state C_2 or C_3.

It is easy to see that for chains with transition matrices (9-25) and (9-26), the process will reach an absorbing state with probability 1. In each case the probability of reaching an absorbing state in one step is $\frac{2}{3}$ and the probability of reaching an absorbing state if we continue to repeat the process is

$$(\tfrac{2}{3}) + (\tfrac{1}{3})(\tfrac{2}{3}) + (\tfrac{1}{3})^2(\tfrac{2}{3}) + (\tfrac{1}{3})^3(\tfrac{2}{3}) + \cdots = \frac{\tfrac{2}{3}}{1 - (\tfrac{1}{3})} = 1$$

Although we have used specific examples with only one nonabsorbing state, the result is true in general. Thus, for any absorbing Markov chain, the probability of reaching an absorbing state is 1.

Some of the interesting properties of absorbing Markov chains include†

1 The average number of times the process will be in each nonabsorbing state
2 The average number of steps required for the process to reach an absorbing state
3 The probability that a process with more than one absorbing state will reach a particular absorbing state

In Sec. 9-2 we proposed finding the probability of x_1 outcomes in state C_1, x_2 outcomes in state C_2, . . . , x_k outcomes in state C_k, if the chain starts in a given state and is repeated n times. We could, of course, evaluate this probability by enumerating all cases with the aid of a tree diagram. No nice answer seems to be available and we shall not pursue this topic.

REFERENCES

1 FELLER, W.: "An Introduction to Probability Theory and its Application," 2d ed., John Wiley & Sons, Inc., New York, 1957.
2 KEMENY, JOHN G., SNELL, J. L., and THOMPSON, G. L.: "Introduction to Finite Mathematics," 2d ed., Prentice-Hall, Inc., Englewood Cliffs, N.J., 1966.
3 PARZEN, E.: "Stochastic Processes," Holden-Day, Inc., San Francisco, 1962.

† For a good discussion at an elementary level see Kemeny, Snell, and Thompson [2], pages 282–291. Chapter 7 of that book includes a number of interesting applications of Markov chains (both regular and absorbing) to behavioral science problems.

9-1 A sequence of random experiments is referred to as a stochastic process. If the outcome of each experiment in a sequence depends at most upon the result of the immediately preceding experiment, then the stochastic process is called a Markov chain.

9-2 The possible categories into which the results of an experiment can be classified are called states. If p_{ij} is the probability that state C_j follows state C_i, then the square matrix P whose elements are p_{ij} is called the transition matrix of the Markov chain.

9-3 If $p_{ij}^{(n)}$ is the probability that a process in state C_i will be in state C_j after n performances of the experiment, then the matrix of the $p_{ij}^{(n)}$, say, $P^{(n)}$, is related to P by

$$P^{(n)} = P^n \tag{9-6}$$

Let the process start in state C_i with probability $p_i^{(0)}$. Then

$$p^{(0)} = [p_1^{(0)}, p_2^{(0)}, \ldots, p_k^{(0)}] \tag{9-7}$$

is called the vector of initial probabilities. Any row vector with nonnegative elements whose sum is 1 is called a probability vector. If $p_i^{(n)}$ is the probability that a process will be in state C_i after n repetitions and

$$p^{(n)} = [p_1^{(n)}, p_2^{(n)}, \ldots, p_k^{(n)}] \tag{9-8}$$

is the vector of these probabilities, then

$$p^{(n)} = p^{(0)}P^n \tag{9-9}$$

A probability vector $a = [a_1, a_2, \ldots, a_k]$ is called a stationary probability vector if

$$aP = a \tag{9-10}$$

If some power of a transition matrix has only positive entries, then it is called a regular matrix. If P is regular, then (9-10) has a unique solution. If a Markov chain has a regular transition matrix P, then

(a) As n increases, each row of P^n approaches the same row vector a.

(b) Each entry a_i, $i = 1, 2, \ldots, k$ of a is positive.

(c) The vector a is the unique stationary probability vector of P.

(d) If $p^{(0)}$ is any vector of initial probabilities, then as n increases the entries of $p^{(n)} = p^{(0)}P^n$ approach the corresponding elements of a. That is, $p_i^{(n)}$ approaches a_i, $i = 1, 2, \ldots, k$.

9-4 Consider a Markov chain with regular transition matrix P. Let n_{ij} be the average numbers of steps required to go from state C_i to state C_j and N be the matrix of the n_{ij}. If the chain has only two states, then

$$N = \begin{bmatrix} n_{11} & n_{12} \\ n_{21} & n_{22} \end{bmatrix} = \begin{bmatrix} 1/a_1 & 1/p_{12} \\ 1/p_{21} & 1/a_2 \end{bmatrix} \qquad (9\text{-}16)$$

If the chain has k states, then the n_{ij} satisfy the equation

$$N = PN_d + J \qquad (9\text{-}22)$$

where N_d is obtained from the N matrix by replacing the diagonal elements with 0's, and J is a square matrix with all elements equal to 1. Using (9-22) and the fact that

$$n_{ij} = \frac{1}{a_i} \qquad (9\text{-}24)$$

all n_{ij} can be found.

APPENDIXES

Appendix A
Some Mathematical Results

Summation Notation

A1-1   SINGLE SUMMATION

Summation signs are used to abbreviate lengthy mathematical expressions.   By definition we have

$$\sum_{i=1}^{n} u(x_i) = u(x_1) + u(x_2) + \cdots + u(x_n) \tag{A1-1}$$

which is read as "the sum of $u(x_i)$, i going from 1 to n."   Each $u(x_i)$ represents a number that may or may not be known later in the discussion.   It is obviously more convenient to write the left-hand side of (A1-1) than to write out the right-hand side of that equation.   Sometimes the limits on the Σ are omitted when it is clear which values of $u(x_i)$ are to be included in the sum.   The upper limit on the sum can be infinity, in

331

which case we write

$$\sum_{i=1}^{\infty} u(x_i) = u(x_1) + u(x_2) + u(x_3) + \cdots \qquad \text{(A1-2)}$$

If the possible values of x under consideration are $x = 0, 1, 2, \ldots, n$ or $x = 0, 1, 2, 3, \ldots$ (all nonnegative integers), then

$$\sum_{x=0}^{n} u(x) = u(0) + u(1) + u(2) + \cdots + u(n) \qquad \text{(A1-3)}$$

and

$$\sum_{x=0}^{\infty} u(x) = u(0) + u(1) + u(2) + \cdots \qquad \text{(A1-4)}$$

are sometimes used. The expression

$$\sum_{x} u(x) \qquad \text{(A1-5)}$$

means that $u(x)$ is summed over all values of x under consideration.
In particular if $f(x)$ is a probability function

$$\sum_{x} f(x) = 1$$

$$\sum_{x} x f(x) = \mu = E(X) = \text{expected value of } X$$

$$\sum_{x} (x - \mu)^2 f(x) = E[(X - \mu)^2] = \text{variance of } X$$

Suppose a random variable X can assume values $x_1 < x_2 < x_3 < \cdots$. Then another notation sometimes used (although, perhaps, it is not too satisfactory) is

$$\sum_{X \leq x} f(x) = f(x_1) + f(x_2) + \cdots + f(x)$$
$$= \Pr(X = x_1) + \Pr(X = x_2) + \cdots + \Pr(X = x)$$

where x is some particular value of X.

EXAMPLE A1-1

Write out $\sum_{i=1}^{6} x_i$.

Solution

By definition (A1-1) we get

$$\sum_{i=1}^{6} x_i = x_1 + x_2 + x_3 + x_4 + x_5 + x_6$$

EXAMPLE A1-2

Write out $\sum_{i=1}^{4} (x_i - \mu)^2 f(x_i)$.

Solution

Definition (A1-1) yields

$$\sum_{i=1}^{4} (x_i - \mu)^2 f(x_i) = (x_1 - \mu)^2 f(x_1) + (x_2 - \mu)^2 f(x_2) + (x_3 - \mu)^2 f(x_3)$$
$$+ (x_4 - \mu)^2 f(x_4)$$

EXAMPLE A1-3

If $x_1 = 3$, $x_2 = 5$, $x_3 = 11$, find $\sum_{i=1}^{3} x_i$, $\left(\sum_{i=1}^{3} x_i\right)^2$, $\sum_{i=1}^{3} x_i^2$.

Solution

$$\sum_{i=1}^{3} x_i = 3 + 5 + 11 = 19 \qquad \left(\sum_{i=1}^{3} x_i\right)^2 = 19^2 = 361$$

$$\sum_{i=1}^{3} x_i^2 = 3^2 + 5^2 + 11^2 = 9 + 25 + 121 = 155$$

Thus, the x_i being known numbers, the summations can be evaluated and yield specific numbers. In addition, if no other values of x are under consideration, then $\sum_x x = 19$, $\sum_x x^2 = 155$, $\left(\sum_x x\right)^2 = 361$.

EXAMPLE A1-4

Write out $\sum_{i=1}^{5} (x_i + y_i)$ and show that this is equivalent to $\sum_{i=1}^{5} x_i + \sum_{i=1}^{5} y_i$.

Solution

Definition (A1-1) yields

$$\sum_{i=1}^{5} (x_i + y_i) = (x_1 + y_1) + (x_2 + y_2) + (x_3 + y_3) + (x_4 + x_4)$$
$$+ (x_5 + y_5)$$

The right-hand side of the last equation can be written as

$$(x_1 + x_2 + x_3 + x_4 + x_5) + (y_1 + y_2 + y_3 + y_4 + y_5) = \sum_{i=1}^{5} x_i + \sum_{i=1}^{5} y_i$$

EXAMPLE A1-5

Write out $\displaystyle\sum_{i=1}^{4} C.$

Solution

By definition (A1-1) we get four terms in the sum. Since there is no subscript to change from term to term, each term is C. Thus

$$\sum_{i=1}^{4} C = C + C + C + C = 4C$$

EXAMPLE A1-6

Write out $\displaystyle\sum_{i=1}^{5} x_i y_i.$

Solution

By Definition (A1-1) we get

$$\sum_{i=1}^{5} x_i y_i = x_1 y_1 + x_2 y_2 + x_3 y_3 + x_4 y_4 + x_5 y_5$$

EXAMPLE A1-7

Write out $\displaystyle\sum_{i=1}^{3} (x_i + y_i)^2.$

Solution

Definition (A1-1) yields

$$\sum_{i=1}^{3} (x_i + y_i)^2 = (x_1 + y_1)^2 + (x_2 + y_2)^2 + (x_3 + y_3)^2$$

If we so desire, each of the binomial terms can be expanded, giving

$$x_1^2 + 2x_1 y_1 + y_1^2 + x_2^2 + 2x_2 y_2 + y_2^2 + x_3^2 + 2x_3 y_3 + y_3^2$$
$$= (x_1^2 + x_2^2 + x_3^2) + 2(x_1 y_1 + x_2 y_2 + x_3 y_3) + (y_1^2 + y_2^2 + y_3^2)$$
$$= \sum_{i=1}^{3} x_i^2 + 2 \sum_{i=1}^{3} x_i y_i + \sum_{i=1}^{3} y_i^2$$

EXAMPLE A1-8

Write out $\left(\sum_{i=1}^{4} x_i\right)^2$, square the expression in the parenthesis, and express as two sums.

Solution

First, by (A1-1) we get

$$\left(\sum_{i=1}^{4} x_i\right)^2 = (x_1 + x_2 + x_3 + x_4)^2$$

Then, squaring the right-hand side of the equation we get four squared terms and six (4 things taken 2 at a time) cross-product terms. These are

$$x_1^2 + x_2^2 + x_3^2 + x_4^2 + 2(x_1x_2 + x_1x_3 + x_1x_4 + x_2x_3 + x_2x_4 + x_3x_4)$$

Of course,

$$x_1^2 + x_2^2 + x_3^2 + x_4^2 = \sum_{i=1}^{4} x_i^2$$

but the cross-product term requires new notation. Two forms are in common usage. One is

$$2(x_1x_2 + x_1x_3 + x_1x_4 + x_2x_3 + x_2x_4 + x_3x_4) = 2 \sum_{\substack{i,j=1 \\ i<j}}^{4} x_i x_j$$

Since we can also write the sum of cross products as

$$
\begin{array}{llll}
& x_1x_2 & + x_1x_3 & + x_1x_4 \\
+ x_2x_1 & & + x_2x_3 & + x_2x_4 \\
+ x_3x_1 & + x_3x_2 & & + x_3x_4 \\
+ x_4x_1 & + x_4x_2 & + x_4x_3 &
\end{array}
$$

another notation frequently used for the sum is

$$\sum_{\substack{i,j=1 \\ i \neq j}}^{4} x_i x_j$$

Since all terms are produced twice (the second time with subscripts reversed from the first time), no 2 multiplier is necessary.

EXAMPLE A1-9

Show that $\displaystyle\sum_{i=1}^{4} 3x_i = 3 \sum_{i=1}^{4} x_i$.

Solution

Definition (A1-1) yields

$$\sum_{i=1}^{4} 3x_i = 3x_1 + 3x_2 + 3x_3 + 3x_4$$

$$= 3(x_1 + x_2 + x_3 + x_4)$$

$$= 3 \sum_{i=1}^{4} x_i$$

The preceding examples illustrate four useful theorems. All are proved by using definition (A1-1). These are:

THEOREM I

$$\sum_{i=1}^{n} C = nC,$$ where C is any quantity that does not have a summation subscript.

Proof

By definition (A1-1) we have

$$\sum_{i=1}^{n} C = \underbrace{C + C + \cdots + C}_{n \text{ terms}} = nC$$

THEOREM II

$$\sum_{i=1}^{n} Cu(x_i) = C \sum_{i=1}^{n} u(x_i)$$

Proof

Definition (A1-1) yields

$$\sum_{i=1}^{n} Cu(x_i) = Cu(x_1) + Cu(x_2) + \cdots + Cu(x_n)$$

$$= C[u(x_1) + u(x_2) + \cdots + u(x_n)]$$

$$= C \sum_{i=1}^{n} u(x_i)$$

THEOREM III

$$\sum_{i=1}^{n} (x_i + y_i - z_i) = \sum_{i=1}^{n} x_i + \sum_{i=1}^{n} y_i - \sum_{i=1}^{n} z_i$$

336

Proof

Definition (A1-1) yields

$$\sum_{i=1}^{n} (x_i + y_i - z_i) = (x_1 + y_1 - z_1) + (x_2 + y_2 - z_2)$$
$$+ \cdots + (x_n + y_n - z_n)$$
$$= (x_1 + x_2 + \cdots + x_n) + (y_1 + y_2 + \cdots + y_n)$$
$$- (z_1 + z_2 + \cdots + z_n)$$
$$= \sum_{i=1}^{n} x_i + \sum_{i=1}^{n} y_i - \sum_{i=1}^{n} z_i$$

Theorem III generalizes in the obvious way.

THEOREM IV

$$\left(\sum_{i=1}^{n} x_i\right)^2 = \sum_{i=1}^{n} x_i^2 + 2 \sum_{\substack{i,j=1 \\ i<j}}^{n} x_i x_j$$
$$= \sum_{i=1}^{n} x_i^2 + \sum_{\substack{i,j=1 \\ i \neq j}}^{n} x_i x_j$$

Proof

A well-known result from algebra states that

$$\left(\sum_{i=1}^{n} x_i\right)^2 = (x_1 + x_2 + \cdots + x_n)^2$$

yields the sum of squares of each term plus twice the sum of all possible cross products. This is easily seen to be true since

$$(x_1 + x_2 + \cdots + x_n)^2 = (x_1 + x_2 + \cdots + x_n)(x_1 + x_2 + \cdots + x_n)$$

and by the definition of multiplication each term in the first parenthesis must be multiplied by each term in the second parenthesis and all results summed.

A1-2 DOUBLE SUMMATION

Double summation does not require any new concepts. It can be regarded as a single summation on one subscript followed by a single summation on a second subscript. Double summation is often used as a matter of convenience in situations where single summation could also be used (by relabeling terms and subscripts).

EXAMPLE A1-10

Evaluate $\displaystyle\sum_{j=1}^{2}\sum_{i=1}^{3} x_{ij}$ and $\displaystyle\sum_{i=1}^{3}\sum_{j=1}^{2} x_{ij}$.

Solution

Using definition (A1-1) twice, we get

$$\sum_{j=1}^{2}\sum_{i=1}^{3} x_{ij} = \sum_{j=1}^{2}(x_{1j} + x_{2j} + x_{3j})$$

$$= (x_{11} + x_{21} + x_{31}) + (x_{12} + x_{22} + x_{32})$$

$$\sum_{i=1}^{3}\sum_{j=1}^{2} x_{ij} = \sum_{i=1}^{3}(x_{i1} + x_{i2})$$

$$= (x_{11} + x_{12}) + (x_{21} + x_{22}) + (x_{31} + x_{32})$$

We note that

$$\sum_{j=1}^{2}\sum_{i=1}^{3} x_{ij} = \sum_{i=1}^{3}\sum_{j=1}^{2} x_{ij}$$

Example A1-10 illustrates the following theorem:

THEOREM V

$$\sum_{j=1}^{b}\sum_{i=1}^{a} x_{ij} = \sum_{i=1}^{a}\sum_{j=1}^{b} x_{ij}$$

That is, if both limits on the summation signs are constants, then the order of summation may be interchanged.

Proof

The theorem can be proved by following the scheme outlined in Example A1-10.

EXAMPLE A1-11

Write out $\displaystyle\sum_{i=1}^{L}\sum_{j=1}^{N_i} v_{ij}$.

Solution

By definition (A1-1) we get

$$\sum_{j=1}^{N_i} v_{ij} = v_{i1} + v_{i2} + v_{i3} + \cdots + v_{iN_i}$$

338

so that

$$\sum_{i=1}^{L} \sum_{j=1}^{N_i} = \sum_{i=1}^{L} (v_{i1} + v_{i2} + \cdots + v_{iN_i})$$

$$= (v_{11} + v_{12} + \cdots + v_{1N_1}) + (v_{21} + v_{22} + \cdots + v_{2N_2})$$
$$+ \cdots + (v_{L1} + v_{L2} + \cdots + v_{LN_L})$$

Corresponding to Theorem I, we have

THEOREM VI

$$\sum_{i=1}^{L} \sum_{j=1}^{N_i} C = C \sum_{i=1}^{L} N_i = CN$$

Proof

By Theorem I

$$\sum_{j=1}^{N_i} C = CN_i$$

Hence

$$\sum_{i=1}^{L} \sum_{j=1}^{N_i} C = \sum_{i=1}^{L} CN_i$$

$$= C \sum_{i=1}^{L} N_i \qquad \text{by Theorem II}$$

$$= CN \qquad \text{by definition of } N$$

The counterpart of Theorem II is

THEOREM VII

$$\sum_{i=1}^{L} \sum_{j=1}^{N_i} Cv_{ij} = C \sum_{i=1}^{L} \sum_{j=1}^{N_i} v_{ij}$$

a result that follows easily, since C will appear in every term of the sum and can be factored out.

Similar to Theorem II is

$$\sum_{i=1}^{L} \sum_{j=1}^{N_i} y_i v_{ij} = \sum_{i=1}^{L} y_i \sum_{j=1}^{N_i} v_{ij}$$

Thus, if part of a product that is being summed involves only the outside index of summation, this part can be factored out of the inside summation sign.

Proof

By Theorem II

$$\sum_{j=1}^{N_i} y_i v_{ij} = y_i \sum_{j=1}^{N_i} v_{ij}$$

Hence the result.

The counterpart of Theorem III is

THEOREM IX ━━━

$$\sum_{i=1}^{L} \sum_{j=1}^{N_i} (x_{ij} + y_{ij} - z_{ij}) = \sum_{i=1}^{L} \sum_{j=1}^{N_i} x_{ij} + \sum_{i=1}^{L} \sum_{j=1}^{N_i} y_{ij} - \sum_{i=1}^{L} \sum_{j=1}^{N_i} z_{ij}$$

Proof

The proof follows easily by expanding the sum as in Example A1-11 followed by regrouping terms.

Binomial Expansions

A2-1   BINOMIAL EXPANSION WITH POSITIVE INTEGRAL EXPONENT

By actual multiplication we can verify that

$(a + b)^2 = (a + b)(a + b) = a^2 + 2ab + b^2$
$(a + b)^3 = (a + b)^2(a + b) = a^3 + 3a^2b + 3ab^2 + b^3$
$(a + b)^4 = (a + b)^3(a + b) = a^4 + 4a^3b + 6a^2b^2 + 4ab^3 + b^4$

Letting n represent the exponent for the three cases, we observe that

(a) *The first term is a^n (and the last is b^n).*
(b) *As we proceed from term to term moving left to right, the exponent on a decreases by 1, the exponent on b increases by 1, and the sum of the two exponents is n.*

(c) *The coefficient of any term is obtained by multiplying the* (A2-1)
exponent in the preceding term by the coefficient in the pre-
ceding term followed by dividing the product by the number
of the preceding term (that is, the first term is number 1, the
second number 2, etc.).

It can be shown that (A2-1) can be used to expand $(a + b)^n$ where n is any positive integer.

By counting permutations we are also able to obtain the coefficient for any term. Consider the coefficient of a^2b in $(a + b)^3 = (a + b)(a + b)$ $(a + b)$. Terms of this type result by selecting

1 a from the first term of the product, a from the second, and b from the third, or aab.
2 a from the first term, b from the second, and a from the third, or aba
3 b from the first term, a from the second, and a from the third, or baa

Hence a^2b terms can be formed in the number of ways three things taken three at a time can be permuted if two are alike and one is different. That is, $3!/(2!)(1!) = 3$. In general, from

$$(a + b)^n = (a + b)(a + b) \cdots (a + b)$$

terms of the type $b^x a^{n-x}$ can be formed in the number of ways n things taken n at time can be permuted if x are alike and $n - x$ are alike. This number is $n!/x!(n - x)! = \binom{n}{x}$. Hence

$$(a + b)^n = \binom{n}{0} a^n + \binom{n}{1} ba^{n-1} + \binom{n}{2} b^2 a^{n-2} + \cdots + \binom{n}{n} b^n \quad (A2\text{-}2)$$

or, in summation form,

$$(a + b)^n = \sum_{x=0}^{n} \binom{n}{x} b^x a^{n-x} \quad (A2\text{-}3)$$

In particular, if $p > 0$, $q > 0$, and $p + q = 1$ we have

$$\sum_{x=0}^{n} \binom{n}{x} p^x q^{n-x} = (q + p)^n = 1 \quad (A2\text{-}4)$$

342

A2-2 BINOMIAL EXPANSION WITH NEGATIVE INTEGRAL EXPONENT

Let n be a positive integer and consider $1/(1 - a)^n = (1 - a)^{-n}$. It seems natural to attempt to expand $(1 - a)^{-n}$ according to the procedure given by (A2-1). If we do this we get

$$(1 - a)^{-n} = 1^{-n} + (-n)1^{-n-1}(-a) + \frac{(-n)(-n - 1)}{2!} 1^{-n-2}(-a)^2$$

$$+ \frac{(-n)(-n - 1)(-n - 2)}{3!} 1^{-n-3}(-a)^3 + \cdots$$

$$= 1 + na + \frac{n(n + 1)}{2!} a^2 + \frac{n(n + 1)(n + 2)}{3!} a^3 + \cdots$$

If we multiply every coefficient in the last expression by $(n - 1)!/(n - 1)!$, then each coefficient represents the value of a combination symbol and we can write

$$(1 - a)^{-n} = \binom{n - 1}{0} + \binom{n}{1} a + \binom{n + 1}{2} a^2 + \binom{n + 2}{3} a^3 + \cdots \tag{A2-5}$$

$$= \sum_{x=0}^{\infty} \binom{n - 1 + x}{x} a^x \tag{A2-6}$$

$$= \sum_{x=0}^{\infty} \binom{x - 1 + n}{n - 1} a^x \tag{A2-7}$$

Since procedure (A2-1) was derived for positive integral exponent, it is natural to inquire as to whether or not (A2-5) is a valid formula. An additional complication arises from the fact that the expansion now contains an infinite number of terms. It can be shown that (A2-5) is correct provided $-1 < a < 1$.

It is easy to verify the following special cases:

$$(1 - a)^{-1} = 1 + a + a^2 + a^3 + \cdots \tag{A2-8}$$
$$(1 - a)^{-2} = 1 + 2a + 3a^2 + 4a^3 + \cdots \tag{A2-9}$$
$$(1 - a)^{-3} = 1 + 3a + 6a^2 + 10a^3 + \cdots \tag{A2-10}$$

By writing out both sums it is easy to verify that

$$\sum_{x=0}^{\infty} \binom{x - 1 + n}{n - 1} a^x = \sum_{y=n}^{\infty} \binom{y - 1}{n - 1} a^{y-n} \tag{A2-11}$$

343

If, in the right side of (A2-11) we replace y by n, n by c, and a by q we have

$$(1 - q)^{-c} = \sum_{n=c}^{\infty} \binom{n-1}{c-1} q^{n-c} \qquad \text{(A2-12)}$$

Multiplying both sides of (A2-12) by p^c gives

$$p^c(1 - q)^{-c} = \sum_{n=c}^{\infty} \binom{n-1}{c-1} p^c q^{n-c} \qquad \text{(A2-13)}$$

Now, if $p > 0$, $q > 0$, and $p = 1 - q$, then (A2-13) yields

$$\sum_{n=c}^{\infty} \binom{n-1}{c-1} p^c q^{n-c} = p^c p^{-c} = 1$$

which agrees with Eq. (4-11).

Matrices

A3-1 DEFINITIONS

A matrix is a rectangular array of numbers or elements arranged in r rows and c columns as follows:

$$
A = \begin{bmatrix} a_{11} & a_{12} & \cdots & a_{1c} \\ a_{21} & a_{22} & \cdots & a_{2c} \\ \cdots & \cdots & \cdots & \cdots \\ a_{r1} & a_{r2} & \cdots & a_{rc} \end{bmatrix}
\tag{A3-1}
$$

The matrix A is sometimes called an $r \times c$ matrix. If $r = c$, then the matrix is also called a square matrix. If $c = 1$, then the matrix consists of a single column and is called a column vector. Similarly, if $r = 1$, the matrix consists of a single row and is called a row vector. Hence

$$x = \begin{bmatrix} a_{11} \\ a_{21} \\ \cdot \\ \cdot \\ \cdot \\ a_{r1} \end{bmatrix}$$

(A3-2)

and

$$y = [a_{11}, a_{12}, \ldots , a_{1c}]$$

(A3-3)

are, respectively, column and row vectors.

Some specific numerical examples are the following:

$$\begin{bmatrix} 3 & 2 & -5 \\ 10 & 0 & 4 \end{bmatrix} \quad \text{is a } 2 \times 3 \text{ matrix}$$

$$\begin{bmatrix} 10 & 16 \\ -1 & 6 \\ 3 & 2 \end{bmatrix} \quad \text{is a } 3 \times 2 \text{ matrix}$$

$$[.4,.2,.4] \quad \text{is a } 1 \times 3 \text{ matrix (row vector)}$$

$$\begin{bmatrix} .6 \\ 5.4 \\ 3.0 \\ 2.1 \end{bmatrix} \quad \text{is a } 4 \times 1 \text{ matrix (column vector)}$$

Two $r \times c$ matrices are said to be equal if and only if all corresponding elements are equal. Thus if

$$\begin{bmatrix} a_{11} & a_{12} & a_{13} \\ a_{21} & a_{22} & a_{23} \end{bmatrix} = \begin{bmatrix} b_{11} & b_{12} & b_{13} \\ b_{21} & b_{22} & b_{23} \end{bmatrix}$$

then we must have $a_{11} = b_{11}$, $a_{12} = b_{12}$, $a_{13} = b_{13}$, $a_{21} = b_{21}$, $a_{22} = b_{22}$, $a_{23} = b_{23}$. Thus, the definition of equality makes sense only if the two matrices have the same number of rows and columns (in which case they are said to have the same order).

A square matrix with diagonal elements $a_{ii} = 1$ and all other elements equal to 0 is called the identity matrix and is denoted by I. Thus

$$I = \begin{bmatrix} 1 & 0 & \cdots & 0 \\ 0 & 1 & \cdots & 0 \\ \cdot & \cdot & \cdots & \cdot \\ 0 & 0 & \cdots & 1 \end{bmatrix}$$

The matrix I plays the same role in the algebra of matrices as 1 does in ordinary arithmetic.

In Chap. 9 we needed a square matrix consisting entirely of 1's. We chose to denote this by J. Thus

$$J = \begin{bmatrix} 1 & 1 & \cdots & 1 \\ 1 & 1 & \cdots & 1 \\ \cdot & \cdot & \cdots & \cdot \\ 1 & 1 & \cdots & 1 \end{bmatrix}$$

A3-2   SOME ALGEBRA OF MATRICES

For two matrices of the same order the sum is defined to be the new matrix of the same order obtained by adding corresponding elements. Thus, if

$$A = \begin{bmatrix} a_{11} & a_{12} & a_{13} \\ a_{21} & a_{22} & a_{23} \end{bmatrix} \qquad B = \begin{bmatrix} b_{11} & b_{12} & b_{13} \\ b_{21} & b_{22} & b_{23} \end{bmatrix}$$

then

$$A + B = \begin{bmatrix} a_{11} + b_{11} & a_{12} + b_{12} & a_{13} + b_{13} \\ a_{21} + b_{21} & a_{22} + b_{22} & a_{23} + b_{23} \end{bmatrix}$$

Similarly, for two matrices of the same order, the difference is a new matrix of the same order obtained by subtracting corresponding elements. Thus

$$A - B = \begin{bmatrix} a_{11} - b_{11} & a_{12} - b_{12} & a_{13} - b_{13} \\ a_{21} - b_{21} & a_{22} - b_{22} & a_{23} - b_{23} \end{bmatrix}$$

Specifically

$$\begin{bmatrix} 7 & 5 & 2 \\ 8 & 0 & 3 \end{bmatrix} + \begin{bmatrix} 6 & 2 & 8 \\ 4 & 7 & 1 \end{bmatrix} = \begin{bmatrix} 13 & 7 & 10 \\ 12 & 7 & 4 \end{bmatrix}$$

$$\begin{bmatrix} 7 & 5 & 2 \\ 8 & 0 & 3 \end{bmatrix} - \begin{bmatrix} 6 & 2 & 8 \\ 4 & 7 & 1 \end{bmatrix} = \begin{bmatrix} 1 & 3 & -6 \\ 4 & -7 & 2 \end{bmatrix}$$

The product AB of two matrices is defined only if A has the same number of columns as B has rows. Suppose A is of order $r \times c$ and B is of order $c \times m$. Then the product AB is defined to be the matrix of order $r \times m$ with elements

$$d_{ij} = a_{i1}b_{1j} + a_{i2}b_{2j} + \cdots + a_{ic}b_{cj} = \sum_{k=1}^{c} a_{ik}b_{kj}$$

Let us illustrate with numerical examples. Suppose that

$$A = \begin{bmatrix} 1 & 5 \\ 2 & 6 \\ 3 & 8 \end{bmatrix} \qquad B = \begin{bmatrix} 0 & 7 & -1 & -3 \\ 4 & 9 & -2 & -4 \end{bmatrix}$$

Then

$$AB = \begin{bmatrix} 1(0)+5(4) & 1(7)+5(9) & 1(-1)+5(-2) & 1(-3)+5(-4) \\ 2(0)+6(4) & 2(7)+6(9) & 2(-1)+6(-2) & 2(-3)+6(-4) \\ 3(0)+8(4) & 3(7)+8(9) & 3(-1)+8(-2) & 3(-3)+8(-4) \end{bmatrix}$$

$$= \begin{bmatrix} 20 & 52 & -11 & -23 \\ 24 & 68 & -14 & -30 \\ 32 & 93 & -19 & -41 \end{bmatrix}$$

In particular (1) the product of a square matrix of order $k \times k$ and another square matrix of order $k \times k$ is a square matrix of order $k \times k$, (2) the product of a matrix of order $1 \times k$ (row vector with k elements) by a square matrix of order $k \times k$ is a matrix of order $1 \times k$ (another row vector). Thus, for example,

$$[2,3,5] \begin{bmatrix} 1 & -2 & 7 \\ 4 & 6 & 8 \\ -1 & 0 & -1 \end{bmatrix}$$

$$= [2(1)+3(4)+5(-1),\ 2(-2)+3(6)+5(0),\ 2(7)+3(8)+5(-1)]$$
$$= [9,14,33]$$

Chapter 9 provides a number of examples of the great simplification achieved by expressing certain mathematical results in matrix notation.

Appendix B

Tables

Appendix B1   The cumulative binomial distribution

Entry $= B(r;n,p) = \sum\limits_{x=0}^{r} b(x;n,p)$ [see Eqs. (4-2) and (4-3)]

n	r	p = .10	p = .20	p = .25	p = .30	p = .40	p = .50
5	0	.59049	.32768	.23730	.16807	.07776	.03125
	1	.91854	.73728	.63281	.52822	.33696	.18750
	2	.99144	.94208	.89648	.83692	.68256	.50000
	3	.99954	.99328	.98437	.96922	.91296	.81250
	4	.99999	.99968	.99902	.99757	.98976	.96875
	5	1.00000	1.00000	1.00000	1.00000	1.00000	1.00000
10	0	.34868	.10737	.05631	.02825	.00605	.00098
	1	.73610	.37581	.24403	.14931	.04636	.01074
	2	.92981	.67780	.52559	.38278	.16729	.05469
	3	.98720	.87913	.77588	.64961	.38228	.17187
	4	.99837	.96721	.92187	.84973	.63310	.37695
	5	.99985	.99363	.98027	.95265	.83376	.62305
	6	.99999	.99914	.99649	.98941	.94524	.82812
	7	1.00000	.99992	.99958	.99841	.98771	.94531
	8		1.00000	.99997	.99986	.99832	.98926
	9			1.00000	.99999	.99990	.99902
	10				1.00000	1.00000	1.00000
15	0	.20589	.03518	.01336	.00475	.00047	.00003
	1	.54904	.16713	.08018	.03527	.00517	.00049
	2	.81594	.39802	.23609	.12683	.02711	.00369
	3	.94444	.64816	.46129	.29687	.09050	.01758
	4	.98728	.83577	.68649	.51549	.21728	.05923
	5	.99775	.93895	.85163	.72162	.40322	.15088
	6	.99969	.98194	.94338	.86886	.60981	.30362
	7	.99997	.99576	.98270	.94999	.78690	.50000
	8	1.00000	.99921	.99581	.98476	.90495	.69638
	9		.99989	.99921	.99635	.96617	.84912
	10		.99999	.99988	.99933	.99065	.94077
	11		1.00000	.99999	.99991	.99807	.98242
	12			1.00000	.99999	.99972	.99631
	13				1.00000	.99997	.99951
	14					1.00000	.99997
	15						1.00000
20	0	.12158	.01153	.00317	.00080	.00004	.00000
	1	.39175	.06918	.02431	.00764	.00052	.00002
	2	.67693	.20608	.09126	.03548	.00361	.00020
	3	.86705	.41145	.22516	.10709	.01596	.00129
	4	.95683	.62965	.41484	.23751	.05095	.00591
	5	.98875	.80421	.61717	.41637	.12560	.02069
	6	.99761	.91331	.78578	.60801	.25001	.05766

n	r	p = .10	p = .20	p = .25	p = .30	p = .40	p = .50
20	7	.99958	.96786	.89819	.77227	.41589	.13159
	8	.99994	.99002	.95907	.88667	.59560	.25172
	9	.99999	.99741	.98614	.95204	.75534	.41190
	10	1.00000	.99944	.99606	.98286	.87248	.58810
	11		.99990	.99906	.99486	.94347	.74828
	12		.99998	.99982	.99872	.97897	.86841
	13		1.00000	.99997	.99974	.99353	.94234
	14			1.00000	.99996	.99839	.97931
	15				.99999	.99968	.99409
	16				1.00000	.99995	.99871
	17					.99999	.99980
	18					1.00000	.99998
	19						1.00000
25	0	.07179	.00378	.00075	.00013	.00000	.00000
	1	.27121	.02739	.00702	.00157	.00005	.00000
	2	.53709	.09823	.03211	.00896	.00043	.00001
	3	.76359	.23399	.09621	.03324	.00237	.00008
	4	.90201	.42067	.21374	.09047	.00947	.00046
	5	.96660	.61669	.37828	.19349	.02936	.00204
	6	.99052	.78004	.56110	.34065	.07357	.00732
	7	.99774	.89088	.72651	.51185	.15355	.02164
	8	.99954	.95323	.85056	.67693	.27353	.05388
	9	.99992	.98267	.92867	.81056	.42462	.11476
	10	.99999	.99445	.97033	.90220	.58577	.21218
	11	1.00000	.99846	.98027	.95575	.73228	.34502
	12		.99963	.99663	.98253	.84623	.50000
	13		.99992	.99908	.99401	.92229	.65498
	14		.99999	.99979	.99822	.96561	.78782
	15		1.00000	.99996	.99955	.98683	.88524
	16			.99999	.99990	.99567	.94612
	17			1.00000	.99998	.99879	.97836
	18				1.00000	.99972	.99268
	19					.99995	.99796
	20					.99999	.99954
	21					1.00000	.99992
	22						.99999
	23						1.00000
50	0	.00515	.00001	.00000	.00000		
	1	.03379	.00019	.00001	.00000		
	2	.11173	.00129	.00009	.00000		
	3	.25029	.00566	.00050	.00003		

n	r	p = .10	p = .20	p = .25	p = .30	p = .40	p = .50
50	4	.43120	.01850	.00211	.00017		
	5	.61612	.04803	.00705	.00072	.00000	
	6	.77023	.10340	.01939	.00249	.00001	
	7	.87785	.19041	.04526	.00726	.00006	
	8	.94213	.30733	.09160	.01825	.00023	
	9	.97546	.44374	.16368	.04023	.00076	.00000
	10	.99065	.58356	.26220	.07885	.00220	.00001
	11	.99678	.71067	.38162	.13904	.00569	.00005
	12	.99900	.81394	.51099	.22287	.01325	.00015
	13	.99971	.88941	.63704	.32788	.02799	.00047
	14	.99993	.93928	.74808	.44683	.05396	.00130
	15	.99998	.96920	.83692	.56918	.09550	.00330
	16	1.00000	.98556	.90169	.68388	.15609	.00767
	17		.99374	.94488	.78219	.23688	.01642
	18		.99749	.97127	.85944	.33561	.03245
	19		.99907	.98608	.91520	.44648	.05946
	20		.99968	.99374	.95224	.56103	.10132
	21		.99990	.99738	.97491	.67014	.16112
	22		.99997	.99898	.98772	.76602	.23994
	23		.99999	.99963	.99441	.84383	.33591
	24		1.00000	.99988	.99763	.90219	.44386
	25			.99996	.99907	.94266	.55614
	26			.99999	.99966	.96859	.66409
	27			1.00000	.99988	.98397	.76006
	28				.99996	.99238	.83888
	29				.99999	.99664	.89868
	30				1.00000	.99863	.94054
	31					.99948	.96755
	32					.99982	.98358
	33					.99994	.99233
	34					.99998	.99670
	35					1.00000	.99870
	36						.99953
	37						.99985
	38						.99995
	39						.99999
	40						1.00000
100	0	.00003					
	1	.00032					
	2	.00194					
	3	.00784					
	4	.02371	.00000				

n	r	p = .10	p = .20	p = .25	p = .30	p = .40	p = .50
100	5	.05758	.00002				
	6	.11716	.00008				
	7	.20605	.00028	.00000			
	8	.32087	.00086	.00001			
	9	.45129	.00233	.00004			
	10	.58316	.00570	.00014	.00000		
	11	.70303	.01257	.00039	.00001		
	12	.80182	.02533	.00103	.00002		
	13	.87612	.04691	.00246	.00006		
	14	.92743	.08044	.00542	.00016		
	15	.96011	.12851	.01108	.00040		
	16	.97940	.19234	.02111	.00097		
	17	.98999	.27119	.03763	.00216		
	18	.99542	.36209	.06301	.00452	.00000	
	19	.99802	.46016	.09953	.00889	.00001	
	20	.99919	.55946	.14883	.01646	.00002	
	21	.99969	.65403	.21144	.02883	.00004	
	22	.99989	.73893	.28637	.04787	.00011	
	23	.99996	.81091	.37018	.07553	.00025	
	24	.99999	.86865	.46167	.11357	.00056	
	25	1.00000	.91252	.55347	.16313	.00119	
	26		.94417	.64174	.22440	.00240	
	27		.96585	.72238	.29637	.00460	.00000
	28		.97998	.79246	.37678	.00843	.00001
	29		.98875	.85046	.46234	.01478	.00002
	30		.99394	.89621	.54912	.02478	.00004
	31		.99687	.93065	.63311	.03985	.00009
	32		.99845	.95540	.71072	.06150	.00020
	33		.99926	.97241	.77926	.09125	.00044
	34		.99966	.98357	.83714	.13034	.00089
	35		.99985	.99059	.88392	.17947	.00176
	36		.99994	.99482	.92012	.23861	.00332
	37		.99998	.99725	.94695	.30681	.00602
	38		.99999	.99860	.96602	.38219	.01049
	39		1.00000	.99931	.97901	.46208	.01760
	40			.99968	.98750	.54329	.02844
	41			.99985	.99283	.62253	.04431
	42			.99994	.99603	.69674	.06661
	43			.99997	.99789	.76347	.09667
	44			.99999	.99891	.82110	.13563
	45			1.00000	.99946	.86891	.18410
	46				.99974	.90702	.24206
	47				.99988	.93621	.30865

n	r	p = .10	p = .20	p = .25	p = .30	p = .40	p = .50
100	48				.99995	.95770	.38218
	49				.99998	.97290	.46021
	50				.99999	.98324	.53979
	51				1.00000	.98999	.61782
	52					.99424	.69135
	53					.99680	.75794
	54					.99829	.81590
	55					.99912	.86437
	56					.99956	.90333
	57					.99979	.93339
	58					.99990	.95569
	59					.99996	.97156
	60					.99998	.98240
	61					.99999	.98951
	62					1.00000	.99398
	63						.99668
	64						.99824
	65						.99911
	66						.99956
	67						.99980
	68						.99991
	69						.99996
	70						.99998
	71						.99999
	72						1.00000

Appendix B2   Table of random digits†
(See Sec. 4-4 for explanation of use)

Row
number

00000	10097 32533	76520 13586	34673 54876	80959 09117	39292 74945
00001	37542 04805	64894 74296	24805 24037	20636 10402	00822 91665
00002	08422 68953	19645 09303	23209 02560	15953 34764	35080 33606
00003	99019 02529	09376 70715	38311 31165	88676 74397	04436 27659
00004	12807 99970	80157 36147	64032 36653	98951 16877	12171 76833
00005	66065 74717	34072 76850	36697 36170	65813 39885	11199 29170
00006	31060 10805	45571 82406	35303 42614	86799 07439	23403 09732
00007	85269 77602	02051 65692	68665 74818	73053 85247	18623 88579
00008	63573 32135	05325 47048	90553 57548	28468 28709	83491 25624
00009	73796 45753	03529 64778	35808 34282	60935 20344	35273 88435
00010	98520 17767	14905 68607	22109 40558	60970 93433	50500 73998
00011	11805 05431	39808 27732	50725 68248	29405 24201	52775 67851
00012	83452 99634	06288 98033	13746 70078	18475 40610	68711 77817
00013	88685 40200	86507 58401	36766 67951	90364 76493	29609 11062
00014	99594 67348	87517 64969	91826 08928	93785 61368	23478 34113
00015	65481 17674	17468 50950	58047 76974	73039 57186	40218 16544
00016	80124 35635	17727 08015	45318 22374	21115 78253	14385 53763
00017	74350 99817	77402 77214	43236 00210	45521 64237	96286 02655
00018	69916 26803	66252 29148	36936 87203	76621 13990	94400 56418
00019	09893 20505	14225 68514	46427 56788	96297 78822	54382 14598
00020	91499 14523	68479 27686	46162 83554	94750 89923	37089 20048
00021	80336 94598	26940 36858	70297 34135	53140 33340	42050 82341
00022	44104 81949	85157 47954	32979 26575	57600 40881	22222 06413
00023	12550 73742	11100 02040	12860 74697	96644 89439	28707 25815
00024	63606 49329	16505 34484	40219 52563	43651 77082	07207 31790
00025	61196 90446	26457 47774	51924 33729	65394 59593	42582 60527
00026	15474 45266	95270 79953	59367 83848	82396 10118	33211 59466
00027	94557 28573	67897 54387	54622 44431	91190 42592	92927 45973
00028	42481 16213	97344 08721	16868 48767	03071 12059	25701 46670
00029	23523 78317	73208 89837	68935 91416	26252 29663	05522 82562
00030	04493 52494	75246 33824	45862 51025	61962 79335	65337 12472
00031	00549 97654	64051 88159	96119 63896	54692 82391	23287 29529
00032	35963 15307	26898 09354	33351 35462	77974 50024	90103 39333
00033	59808 08391	45427 26842	83609 49700	13021 24892	78565 20106
00034	46058 85236	01390 92286	77281 44077	93910 83647	70617 42941

† Extracted with permission from the Rand Corporation publication "A Million Random Digits," The Free Press of Glencoe, New York, 1955.

**Row
number**

00035	32179	00597	87379	25241	05567	07007	86743	17157	85394	11838
00036	69234	61406	20117	45204	15956	60000	18743	92423	97118	96338
00037	19565	41430	01758	75379	40419	21585	66674	36806	84962	85207
00038	45155	14938	19476	07246	43667	94543	59047	90033	20826	69541
00039	94864	31994	36168	10851	34888	81553	01540	35456	05014	51176
00040	98086	24826	45240	28404	44999	08896	39094	73407	35441	31880
00041	33185	16232	41941	50949	89435	48581	88695	41994	37548	73043
00042	80951	00406	96382	70774	20151	23387	25016	25298	94624	61171
00043	79752	49140	71961	28296	69861	02591	74852	20539	00387	59579
00044	18633	32537	98145	06571	31010	24674	05455	61427	77938	91936
00045	74029	43902	77557	32270	97790	17119	52527	58021	80814	51748
00046	54178	45611	80993	37143	05335	12969	56127	19255	36040	90324
00047	11664	49883	52079	84827	59381	71539	09973	33440	88461	23356
00048	48324	77928	31249	64710	02295	36870	32307	57546	15020	09994
00049	69074	94138	87637	91976	35584	04401	10518	21615	01848	76938
00050	09188	20097	32825	39527	04220	86304	83389	87374	64278	58044
00051	90045	85497	51981	50654	94938	81997	91870	76150	68476	64659
00052	73189	50207	47677	26269	62290	64464	27124	67018	41361	82760
00053	75768	76490	20971	87749	90429	12272	95375	05871	93823	43178
00054	54016	44056	66281	31003	00682	27398	20714	53295	07706	17813
00055	08358	69910	78542	42785	13661	58873	04618	97553	31223	08420
00056	28306	03264	81333	10591	40510	07893	32604	60475	94119	01840
00057	53840	86233	81594	13628	51215	90290	28466	68795	77762	20791
00058	91757	53741	61613	62269	50263	90212	55781	76514	83483	47055
00059	89415	92694	00397	58391	12607	17646	48949	72306	94541	37408
00060	77513	03820	86864	29901	68414	82774	51908	13980	72893	55507
00061	19502	37174	69979	20288	55210	29773	74287	75251	65344	67415
00062	21818	59313	93278	81757	05686	73156	07082	85046	31853	38452
00063	51474	66499	68107	23621	94049	91345	42836	09191	08007	45449
00064	99559	68331	62535	24170	69777	12830	74819	78142	43860	72834
00065	33713	48007	93584	72869	51926	64721	58303	29822	93174	93972
00066	85274	86893	11303	22970	28834	34137	73515	90400	71148	43643
00067	84133	89640	44035	52166	73852	70091	61222	60561	62327	18423
00068	56732	16234	17395	96131	10123	91622	85496	57560	81604	18880
00069	65138	56806	87648	85261	34313	65861	45875	21069	85644	47277

Row
number

00070	38001	02176	81719	11711	71602	92937	74219	64049	65584	49698
00071	37402	96397	01304	77586	56271	10086	47324	62605	40030	37438
00072	97125	40348	87083	31417	21815	39250	75237	62047	15501	29578
00073	21826	41134	47143	34072	64638	85902	49139	06441	03856	54552
00074	73135	42742	95719	09035	85794	74296	08789	88156	64691	19202
00075	07638	77929	03061	18072	96207	44156	23821	99538	04713	66994
00076	60528	83441	07954	19814	59175	20695	05533	52139	61212	06455
00077	83596	35655	06958	92983	05128	09719	77433	53783	92301	50498
00078	10850	62746	99599	10507	13499	06319	53075	71839	06410	19362
00079	39820	98952	43622	63147	64421	80814	43800	09351	31024	73167
00080	59580	06478	75569	78800	88835	54486	23768	06156	04111	08408
00081	38508	07341	23793	48763	90822	97022	17719	04207	95954	49953
00082	30692	70668	94688	16127	56196	80091	82067	63400	05462	69200
00083	65443	95659	18238	27437	49632	24041	08337	65676	96299	90836
00084	27267	50264	13192	72294	07477	44606	17985	48911	97341	30358
00085	91307	06991	19072	24210	36699	53728	28825	35793	28976	66252
00086	68434	94688	84473	13622	62126	98408	12843	82590	09815	93146
00087	48908	15877	54745	24591	35700	04754	83824	52692	54130	55160
00088	06913	45197	42672	78601	11883	09528	63011	98901	14974	40344
00089	10455	16019	14210	33712	91342	37821	88325	80851	43667	70883
00090	12883	97343	65027	61184	04285	01392	17974	15077	90712	26769
00091	21778	30976	38807	36961	31649	42096	63281	02023	08816	47449
00092	19523	59515	65122	59659	86283	68258	69572	13798	16435	91529
00093	67245	52670	35583	16563	79246	86686	76463	34222	26655	90802
00094	60584	47377	07500	37992	45134	26529	26760	83637	41326	44344
00095	53853	41377	36066	94850	58838	73859	49364	73331	96240	43642
00096	24637	38736	74384	89342	52623	07992	12369	18601	03742	83873
00097	83080	12451	38992	22815	07759	51777	97377	27585	51972	37867
00098	16444	24334	36151	99073	27493	70939	85130	32552	54846	54759
00099	60790	18157	57178	65762	11161	78576	45819	52979	65130	04860
00100	03991	10461	93716	16894	66083	24653	84609	58232	88618	19161
00101	38555	95554	32886	59780	08355	60860	29735	47762	71299	23853
00102	17546	73704	92052	46215	55121	29281	59076	07936	27954	58909
00103	32643	52861	95819	06831	00911	98936	76355	93779	80863	00514
00104	69572	68777	39510	35905	14060	40619	29549	69616	33564	60780

Row
number

00105	24122	66591	27699	06494	14845	46672	61958	77100	90899	75754
00106	61196	30231	92962	61773	41839	55382	17267	70943	78038	70267
00107	30532	21704	10274	12202	39685	23309	10061	68829	55986	66485
00108	03788	97599	75867	20717	74416	53166	35208	33374	87539	08823
00109	48228	63379	85783	47619	53152	67433	35663	52972	16818	60311
00110	60365	94653	35075	33949	42614	29297	01918	28316	98953	73231
00111	83799	42402	56623	34442	34994	41374	70071	14736	09958	18065
00112	32960	07405	36409	83232	99385	41600	11133	07586	15917	06253
00113	19322	53845	57620	52606	66497	68646	78138	66559	19640	99413
00114	11220	94747	07399	37408	48509	23929	27482	45476	85244	35159
00115	31751	57260	68980	05339	15470	48355	88651	22596	03152	19121
00116	88492	99382	14454	04504	20094	98977	74843	93413	22109	78508
00117	30934	47744	07481	83828	73788	06533	28597	20405	94205	20380
00118	22888	48893	27499	98748	60530	45128	74022	84617	82037	10268
00119	78212	16993	35902	91386	44372	15486	65741	14014	87481	37220
00120	41849	84547	46850	52326	34677	58300	74910	64345	19325	81549
00121	46352	33049	69248	93460	45305	07521	61318	31855	14413	70951
00122	11087	96294	14013	31792	59747	67277	76503	34513	39663	77544
00123	52701	08337	56303	87315	16520	69676	11654	99893	02181	68161
00124	57275	36898	81304	48585	68652	27376	92852	55866	88448	03584
00125	20857	73156	70284	24326	79375	95220	01159	63267	10622	48391
00126	15633	84924	90415	93614	33521	26665	55823	47641	86225	31704
00127	92694	48297	39904	02115	59589	49067	66821	41575	49767	04037
00128	77613	19019	88152	00080	20554	91409	96277	48257	50816	97616
00129	38688	32486	45134	63545	59404	72059	43947	51680	43852	59693
00130	25163	01889	70014	15021	41290	67312	71857	15957	68971	11403
00131	65251	07629	37239	33295	05870	01119	92784	26340	18477	65622
00132	36815	43625	18637	37509	82444	99005	04921	73701	14707	93997
00133	64397	11692	05327	82162	20247	81759	45197	25332	83745	22567
00134	04515	25624	95096	67946	48460	85558	15191	18782	16930	33361
00135	83761	60873	43253	84145	60833	25983	01291	41349	20368	07126
00136	14387	06345	80854	09279	43529	06318	38384	74761	41196	37480
00137	51321	92246	80088	77074	88722	56736	66164	49431	66919	31678
00138	72472	00008	80890	18002	94813	31900	54155	83436	35352	54131
00139	05466	55306	93128	18464	74457	90561	72848	11834	79982	68416

**Row
number**

00140	39528	72484	82474	25593	48545	35247	18619	13674	18611	19241
00141	81616	18711	53342	44276	75122	11724	74627	73707	58319	15997
00142	07586	16120	82641	22820	92904	13141	32392	19763	61199	67940
00143	90767	04235	13574	17200	69902	63742	78464	22501	18627	90872
00144	40188	28193	29593	88627	94972	11598	62095	36787	00441	58997
00145	34414	82157	86887	55087	19152	00023	12302	80783	32624	68691
00146	63439	75363	44989	16822	36024	00867	76378	41605	65961	73488
00147	67049	09070	93399	45547	94458	74284	05041	49807	20288	34060
00148	79495	04146	52162	90286	54158	34243	46978	35482	59362	95938
00149	91704	30552	04737	21031	75051	93029	47665	64382	99782	93478
00150	94015	46874	32444	48277	59820	96163	64654	25843	41145	42820
00151	74108	88222	88570	74015	25704	91035	01755	14750	48968	38603
00152	62880	87873	95160	59221	22304	90314	72877	17334	39283	04149
00153	11748	12102	80580	41867	17710	59621	06554	07850	73950	79552
00154	17944	05600	60478	03343	25852	58905	57216	39618	49856	99326
00155	66067	42792	95043	52680	46780	56487	09971	59481	37006	22186
00156	54244	91030	45547	70818	59849	96169	61459	21647	87417	17198
00157	30945	57589	31732	57260	47670	07654	46376	25366	94746	49580
00158	69170	37403	86995	90307	94304	71803	26825	05511	12459	91314
00159	08345	88975	35841	85771	08105	59987	87112	21476	14713	71181
00160	27767	43584	85301	88977	29490	69714	73035	41207	74699	09310
00161	13025	14338	54066	15243	47724	66733	47431	43905	31048	56699
00162	80217	36292	98525	24335	24432	24896	43277	58874	11466	16082
00163	10875	62004	90391	61105	57411	06368	53856	30743	08670	84741
00164	54127	57326	26629	19087	24472	88779	30540	27886	61732	75454
00165	60311	42824	37301	42678	45990	43242	17374	52003	70707	70214
00166	49739	71484	92003	98086	76668	73209	59202	11973	02902	33250
00167	78626	51594	16453	94614	39014	97066	83012	09832	25571	77628
00168	66692	13986	99837	00582	81232	44987	09504	96412	90193	79568
00169	44071	28091	07362	97703	76447	42537	98524	97831	65704	09514
00170	41468	85149	49554	17994	14924	39650	95294	00556	70481	06905
00171	94559	37559	49678	53119	70312	05682	66986	34099	74474	20740
00172	41615	70360	64114	58660	90850	64618	80620	51790	11436	38072
00173	50273	93113	41794	86861	24781	89683	55411	85667	77535	99892
00174	41396	80504	90670	08289	40902	05069	95083	06783	28102	57816

Row
number

00175	25807 24260	71529 78920	72682 07385	90726 57166	98884 08583
00176	06170 97965	88302 98041	21443 41808	68984 83620	89747 98882
00177	60808 54444	74412 81105	01176 28838	36421 16489	18059 51061
00178	80940 44893	10408 36222	80582 71944	92638 40333	67054 16067
00179	19516 90120	46759 71643	13177 55292	21036 82808	77501 97427
00180	49386 54480	23604 23554	21785 41101	91178 10174	29420 90438
00181	06312 88940	15995 69321	47458 64809	98189 81851	29651 84215
00182	60942 00307	11897 92674	40405 68032	96717 54244	10701 41393
00183	92329 98932	78284 46347	71209 92061	39448 93136	25722 08564
00184	77936 63574	31384 51924	85561 29671	58137 17820	22751 36518
00185	38101 77756	11657 13897	95889 57067	47648 13885	70669 93406
00186	39641 69457	91339 22502	92613 89719	11947 56203	19324 20504
00187	84054 40455	99396 63680	67667 60631	69181 96845	38525 11600
00188	47468 03577	57649 63266	24700 71594	14004 23153	69249 05747
00189	43321 31370	28977 23896	76479 68562	62342 07589	08899 05985
00190	64281 61826	18555 64937	13173 33365	78851 16499	87064 13075
00191	66847 70495	32350 02985	86716 38746	26313 77463	55387 72681
00192	72461 33230	21529 53424	92581 02262	78438 66276	18396 73538
00193	21032 91050	13058 16218	12470 56500	15292 76139	59526 52113
00194	95362 67011	06651 16136	01016 00857	55018 56374	35824 71708
00195	49712 97380	10404 55452	34030 60726	75211 10271	36633 68424
00196	58275 61764	97586 54716	50259 46345	87195 46092	26787 60939
00197	89514 11788	68224 23417	73959 76145	30342 40277	11049 72049
00198	15472 50669	48139 36732	46874 37088	73465 09819	58869 35220
00199	12120 86124	51247 44302	60883 52109	21437 36786	49226 77837

Entries are $p(x) = p(N,n,k,x)$ and $P(r) = P(N,n,k,r)$ [see Eqs. (4-24) and (4-26)]

N	n	k	r or x	P(r)	p(x)	N	n	k	r or x	P(r)	p(x)
10	1	1	0	0.900000	0.900000	10	5	3	0	0.083333	0.083333
10	1	1	1	1.000000	0.100000	10	5	3	1	0.500000	0.416667
10	2	1	0	0.800000	0.800000	10	5	3	2	0.916667	0.416667
10	2	1	1	1.000000	0.200000	10	5	3	3	1.000000	0.083333
10	2	2	0	0.622222	0.622222	10	5	4	0	0.023810	0.023810
10	2	2	1	0.977778	0.355556	10	5	4	1	0.261905	0.238095
10	2	2	2	1.000000	0.022222	10	5	4	2	0.738095	0.476190
10	3	1	0	0.700000	0.700000	10	5	4	3	0.976190	0.238095
10	3	2	1	1.000000	0.300000	10	5	4	4	1.000000	0.023810
10	3	2	0	0.466667	0.466667	10	5	5	0	0.003968	0.003968
10	3	2	1	0.933333	0.466667	10	5	5	1	0.103175	0.099206
10	3	2	2	1.000000	0.066667	10	5	5	2	0.500000	0.396825
10	3	3	0	0.291667	0.291667	10	5	5	3	0.896825	0.396825
10	3	3	1	0.816667	0.525000	10	5	5	4	0.996032	0.099206
10	3	3	2	0.991667	0.175000	10	5	5	5	1.000000	0.003968
10	3	3	3	1.000000	0.008333	10	6	1	0	0.400000	0.400000
10	4	1	0	0.600000	0.600000	10	6	1	1	1.000000	0.600000
10	4	1	1	1.000000	0.400000	10	6	2	0	0.133333	0.133333
10	4	2	0	0.333333	0.333333	10	6	2	1	0.666667	0.533333
10	4	2	1	0.866667	0.533333	10	6	2	2	1.000000	0.333333
10	4	2	2	1.000000	0.133333	10	6	3	0	0.033333	0.033333
10	4	3	0	0.166667	0.166667	10	6	3	1	0.333333	0.300000
10	4	3	1	0.666667	0.500000	10	6	3	2	0.833333	0.500000
10	4	3	2	0.966667	0.300000	10	6	3	3	1.000000	0.166667
10	4	3	3	1.000000	0.033333	10	6	4	0	0.004762	0.004762
10	4	4	0	0.071429	0.071429	10	6	4	1	0.119048	0.114286
10	4	4	1	0.452381	0.380952	10	6	4	2	0.547619	0.428571
10	4	4	2	0.880952	0.428571	10	6	4	3	0.928571	0.380952
10	4	4	3	0.995238	0.114286	10	6	4	4	1.000000	0.071429
10	4	4	4	1.000000	0.004762	10	6	5	1	0.023810	0.023810
10	5	1	0	0.500000	0.500000	10	6	5	2	0.261905	0.238095
10	5	1	1	1.000000	0.500000	10	6	5	3	0.738095	0.476190
10	5	2	0	0.222222	0.222222	10	6	5	4	0.976190	0.238095
10	5	2	1	0.777778	0.555556	10	6	5	5	1.000000	0.023810
10	5	2	2	1.000000	0.222222	10	6	6	2	0.071429	0.071429

† Extracted with permission from Gerald J. Lieberman and Donald B. Owen, "Tables of the Hypergeometric Probability Distribution," Stanford University Press, Stanford, Calif., 1961.

N	n	k	r or x	P(r)	p(x)	N	n	k	r or x	P(r)	p(x)
10	6	6	3	0.452381	0.380952	10	8	3	2	0.533333	0.466667
10	6	6	4	0.880952	0.428571	10	8	3	3	1.000000	0.466667
10	6	6	5	0.995238	0.114286	10	8	4	2	0.133333	0.133333
10	6	6	6	1.000000	0.004762	10	8	4	3	0.666667	0.533333
10	7	1	0	0.300000	0.300000	10	8	4	4	1.000000	0.333333
10	7	1	1	1.000000	0.700000	10	8	5	3	0.222222	0.222222
10	7	2	0	0.066667	0.066667	10	8	5	4	0.777778	0.555556
10	7	2	1	0.533333	0.466667	10	8	5	5	1.000000	0.222222
10	7	2	2	1.000000	0.466667	10	8	6	4	0.333333	0.333333
10	7	3	0	0.008333	0.008333	10	8	6	5	0.866667	0.533333
10	7	3	1	0.183333	0.175000	10	8	6	6	1.000000	0.133333
10	7	3	2	0.708333	0.525000	10	8	7	5	0.466667	0.466667
10	7	3	3	1.000000	0.291667	10	8	7	6	0.933333	0.466667
10	7	4	1	0.033333	0.033333	10	8	7	7	1.000000	0.066667
10	7	4	2	0.333333	0.300000	10	8	8	6	0.622222	0.622222
10	7	4	3	0.833333	0.500000	10	8	8	7	0.977778	0.355556
10	7	4	4	1.000000	0.166667	10	8	8	8	1.000000	0.022222
10	7	5	2	0.083333	0.083333	10	9	1	0	0.100000	0.100000
10	7	5	3	0.500000	0.416667	10	9	1	1	1.000000	0.900000
10	7	5	4	0.916667	0.416667	10	9	2	1	0.200000	0.200000
10	7	5	5	1.000000	0.083333	10	9	2	2	1.000000	0.800000
10	7	6	3	0.166667	0.166667	10	9	3	2	0.300000	0.300000
10	7	6	4	0.666667	0.500000	10	9	3	3	1.000000	0.700000
10	7	6	5	0.966667	0.300000	10	9	4	3	0.400000	0.400000
10	7	6	6	1.000000	0.033333	10	9	4	4	1.000000	0.600000
10	7	7	4	0.291667	0.291667	10	9	5	4	0.500000	0.500000
10	7	7	5	0.816667	0.525000	10	9	5	5	1.000000	0.500000
10	7	7	6	0.991667	0.175000	10	9	6	5	0.600000	0.600000
10	7	7	7	1.000000	0.008333	10	9	6	6	1.000000	0.400000
10	8	1	0	0.200000	0.200000	10	9	7	6	0.700000	0.700000
10	8	1	1	1.000000	0.800000	10	9	7	7	1.000000	0.300000
10	8	2	0	0.022222	0.022222	10	9	8	7	0.800000	0.800000
10	8	2	1	0.377778	0.355556	10	9	8	8	1.000000	0.200000
10	8	2	2	1.000000	0.622222	10	9	9	8	0.900000	0.900000
10	8	3	1	0.066667	0.066667	10	9	9	9	1.000000	0.100000

Entry $= P(r;\mu) = \sum_{x=0}^{r} p(x;\mu)$ [see Eqs. (4-33) and (4-34)]

r	$\mu = .1$	$\mu = .2$	$\mu = .3$	$\mu = .4$	$\mu = .5$
0	.90484	.81873	.74082	.67302	.60653
1	.99532	.98248	.96306	.93845	.90980
2	.99985	.99885	.99640	.99207	.98561
3	1.00000	.99994	.99973	.99922	.99825
4		1.00000	.99998	.99994	.99983
5			1.00000	1.00000	.99999
6					1.00000

r	$\mu = .6$	$\mu = .7$	$\mu = .8$	$\mu = .9$	$\mu = 1.0$
0	.54881	.49658	.44933	.40657	.36788
1	.87810	.84419	.80879	.77248	.73576
2	.97688	.96586	.95258	.93714	.91970
3	.99664	.99425	.99092	.98654	.98101
4	.99961	.99921	.99859	.99766	.99634
5	.99996	.99991	.99982	.99966	.99941
6	1.00000	.99999	.99998	.99996	.99992
7		1.00000	1.00000	1.00000	.99999
8					1000000

† Extracted with permission from D. B. Owen, "Handbook of Statistical Tables," Addison-Wesley Publishing Company, Inc., Reading, Mass., 1962.

r	μ = 2	μ = 3	μ = 4	μ = 5	μ = 6
0	.13534	.04979	.01832	.00674	.00248
1	.40601	.19915	.09158	.04043	.01735
2	.67668	.42319	.23810	.12465	.06197
3	.85712	.64723	.43347	.26503	.15120
4	.94735	.81526	.62884	.44049	.28506
5	.98344	.91608	.78513	.61596	.44568
6	.99547	.96649	.88933	.76218	.60630
7	.99890	.98810	.94887	.86663	.74398
8	.99976	.99620	.97864	.93191	.84724
9	.99995	.99890	.99187	.96817	.91608
10	.99999	.99971	.99716	.98630	.95738
11	1.00000	.99993	.99908	.99455	.97991
12		.99998	.99973	.99798	.99117
13		1.00000	.99992	.99930	.99637
14			.99998	.99977	.99860
15			1.00000	.99993	.99949
16				.99998	.99982
17				1.00000	.99994
18					.99998
19					1.00000

r	$\mu = 7$	$\mu = 8$	$\mu = 9$	$\mu = 10$
0	.00091	.00033	.00012	.00004
1	.00730	.00302	.00123	.00050
2	.02964	.01375	.00623	.00277
3	.08176	.04238	.02123	.01034
4	.17299	.09963	.05496	.02925
5	.30071	.19124	.11569	.06709
6	.44971	.31337	.20678	.13014
7	.59871	.45296	.32390	.22022
8	.72909	.59255	.45565	.33282
9	.83050	.71662	.58741	.45793
10	.90148	.81589	.70599	.58304
11	.94665	.88808	.80301	.69678
12	.97300	.93620	.87577	.79156
13	.98719	.96582	.92615	.86446
14	.99428	.98274	.95853	.91654
15	.99759	.99177	.97796	.95126
16	.99904	.99628	.98889	.97296
17	.99964	.99841	.99468	.98572
18	.99987	.99935	.99757	.99281
19	.99996	.99975	.99894	.99655
20	.99999	.99991	.99956	.99841
21	1.00000	.99997	.99982	.99930
22		.99999	.99993	.99970
23		1.00000	.99998	.99988
24			.99999	.99995
25			1.00000	.99998
26				.99999
27				1.00000

Appendix B5 The cumulative standardized normal distribution function† Entry $= \Pr[Z < z_p] = p$

z_p	.00	.01	.02	.03	.04	.05	.06	.07	.08	.09
−.0	.5000	.4960	.4920	.4880	.4840	.4801	.4761	.4721	.4681	.4641
−.1	.4602	.4562	.4522	.4483	.4443	.4404	.4364	.4325	.4286	.4247
−.2	.4207	.4168	.4129	.4090	.4052	.4013	.3974	.3936	.3897	.3859
−.3	.3821	.3783	.3745	.3707	.3669	.3632	.3594	.3557	.3520	.3483
−.4	.3446	.3409	.3372	.3336	.3300	.3264	.3228	.3192	.3156	.3121
−.5	.3085	.3050	.3015	.2981	.2946	.2912	.2877	.2843	.2810	.2776
−.6	.2743	.2709	.2676	.2643	.2611	.2578	.2546	.2514	.2483	.2451
−.7	.2420	.2389	.2358	.2327	.2297	.2266	.2236	.2206	.2177	.2148
−.8	.2119	.2090	.2061	.2063	.2005	.1977	.1949	.1922	.1894	.1867
−.9	.1841	.1814	.1788	.1762	.1736	.1711	.1685	.1660	.1635	.1611
−1.0	.1587	.1562	.1539	.1515	.1492	.1469	.1446	.1423	.1401	.1379
−1.1	.1357	.1335	.1314	.1292	.1271	.1251	.1230	.1210	.1190	.1170
−1.2	.1151	.1131	.1112	.1093	.1075	.1056	.1038	.1020	.1003	.09853
−1.3	.09680	.09510	.09342	.09176	.09012	.08851	.08691	.08534	.08379	.08226
−1.4	.08076	.07927	.07780	.07636	.07493	.07353	.07215	.07078	.06944	.06811
−1.5	.06681	.06552	.06426	.06301	.06178	.06057	.05938	.05821	.05705	.05592
−1.6	.05480	.05370	.05262	.05155	.05050	.04947	.04846	.04746	.04648	.04551
−1.7	.04457	.04363	.04272	.04182	.04093	.04006	.03920	.03836	.03754	.03673
−1.8	.03593	.03515	.03438	.03362	.03288	.03216	.03144	.03074	.03005	.02938
−1.9	.02872	.02807	.02743	.02680	.02619	.02559	.02500	.02442	.02385	.02330
−2.0	.02275	.02222	.02169	.02118	.02068	.02018	.01970	.01923	.01876	.01831
−2.1	.01786	.01743	.01700	.01659	.01616	.01578	.01539	.01500	.01463	.01426
−2.2	.01390	.01355	.01321	.01287	.01255	.01222	.01191	.01160	.01130	.01101
−2.3	.01072	.01044	.01017	$.0^2 9903$	$.0^2 9642$	$.0^2 9387$	$.0^2 9137$	$.0^2 8894$	$.0^2 8656$	$.0^2 8424$
−2.4	$.0^2 8198$	$.0^2 7976$	$.0^2 7760$	$.0^2 7549$	$.0^2 7344$	$.0^2 7143$	$.0^2 6947$	$.0^2 6756$	$.0^2 6569$	$.0^2 6387$
−2.5	$.0^2 6210$	$.0^2 6037$	$.0^2 5868$	$.0^2 5703$	$.0^2 5543$	$.0^2 5386$	$.0^2 5234$	$.0^2 5085$	$.0^2 4940$	$.0^2 4799$
−2.6	$.0^2 4661$	$.0^2 4527$	$.0^2 4396$	$.0^2 4269$	$.0^2 4145$	$.0^2 4025$	$.0^2 3907$	$.0^2 3793$	$.0^2 3681$	$.0^2 3573$
−2.7	$.0^2 3467$	$.0^2 3364$	$.0^2 3264$	$.0^2 3167$	$.0^2 3072$	$.0^2 2980$	$.0^2 2890$	$.0^2 2803$	$.0^2 2718$	$.0^2 2635$
−2.8	$.0^2 2555$	$.0^2 2477$	$.0^2 2401$	$.0^2 2327$	$.0^2 2256$	$.0^2 2186$	$.0^2 2118$	$.0^2 2052$	$.0^2 1988$	$.0^2 1926$
−2.9	$.0^2 1866$	$.0^2 1807$	$.0^2 1750$	$.0^2 1695$	$.0^2 1641$	$.0^2 1589$	$.0^2 1538$	$.0^2 1489$	$.0^2 1441$	$.0^2 1395$

x	.00	.01	.02	.03	.04	.05	.06	.07	.08	.09
.1	.5398	.5438	.5478	.5517	.5557	.5596	.5636	.5675	.5714	.5753
.2	.5793	.5832	.5871	.5910	.5948	.5987	.6026	.6064	.6103	.6141
.3	.6179	.6217	.6255	.6293	.6331	.6368	.6406	.6443	.6480	.6517
.4	.6554	.6591	.6628	.6664	.6700	.6736	.6772	.6808	.6844	.6879
.5	.6915	.6950	.6985	.7019	.7054	.7088	.7123	.7157	.7190	.7224
.6	.7257	.7291	.7324	.7357	.7389	.7422	.7454	.7486	.7517	.7549
.7	.7580	.7611	.7642	.7673	.7703	.7734	.7764	.7794	.7823	.7852
.8	.7881	.7910	.7939	.7967	.7995	.8023	.8051	.8078	.8106	.8133
.9	.8159	.8186	.8212	.8238	.8264	.8289	.8315	.8340	.8365	.8389
1.0	.8413	.8438	.8461	.8485	.8508	.8531	.8554	.8577	.8599	.8621
1.1	.8643	.8665	.8686	.8708	.8729	.8749	.8770	.8790	.8810	.8830
1.2	.8849	.8869	.8888	.8907	.8925	.8944	.8962	.8980	.8997	.90147
1.3	.90320	.90490	.90658	.90824	.90988	.91149	.91309	.91466	.91621	.91774
1.4	.91924	.92073	.92220	.92364	.92507	.92647	.92785	.92922	.93056	.93189
1.5	.93319	.93448	.93574	.93699	.93822	.93943	.94062	.94179	.94295	.94408
1.6	.94520	.94630	.94738	.94845	.94950	.95053	.95154	.95254	.95352	.95449
1.7	.95543	.95637	.95728	.95818	.95907	.95994	.96080	.96164	.96246	.96327
1.8	.96407	.96485	.96562	.96638	.96712	.96784	.96856	.96926	.96995	.97062
1.9	.97128	.97193	.97257	.97320	.97381	.97441	.97500	.97558	.97615	.97670
2.0	.97725	.97778	.97831	.97882	.97932	.97982	.98030	.98077	.98124	.98169
2.1	.98214	.98257	.98300	.98341	.98382	.98422	.98461	.98500	.98537	.98574
2.2	.98610	.98645	.98679	.98713	.98745	.98778	.98809	.98840	.98870	.98899
2.3	.98928	.98956	.98983	$.9^{2}0097$	$.9^{2}0358$	$.9^{2}0613$	$.9^{2}0863$	$.9^{2}1106$	$.9^{2}1344$	$.9^{2}1576$
2.4	$.9^{2}1802$	$.9^{2}2024$	$.9^{2}2240$	$.9^{2}2451$	$.9^{2}2656$	$.9^{2}2857$	$.9^{2}3053$	$.9^{2}3244$	$.9^{2}3431$	$.9^{2}3613$
2.5	$.9^{2}3790$	$.9^{2}3963$	$.9^{2}4132$	$.9^{2}4297$	$.9^{2}4457$	$.9^{2}4614$	$.9^{2}4766$	$.9^{2}4915$	$.9^{2}5060$	$.9^{2}5201$
2.6	$.9^{2}5339$	$.9^{2}5473$	$.9^{2}5604$	$.9^{2}5731$	$.9^{2}5855$	$.9^{2}5975$	$.9^{2}6093$	$.9^{2}6207$	$.9^{2}6319$	$.9^{2}6427$
2.7	$.9^{2}6533$	$.9^{2}6636$	$.9^{2}6736$	$.9^{2}6833$	$.9^{2}6928$	$.9^{2}7020$	$.9^{2}7110$	$.9^{2}7197$	$.9^{2}7282$	$.9^{2}7365$
2.8	$.9^{2}7445$	$.9^{2}7523$	$.9^{2}7599$	$.9^{2}7673$	$.9^{2}7744$	$.9^{2}7814$	$.9^{2}7882$	$.9^{2}7948$	$.9^{2}8012$	$.9^{2}8074$
2.9	$.9^{2}8134$	$.9^{2}8193$	$.9^{2}8250$	$.9^{2}8305$	$.9^{2}8359$	$.9^{2}8411$	$.9^{2}8462$	$.9^{2}8511$	$.9^{2}8559$	$.9^{2}8605$
3.0	$.9^{2}8650$	$.9^{2}8694$	$.9^{2}8736$	$.9^{2}8777$	$.9^{2}8817$	$.9^{2}8856$	$.9^{2}8893$	$.9^{2}8930$	$.9^{2}8965$	$.9^{2}8999$

† Reprinted with permission from A. Hald, "Statistical Tables and Formulas," John Wiley & Sons, Inc., New York, 1952.

Note: $.0^{2}1350 = .001350$ $.9^{2}8650 = .998650$.

Answers to Exercises

1-1 $\frac{3}{16}$

1-3 $\frac{2}{6}$

1-5 $\frac{18}{38}$, $\frac{16}{38}$, $\frac{7}{38}$

1-7   .6, .4

1-9   Count the number of July fourths during the past 100 years that were calm, dry, and normal, and divide by 100. We might even consider using all days a week or two on either side of July 4 to get a larger n and a more stable ratio.

1-11   When $n = 100$ the range should be smaller, indicating increased "stability."

1-13   The sample space that would most likely be selected consists of simple events $HHH, HHT, HTH, HTT, THH, THT, TTH, TTT$. The weight $\frac{1}{8}$ for each point is reasonable. Three of the points correspond to obtaining two heads. Thus the probability that two out of three times the result is heads is $\frac{3}{8}$.

1-15   The appropriate sample space consists of six points that could be designated W, W, W, W, R, R where W stands for white, R for red. The weight $\frac{1}{6}$ is reasonable for each point. The probability of drawing a red ball is $\frac{2}{6}$.

1-17 The appropriate sample space consists of 36 points that could be designated

(1,1)	(2,1)	(3,1)	(4,1)	(5,1)	(6,1)
(1,2)	(2,2)	(3,2)	(4,2)	(5,2)	(6,2)
(1,3)	(2,3)	(3,3)	(4,3)	(5,3)	(6,3)
(1,4)	(2,4)	(3,4)	(4,4)	(5,4)	(6,4)
(1,5)	(2,5)	(3,5)	(4,5)	(5,5)	(6,5)
(1,6)	(2,6)	(3,6)	(4,6)	(5,6)	(6,6)

The weight $\frac{1}{36}$ is reasonable for each point. Since six of the points produce a total of seven, the desired probability is $\frac{6}{36}$.

1-19 The results suggest a sample space consisting of three points, which can be designated A, B, and C to stand for winning by the respective individuals. The weights we would use are $\frac{60}{100}$, $\frac{30}{100}$, $\frac{10}{100}$, respectively. The probability that A loses the next race is $\frac{30}{100} + \frac{10}{100} = \frac{40}{100}$.

1-21 The suggested sample space contains five points, which could be designated by W, Y, R, G, and B with weights $\frac{800}{2,000}$, $\frac{500}{2,000}$, $\frac{300}{2,000}$, $\frac{300}{2,000}$, $\frac{100}{2,000}$. The probability that the can contains red, white, or blue paint is $\frac{300}{2,000} + \frac{800}{2,000} + \frac{100}{2,000} = \frac{1,200}{2,000}$.

1-23 The suggested sample space contains 86 points, which could be designated by 10, 11, . . . , 94, 95, where the designation indicates that death occurs during that year of life. The weights we would use are $q_{10}, q_{11}, \ldots, q_{95}$.

1-25 $\frac{1}{10}$, $\frac{2}{10}$, $\frac{3}{10}$, $\frac{4}{10}$

1-27 Let D designate the simple event "a decision is reached" and N the simple event "no decision is reached." An infinite number of sample points are required. We could denote them by D (a decision is reached on the first trial), ND (no decision followed by a decision), NND (no decision twice followed by a decision), $NNND$, etc.

2-1 After eliminating (1,1), (2,2), (3,3), and (4,4) the answers are as given in Example 2-1.

2-3 Yes

2-5 .29

2-7 .85, .70

2-9 Yes, $\frac{15}{36}$

2-11 $\frac{5}{12}$

2-13 $(74,173/100,000)/(75,782/100,000) = 74,173/75,782$

2-15 $\frac{1}{4}$

2-17 $\frac{5}{6}$ since the probability that a student graduates is .48, while the probability that the student is male and graduates is .40.

2-19 $\frac{8}{663}$

2-21 $(\frac{5}{6})^3(\frac{1}{6})$

2-23 $1 - (\frac{1}{2})(\frac{1}{3})(\frac{1}{4}) = \frac{23}{24}$

2-25 $\frac{6}{31}$. Consider a four-point sample space with weights $\frac{1}{6}$, $\frac{5}{36}$, $\frac{25}{216}$, $\frac{125}{216}$.

2-27 $\frac{13}{52}$

2-29 $\frac{3}{4}$, $\frac{11}{16}$

2-31 All pairs are independent, but A_1, A_2, A_3 taken together are not.

2-33 (2-20) holds, but none of the equations of (2-21) hold.

2-35 No, because an intelligent individual would answer both correctly or both incorrectly.

2-37 Nearly everyone's experience would indicate that the events are not independent. Those who get A in the first course probably have a much higher probability of getting A in the second than someone who got less than A in the first.

2-39 .9702. We have no reason to suspect that events are not independent.

2-41 $5! = 120$, $(3 \cdot 2 \cdot 3 \cdot 2 \cdot 1)/120 = \frac{3}{10}$

2-43 6,720, 56

2-45 1,584

2-47 50,400

2-49 $13 \cdot 12 \cdot \binom{4}{3}\binom{4}{2} / \binom{52}{5} = 3,744/2,598,960$

2-51 $\frac{18}{23}$, $\frac{5}{23}$

2-53 $\frac{1}{4}$, $\frac{3}{5}$

2-55 $\frac{4}{5}$, $\frac{3}{4}$

2-57 .52, $\frac{12}{13}$

3-1 Let X = number of heads; then

x	0	1	2
$f(x)$	$\frac{1}{4}$	$\frac{2}{4}$	$\frac{1}{4}$

3-3

x	1	2	3	4	5	6
$F(x)$	$\frac{1}{21}$	$\frac{3}{21}$	$\frac{6}{21}$	$\frac{10}{21}$	$\frac{15}{21}$	1

3-5 With only four brands one might just as well write down all $4! = 24$ orders in which the individual could name the cigarettes. Since no discriminatory powers are possessed each order should be assigned probability $\frac{1}{24}$. Then count the number of matches with the correct order, say, $ABCD$, in each permutation. This yields the probability distribution

x	0	1	2	3	4
$f(x)$	$\frac{9}{24}$	$\frac{8}{24}$	$\frac{6}{24}$	0	$\frac{1}{24}$

3-7 (a) $\Pr(X \leq 2) = .53$, $\Pr(X = 2) = .53 - .27 = .26$, $\Pr(2 \leq X \leq 4)$ $= \Pr(X \leq 4) - \Pr(X \leq 1) = .65$

(b)

x	0	1	2	3	4	5
$f(x)$	.13	.14	.26	.31	.08	.08

3-9 If X is the sum, then the probability distribution is

x	2	3	4	5	6	7	8
$f(x)$	$\frac{1}{16}$	$\frac{2}{16}$	$\frac{3}{16}$	$\frac{4}{16}$	$\frac{3}{16}$	$\frac{2}{16}$	$\frac{1}{16}$

3-11

x	2	3	4	5	6	7	8
$f(x)$	.01	.04	.10	.20	.25	.24	.16
$F(x)$	.01	.05	.15	.35	.60	.84	1

3-13 $f(n) = (\frac{5}{6})^{n-1}(\frac{1}{6})$, $n = 1, 2, 3 \ldots$

3-15 $g(y) = (y + 3)/15$, $y = -2, -1, 0, 1, 2$

z	0	1	2
$h(z)$	$\frac{3}{15}$	$\frac{6}{15}$	$\frac{6}{15}$

$\Pr(Z \leq 1) = \frac{9}{15}$

3-17 $g(y) = (\frac{1}{2})^{\sqrt{y}}$ $y = 1, 4, 9, \ldots$
$h(z) = (\frac{1}{2})^{z^2}$ $z = 1, \sqrt{2}, \sqrt{3}, \ldots$

3-19

z	$-\$2$	$-\$1$	$\$1$	$\$2$
$h(z)$	$\frac{7}{36}$	$\frac{12}{36}$	$\frac{9}{36}$	$\frac{8}{36}$

370

3-21

y	0	4	16
$g(y)$	$17\!\!/_{24}$	$6\!\!/_{24}$	$1\!\!/_{24}$

3-23 $1\frac{1}{3}$

3-25 $-\$\frac{1}{36} \cong -\$.03$

3-27 1

3-29 2.31

3-31 $\frac{5}{3}$

3-33 5.64

3-35 $14\!\!/_9$, $16\!\!/_{15}$

3-37 $2\text{,}915\!\!/_{1\text{,}296} \cong 2.25$

3-39 $35\!\!/_6$

3-41 5, $\frac{5}{2}$

3-43 1

3-45 $(k + 1)/2$, $(k^2 - 1)/12$

3-47 $E(Y) = E(X^2) = 10$, $E(Y^2) = E(X^4) = 130$, $\sigma_Y^2 = 30$

3-49 (a) $\leq \frac{4}{9}$ (b) $\leq \frac{4}{9}$ (c) $\geq \frac{3}{4}$ (d) $\geq \frac{3}{4}$ (e) ≥ 100

3-51 $\Pr(|X| \geq 1) \leq 4p_1^2$. If $p_1 = \frac{1}{2}$, then equality is achieved.

4-1 .16308, .13230. Since some players appear to hit in streaks, we might doubt that p remains constant and successive trials are independent.

4-3 .00163. We are apt to doubt the claim.

4-5 .07257, .03268. Perhaps one should improve with practice, so that p may change from trial to trial. In addition, successive trials may not be independent, since a thrower may tend to follow a poor results by another poor result. $\mu = 10$, $\sigma^2 = 9$, $\sigma = 3$.

4-7 .12560. Since cloudy weather may be associated with fronts lasting several days, there is real doubt that the independence condition is satisfied.

4-9 $\Pr(X \geq 29) = .00004$ if $p = .30$ for the new cure. Consequently, it is reasonable to believe that the new cure is better.

4-11 .36689, $320\!\!/_3$ or about 107, $2\text{,}240\!\!/_9$

4-13 .00001. We are apt to conclude that the probability of a success is less than .90.

4-15 .90874, 12

4-19 We start in row 121, column 20. The first number is 0, which we could let correspond to 10 and use one digit numbers. Below 0 are 2 and 5. Hence individuals with numbers 2, 5, and 10 prepare the report. If we use two-digit numbers, then the first number is 04 and we use individuals 4, 2, and 5.

4-21 2.6, .432.

4-23 $94/54\text{,}145 = .001736$; $95/54\text{,}145 = .001755$ (both calculated without tables). $E(X) = \frac{5}{13}$.

4-25 2, $16\!\!/_{11}$

4-27 $P(32,8,3,1)$. Using the approximation we get $\displaystyle\sum_{x=0}^{1} b(x;3,\frac{1}{4}) = 27\!\!/_{64} + 27\!\!/_{64}$
$= .84375$.

4-31 Conditions (4-32) seem reasonable but adequacy of the model would have to be judged by experience. Using the Poisson with $\mu = 5$ yields .38404.

4-33 .00345. With $\mu = 10$, it is quite unlikely that the number of flats will be 20 or more.

4-35 .08346

5-1

x	1	2	3	4
$f_1(x)$	$\frac{1}{5}$	$\frac{1}{5}$	$\frac{2}{5}$	$\frac{1}{5}$

y	1	2	3
$f_2(y)$	$\frac{1}{6}$	$\frac{2}{6}$	$\frac{3}{6}$

The random variables are independent.

5-3

x_2 \ x_1	1	2	3	4	5	Row totals
1	.03	.06	.12	.06	.03	.3
2	.03	.06	.12	.06	.03	.3
3	.02	.04	.08	.04	.02	.2
4	.02	.04	.08	.04	.02	.2
Column totals	.1	.2	.4	.2	.1	1

5-5 $f(x_1,x_2) = \binom{n_1}{x_1}\binom{n_2}{x_2} p_1^{x_1} p_2^{x_2}(1-p_1)^{n_1-x_1}(1-p_2)^{n_2-x_2}$ $\quad x_1 = 0, 1, \ldots, n_1$
$x_2 = 0, 1, \ldots, n_2$

5-7 $f(x,y) = \dfrac{\binom{3}{x}\binom{2}{y}\binom{5}{5-x-y}}{\binom{10}{5}}$ $\quad \begin{aligned} x &= 0, 1, 2, 3 \\ y &= 0, 1, 2 \end{aligned}$

y \ x	0	1	2	3	Row totals
0	$\frac{1}{252}$	$\frac{15}{252}$	$\frac{30}{252}$	$\frac{10}{252}$	$\frac{56}{252}$
1	$\frac{10}{252}$	$\frac{60}{252}$	$\frac{60}{252}$	$\frac{10}{252}$	$\frac{140}{252}$
2	$\frac{10}{252}$	$\frac{30}{252}$	$\frac{15}{252}$	$\frac{1}{252}$	$\frac{56}{252}$
Column totals	$\frac{21}{252}$	$\frac{105}{252}$	$\frac{105}{252}$	$\frac{21}{252}$	1

5-13 If $Y = 1$, then the conditional probability distribution of X is

x	0	1	2	3
$g(x\mid 1)$	$\frac{2}{40}$	$\frac{18}{40}$	$\frac{18}{40}$	$\frac{2}{40}$

If $X = 2$, the conditional probability distribution of Y is

y	0	1	2
$h(y\mid 2)$	$\frac{9}{30}$	$\frac{18}{30}$	$\frac{3}{30}$

5-15

z	0	1	2	4
$g(z)$	$\frac{111}{144}$	$\frac{20}{144}$	$\frac{12}{144}$	$\frac{1}{144}$

5-17

z_2 \ z_1	0	1	2	4	Row totals
0	$\frac{25}{144}$	0	0	0	$\frac{25}{144}$
1	$\frac{60}{144}$	0	0	0	$\frac{60}{144}$
2	$\frac{26}{144}$	$\frac{20}{144}$	0	0	$\frac{46}{144}$
3	0	0	$\frac{12}{144}$	0	$\frac{12}{144}$
4	0	0	0	$\frac{1}{144}$	$\frac{1}{144}$
Column totals	$\frac{111}{144}$	$\frac{20}{144}$	$\frac{12}{144}$	$\frac{1}{144}$	1

y	0	1	2	3	4	5
$g(y)$	$\frac{4}{72}$	$\frac{16}{72}$	$\frac{25}{72}$	$\frac{19}{72}$	$\frac{7}{72}$	$\frac{1}{72}$

5-19

y	2	3	4	5	6	7	8
$g(y)$	$\frac{1}{16}$	$\frac{2}{16}$	$\frac{3}{16}$	$\frac{4}{16}$	$\frac{3}{16}$	$\frac{2}{16}$	$\frac{1}{16}$

5-21

5-23 $\quad g(y_1,y_2) = \dfrac{n!}{(y_1 - y_2)!y_2!(n - y_1)!} p_1^{y_1-y_2} p_2^{y_2}(1 - p_1 - p_2)^{n-y_1}$

$$y_1 = 0, 1, \ldots, n$$
$$y_2 = 0, 1, \ldots, n$$
$$y_2 \leqq y_1$$

5-25 $\quad .55626$

5-27 $\quad E(X + Y) = E(X) + E(Y) = 7 + (\frac{5}{2}) = \frac{19}{2}$

$\qquad \mathrm{Var}(X + Y) = \mathrm{Var}(X) + \mathrm{Var}(Y) = 14 + (\frac{5}{4}) = 6\frac{1}{4}$

5-29 $\quad E(X + Y) = E(X) + E(Y) = (\frac{1}{3}) + 1 = \frac{4}{3}$

$\qquad \mathrm{Var}(X + Y) = \mathrm{Var}(X) + \mathrm{Var}(Y) = (\frac{5}{18}) + (\frac{1}{2}) = \frac{7}{9}$

5-31 $\quad E(X + Y) = E(X) + E(Y) = (\frac{3}{2}) + 1 = \frac{5}{2}$

$\qquad \mathrm{Var}(X + Y) = \mathrm{Var}(X) + \mathrm{Var}(Y) + 2\mathrm{Cov}(X,Y)$
$\qquad\qquad\qquad\quad = (1\frac{5}{28}) + (\frac{3}{4}) + 2(-\frac{3}{14}) = 1\frac{5}{28}$

$\qquad$ since $\mathrm{Cov}(X,Y) = (\frac{9}{4}) - (\frac{3}{2}) = -\frac{3}{14}$ by (5-22).

5-33 $\quad E(X_1 + X_2) = \mu_1 + \mu_2, \mathrm{Var}(X_1 + X_2) = \mu_1 + \mu_2$

5-35 $\quad \frac{5}{6}, 11\frac{1}{36}$

5-39 $\quad p\mu$

5-43 $\quad f(x_1,x_2,x_3,x_4) = x_1x_2x_3x_4/10^4 \qquad x_i = 1, 2, 3, 4$
$\qquad\qquad\qquad\qquad\qquad\qquad\qquad\qquad\quad i = 1, 2, 3, 4$

$\qquad E(Y_1) = 4E(X) = 12, \mathrm{Var}(Y_1) = 4\mathrm{Var}(X) = 4,$ and $E(X_1X_2X_3X_3) = E(X_1)E(X_2)E(X_3)E(X_4) = 3^4 = 81$

5-45 $\quad f(x_1, \ldots, x_n) = p^n(1 - p)^{x_1+x_2+\cdots+x_n-n} \qquad x_i = 1, 2, 3, \ldots$
$\qquad\qquad\qquad\qquad\qquad\qquad\qquad\qquad\qquad\qquad i = 1, 2, \ldots, n$

5-49 $\quad n = 30$

5-51 $\quad (5!/1!1!3!)/(.5)(.3)(.2)^3 = .024.$ If drawing is done without replacement, probabilities associated with the three categories will change slightly from selection to selection, and successive draws will be dependent. However, with a large number of students neither of these objections is serious. Both can be overcome by drawing with replacement. If $X_1 =$ number on committee from fraternities and sororities, then the marginal distribution of X_1 is binomial with $n = 5$, $p = .20$ and $\mathrm{Pr}(X_1 \geqq 3) = 1 - \sum\limits_{x_1=0}^{2} b(x_1;5,.20) = 1 - .94208$
$\qquad = .05792.$

5-53 $\quad \mathrm{Var}(Y_1) = \frac{9}{20} = .45$

5-55 $\quad$ Let X_1, X_2, X_3, X_4 be the numbers falling in categories C_1, C_2, C_3, C_4 respectively. The multinomial seems to be a reasonable model to describe the behavior of the random variables. One might suspect that p_1 would increase if the roadblock continued to operate and drivers realized that they are liable to be stopped. That is, drivers would be more apt to have their automobile in good operating condition. We have $I = (\$0)X_1 + (\$5)X_2 + (\$8)X_3 + (\$10)X_4.$

Assuming the multinomial (with $n = 100$) is satisfactory, we get $E(I) = \$460$, standard deviation $= \sqrt{\text{Var}(I)} = \$\sqrt{1,164} = \$34.18$.

5-57 $\quad f(x_1, x_2) = \dfrac{\binom{3}{x_1}\binom{2}{x_2}\binom{5}{6 - x_1 - x_2}}{\binom{10}{6}} \qquad \begin{array}{l} x_1 = 0, 1, 2, 3 \\ x_2 = 0, 1, 2 \end{array}$

$$\dfrac{\binom{3}{2}\binom{2}{1}\binom{5}{3}}{\binom{10}{6}} = \dfrac{2}{7}, \qquad g(y) = \dfrac{\binom{5}{y}\binom{5}{6 - y}}{\binom{10}{6}} \qquad y = 1, 2, 3, 4, 5$$

$\Pr(Y \leq 3) = .738095 \qquad$ *from Appendix B3*

5-59 $\quad E(X_i) = n/N,\ \text{Var}(X_i) = p(1 - p)$ where $p = n/N,\ \text{Cov}(X_i, X_j)$
$\quad = -p(1 - p)/(N - 1)$.

5-63 $\quad G_3(y_3) = [F(y_3)]^3 = (y_3/6)^3 \qquad y_3 = 1, 2, 3, 4, 5, 6$

$$g(y_3) = \left(\dfrac{y_3}{6}\right)^3 - \left(\dfrac{y_3 - 1}{6}\right)^3$$

$E(Y_3) = {}^{119}\!\!/_{24}$

5-65 In Example 4-4 we discussed the appropriateness of the binomial model for a score on a single examination. Hence the probability function of X, the score on a single examination, is $f(x) = b(x; 25, .20)$. Because of guessing, the four scores X_1, X_2, X_3, X_4 will be independent. Hence, if $Y_1 = \min X_i$

$\Pr(Y_1 \leq 5) = 1 - [1 - F(5)]^4$

$F(5) = \Pr(X \leq 5) = .61669 \qquad$ *from Appendix B1*

Thus

$\Pr(Y_1 \leq 5) = 1 - (.38331)^4 = 1 - .023 = .977$

5-67 $\quad F(1) = .75 \qquad \Pr(Y_1 \leq 1) = 1 - (1 - .75)^5 = {}^{1,023}\!\!/_{1,024}$

5-69

y_{10}	$g(y_{10})$
1	.00001
2	.00258
3	.04537
4	.19670
5	.34344
6	.28438
7	.11038
8	.01658
9	.00056
10	.00000

6-1 (a) .91466 (b) .6633 (c) .1587 (d) .3174
6-3 (a) 1.645 (b) 2.326 (c) 2.576
6-5 (a) .1587 (b) .02275
6-7 .0304
6-9 With continuity correction $\Pr(-10.1 < Z < .70) = .76$. Without continuity correction $\Pr(-10 < Z < .60) = .73$. The exact answer is .75794.

6-11 With continuity correction $\Pr(-20.025 < Z < 1.775) = .038$. Without continuity correction $\Pr(-20 < Z < 1.8) = .036$.

6-13 $\Pr(-1.97 < Z < 1.21) = .86$

7-1 The best estimate of p is .28. The probabilities are respectively .99161, .98517, .97861.

7-3 According to (7-10) the best estimator of μ is $\bar{X}$. Here $\bar{x} = .30$. $\Pr(.30 \leq \bar{X} \leq .70) = \Pr(6 \leq Y \leq 14) = .84945$ where Y has a Poisson distribution with parameter $20(.5) = 10$.

7-5 (a) $x/n = \frac{2}{25} = .04$ is an unbiased estimate of k/N, and $Nx/n = 40$ is an unbiased estimate of k.

$$\Pr\left(\frac{k}{N} - .01 \leq \frac{X}{n} \leq \frac{k}{N} + .01\right) = \Pr(1 \leq X \leq 1.5) = \Pr(X = 1)$$

$$= \frac{\binom{50}{1}\binom{950}{24}}{\binom{1,000}{25}}$$

(b) An unbiased estimate is $x/n = \frac{110}{200} = .55$.

7-7 $y = 6$, $n = 20$, short interval $(.13,.64)$, shortest unbiased interval $(.105,.59)$, equal-tail interval $(.11,.655)$.

7-9 The binomial is appropriate with $y = 9$, $n = 10$. Hence the interval is $(.5550,.9975)$.

7-11 $\bar{x} = .8$, $z_{1-\alpha/2} = z_{.975} = 1.96$, and the interval is $(.72,.88)$.

7-13 $y = 324$, $n = 200$, $\bar{x} = 1.62$, and the interval is $(1.44,1.80)$.

7-15 .21996, .43890, .65935, .92643, .99268. The power is decreased and β is increased for all $p < .90$.

7-17 .87612, .04693, .05311, .69319. The power is decreased.

7-19 Test H_0: $p = .90$ against H_1: $p < .90$, where p is the fraction of seeds that will germinate. With $p = .90$, $\Pr(Y \leq 40) = .02454$, $\Pr(Y \leq 41) = .05787$. Hence, the critical region is $y \leq 40$ with $\alpha = .02454$. We could also test H_0: $p = .10$ against H_1: $p > .10$ regarding p as the fraction of seeds that will not germinate. With $p = .10$, $\Pr(Y \geq 10) = .02454$, $\Pr(Y \geq 9) = .05787$, and the critical region is $y \geq 10$ with $\alpha = .02454$. With either formulation accept H_0 and do not reject the company's claim.

7-21 Test H_0: $p = .25$ against H_1: $p \neq .25$. The critical region is $y \leq 1$ and $y \geq 10$ with $\alpha = .02431 + .01386 = .03817$. Hence the theory is not contradicted.

7-23 $n = 25$ with critical region $y \geq 9$.

7-25 Test H_0: $p = .30$ against H_1: $p > .30$. The critical region is $y \geq 24$ with $\alpha = .00559$. Hence, H_0 is rejected and the new cure is recommended. We observe that $\Pr(Y \geq 29) = .00004$ if H_0 is true.

7-27 We test H_0: $p = .10$ against H_1: $p > .10$ and reject if $y \geq 10$ with $\alpha = .02454$. If $p = .30$, power $= \Pr(Y \geq 10) = .94977$ so that $n = 50$ is a sufficiently large sample. Accept H_0 (and the examination) with $y = 9$.

7-29 (b) $\sum\limits_{x=y_1}^{n} b^*(x;y_1,p) = \sum\limits_{y=y_1}^{n} b(y;n,p) =$ power of the test.

7-31 The power is $\Pr(Y \geq 10)$ computed with Poisson parameter $5(1.6) = 8$. This yields .28338. If $\mu = 2$, then the same calculation with parameter $5(2) = 10$ yields .54207.

7-33   With parameter $4(2.5) = 10$ we get $\Pr(Y \leq 2) + \Pr(Y \geq 15) = .00277 + .08346 = .08623$. With $4(1) = 4$ we get $\Pr(Y \leq 2) + \Pr(Y \geq 15) = .23810 + .00002 = .23812$. With such a small n we do not expect large values for the power.

7-35   Yes. If H_0 is true, Y has a Poisson distribution with parameter 10, the critical region is $y \leq 4$, and $\alpha = .02925$. If $\mu = 1$, then Y has a Poisson distribution with parameter 2 and $\Pr(Y \leq 4) = .94735$.

7-37   The observed value of the statistic is $z' = -2$. Thus H_0: $p = .90$ is rejected in favor of H_1: $p < .90$.

7-39   The interval for the proportion which fail to germinate is $(.102, .128)$.

7-41   If $p = .23$, power $\cong \Pr(Z < -.37) = .36$. If $p = .27$, power $\cong \Pr(Z < -.35) = .36$.

7-43   As in Exercise 7-28, test H_0: $\mu = 1$ against H_1: $\mu > 1$. The observed value of the statistic is $z' = 1.8$. Hence, H_0 is rejected.

7-45 $(.98, 1.26)$

8-1   The ordered samples are

7,8	8,7	5,7	16,7	14,7
7,5	8,5	5,8	16,8	14,8
7,16	8,16	5,16	16,5	14,5
7,14	8,14	5,14	16,14	14,16

with sample means

7.5	7.5	6	11.5	10.5
6	6.5	6.5	12	11
11.5	12	10.5	10.5	9.5
10.5	11	9.5	15	15

Since every ordered sample has an equal chance of being selected, $\frac{1}{20}$ is the probability to assign to each sample and to each sample mean. Hence

$$E(\bar{Y}) = 7.5(\tfrac{1}{20}) + 6(\tfrac{1}{20}) + \cdots + 15(\tfrac{1}{20}) = 10$$

$$\bar{V} = \frac{7 + 8 + 5 + 16 + 14}{5} = 10$$

$$\sigma'^2 = \frac{(7-10)^2 + (8-10)^2 + \cdots + (14-10)^2}{4} = \frac{45}{2}$$

$$\mathrm{Var}(\bar{Y}) = \frac{(7.5-10)^2 + (6-10)^2 + \cdots + (15-10)^2}{20} = \frac{27}{4}$$

8-3 $\sum_{i=1}^{20} y_i = 34$, $\hat{y} = 4{,}000(34)/20 = 6{,}800$ is the estimate of the number of children attending public school. $s^2 = 2.22$, $s = 1.49$. The confidence interval is $6{,}800 \pm 1.96(1.49)\sqrt{4{,}000(3{,}980)}/\sqrt{20}$ or $6{,}800 \pm 2{,}605$ or $(4{,}195, 9{,}405)$. $\Pr(\hat{Y} < 6{,}800) \cong \Pr(Z < -.15) = .44$.

8-9   The sample variances are

0	.5	2	40.5	24.5
.5	0	4.5	32	18
2	4.5	0	60.5	40.5
40.5	32	60.5	0	2
24.5	18	40.5	2	0

Each of the 25 sample variances has probability $\frac{1}{25}$ of being selected. Hence

$$E(S^2) = 0 \; (\tfrac{1}{25}) + .5(\tfrac{1}{25}) + \cdots + 0 \; (\tfrac{1}{25}) = 18$$

$$\sigma^2 = \frac{(7 - 10)^2 + (8 - 10)^2 + (5 - 10)^2 + (16 - 10)^2 + (14 - 10)^2}{5} = 18$$

8-11 $\operatorname{var}(\bar{y}) = 4.0$, $\operatorname{var}(\hat{y}) = 4,000,000$. Confidence interval for $\bar{V}$ is $(50 \pm 1.645 \sqrt{4})$ or $(\$46.71, \$53.29)$. The confidence interval for $\hat{V}$ is $(\$46,710, \$53,290)$.

8-13 In both cases $\hat{y} = 4,000$. With equal probabilities $\operatorname{var}(\hat{y}) = 1,925,000$. With unequal probabilities $\operatorname{var}(\hat{y}) = 100,000$.

8-15 Since $E(X_i) = p$,

$$E(\bar{Y}) = \frac{1}{Np} \sum_{i=1}^{N} v_i E(X_i) = \bar{V}$$

Also $\operatorname{Var}(X_i) = (1 - p)(p)$. Hence

$$\operatorname{Var}(\bar{Y}) = \sum_{i=1}^{N} \frac{v_i^2}{N^2 p^2} \operatorname{Var}(X_i) = \frac{1 - p}{p N^2} \sum_{i=1}^{N} v_i^2$$

The sample size has a binomial distribution with parameters N and p. Hence expected sample size is Np.

8-17 The possible samples, corresponding probabilities, and corresponding $\bar{y}$ are

Samples	$-1, 0, 1$	$-1, 0$	$-1, 1$	$0, 1$	-1	0	1	nothing
Probability	$(\tfrac{1}{3})^3 = \tfrac{1}{27}$	$(\tfrac{1}{3})^2(\tfrac{2}{3}) = \tfrac{2}{27}$	$\tfrac{2}{27}$	$\tfrac{2}{27}$	$\tfrac{4}{27}$	$\tfrac{4}{27}$	$\tfrac{4}{27}$	$\tfrac{8}{27}$
$\bar{y}$	0	-1	0	1	-1	0	1	0

Hence the probability distribution of $\bar{Y}$ is

$\bar{y}$	-1	0	1
$g(\bar{y})$	$\tfrac{6}{27}$	$\tfrac{15}{27}$	$\tfrac{6}{27}$

$\operatorname{Var}(\bar{Y}) = \frac{2}{9}$. $\operatorname{Var}(\bar{Y})$ with fixed sample size $n = 1$ is $\frac{3}{6}$.

8-21 $n_1 = 20$, $n_2 = 50$, $n_3 = 30$. $\bar{y} = 2.9$, $\operatorname{var}(\bar{y}) = .0625$, $\operatorname{Pr}(\bar{Y} > 3.0) \cong .34$. The interval is $(2.41, 3.39)$.

8-23 The inequality reduces to

$$\sum_{i=1}^{L} \frac{N_i^2 \sigma_i^2}{N^2 n_i} \geqq \frac{1}{N^2 n} \left(\sum_{i=1}^{L} N_i \sigma_i \right)^2$$

the desired result.

9-1 $\begin{bmatrix} 0 & 1 & 0 \\ \frac{1}{4} & \frac{1}{2} & \frac{1}{4} \\ 0 & 1 & 0 \end{bmatrix}$

9-3 $\begin{bmatrix} \frac{1}{4} & \frac{1}{4} & \frac{1}{4} & \frac{1}{4} \\ 0 & \frac{2}{4} & \frac{1}{4} & \frac{1}{4} \\ 0 & 0 & \frac{3}{4} & \frac{1}{4} \\ 0 & 0 & 0 & 1 \end{bmatrix}$

9-5 $\begin{bmatrix} 1/4 & 3/4 \\ 2/3 & 1/3 \end{bmatrix}$

9-7 $\begin{bmatrix} 1/2 & 1/2 & 0 & 0 \\ 1/2 & 0 & 1/2 & 0 \\ 0 & 1/2 & 0 & 1/2 \\ 0 & 0 & 1/2 & 1/2 \end{bmatrix}$

9-9 All entries in P^2 are positive. $a = [1/6, 4/6, 1/6]$. After a sufficiently large number of steps the probabilities of finding 0, 1, or 2 white balls in urn 1 are very close to $1/6$, $4/6$, and $1/6$, respectively.

9-11 $P^{(2)} = \begin{bmatrix} 1/16 & 3/16 & 5/16 & 7/16 \\ 0 & 4/16 & 5/16 & 7/16 \\ 0 & 0 & 9/16 & 7/16 \\ 0 & 0 & 0 & 1 \end{bmatrix} = P^2$

$P^{(4)} = \begin{bmatrix} 1/256 & 15/256 & 65/256 & 175/256 \\ 0 & 16/256 & 65/256 & 175/256 \\ 0 & 0 & 81/256 & 175/256 \\ 0 & 0 & 0 & 1 \end{bmatrix} = P^2 P^2$

It is apparent that the last row remains unchanged for any P^n so that the matrix is not regular. Solving $aP = a$ yields $a_1/4 = a_1$ or $a_1 = 0$, $(a_1/4) + (2a_2/4) = 0$ or $a_2 = 0$, $(a_1/4) + (a_2/4) + (3a_3/4) = a_3$ or $a_3 = 0$, $(a_1/4) + (a_2/4) + (a_3/4) + a_4 = a_4$ or $a_4 = a_4$. Hence $a = [0,0,0,1]$. Although it is not apparent what P^n will approach from looking at $P^{(4)}$, the probability that a 4 will be rolled eventually is 1 (the sum of probabilities in a negative binomial distribution with $p = 1/4$, $c = 1$). Hence p^n will approach

$\begin{bmatrix} 0 & 0 & 0 & 1 \\ 0 & 0 & 0 & 1 \\ 0 & 0 & 0 & 1 \\ 0 & 0 & 0 & 1 \end{bmatrix}$

9-13 $[8/17, 9/17]$

9-15 $P^{(2)} = \begin{bmatrix} 1/2 & 1/4 & 1/4 & 0 \\ 1/4 & 1/2 & 0 & 1/4 \\ 1/4 & 0 & 1/2 & 1/4 \\ 0 & 1/4 & 1/4 & 1/2 \end{bmatrix}$ $P^{(3)} = \begin{bmatrix} 3/8 & 3/8 & 1/8 & 1/8 \\ 3/8 & 1/8 & 3/8 & 1/8 \\ 1/8 & 3/8 & 1/8 & 3/8 \\ 1/8 & 1/8 & 3/8 & 3/8 \end{bmatrix}$

Since all entries of P^3 are positive, P is regular. $[1/4, 1/4, 1/4, 1/4]$. Individual A will be in each of four states about $1/4$ of the time if they continue to toss pennies.

9-17 $\left[\dfrac{p^2}{p^2 + pq + q^2}, \dfrac{pq}{p^2 + pq + q^2}, \dfrac{q^2}{p^2 + pq + q^2} \right]$

9-19 $P = \begin{bmatrix} q & p & 0 & 0 \\ q & 0 & p & 0 \\ 0 & q & 0 & p \\ 0 & 0 & q & p \end{bmatrix}$

$a = [q^3/D, q^2p/D, qp^2/D, p^3/D]$ where $D = q^3 + q^2p + qp^2 + p^3$.

9-21 $P^n = P$ for all n. Hence P is not regular.

9-23 $\begin{bmatrix} 6 & 1 & 6 \\ 5 & 3/2 & 5 \\ 6 & 1 & 6 \end{bmatrix}$ $n_{12} = n_{32} = 1$

If urn 1 contains 0 or 2 white balls, then on the next step it must contain 1 white ball. $n_{23} = 5$.

9-25 $\quad \begin{bmatrix} 11\frac{1}{6} & 70\frac{5}{9} & 25\frac{2}{3} \\ 40\frac{5}{9} & 11\frac{1}{3} & 20\frac{2}{3} \\ 35\frac{5}{9} & 50\frac{5}{9} & 11\frac{1}{2} \end{bmatrix} \qquad n_{13} = 25\frac{2}{3}$

9-27 $\quad \begin{bmatrix} 4 & 2 & 6 & 12 \\ 6 & 4 & 4 & 10 \\ 10 & 4 & 4 & 6 \\ 12 & 6 & 2 & 4 \end{bmatrix} \qquad n_{31} = 10$

9-29 $\quad \begin{bmatrix} 10\frac{2}{3} & 4 & 5\frac{1}{2} & 4 \\ 9\frac{1}{4} & 5 & 9\frac{1}{4} & 5 \\ 5\frac{1}{2} & 4 & 10\frac{2}{3} & 4 \\ 9\frac{1}{4} & 5 & 9\frac{1}{4} & 5 \end{bmatrix} \qquad n_{42} = 5$

9-31 $\quad \begin{bmatrix} 1/p^2 & 1/q & 1/pq & (1+q)/q^2 \\ (1+p)/p^2 & 1/pq & 1/p & 1/q^2 \\ 1/p^2 & 1/q & 1/pq & (1+q)/q^2 \\ (1+p)/p^2 & 1/pq & 1/p & 1/q^2 \end{bmatrix}$

Index